Managing Extreme Climate Change Risks through Insurance

In recent years, the damage caused by natural disasters has increased worldwide; this trend will only continue with the impact of climate change. Despite this, the role for the most common mechanism for managing risk – insurance – has received little attention. This book considers the contribution that insurance arrangements can make to society's management of the risks of natural hazards in a changing climate. It also looks at the potential impacts of climate change on the insurance sector, and insurers' responses to climate change. The author combines theory with evidence from the rich experience of the Netherlands together with examples from around the world. He recognizes the role of the individual in preparing for disasters, as well as the difficulties individuals have in understanding and dealing with infrequent risks. Written in plain language, this book will appeal to researchers and policy makers alike.

W. J. WOUTER BOTZEN is Assistant Professor at the Department of Environmental Economics of the Institute for Environmental Studies, VU University Amsterdam and a visiting scholar at the Risk Management and Decision Processes Center of the Wharton School, University of Pennsylvania.

Managing Extreme Climate Change Risks through Insurance

W. J. WOUTER BOTZEN

CAMBRIDGE
UNIVERSITY PRESS

CAMBRIDGE
UNIVERSITY PRESS

University Printing House, Cambridge CB2 8BS, United Kingdom

Cambridge University Press is part of the University of Cambridge.

It furthers the University's mission by disseminating knowledge in the pursuit of education, learning and research at the highest international levels of excellence.

www.cambridge.org
Information on this title: www.cambridge.org/9781316600887

First published 2013
First paperback edition 2015

A catalogue record for this publication is available from the British Library

Library of Congress Cataloguing in Publication data
Botzen, W. J. Wouter
Managing extreme climate change risks through insurance / Wouter Botzen.
pages cm
Includes bibliographical references and index.
ISBN 978-1-107-03327-6 (hbk.)
1. Risk management. 2. Natural disasters. 3. Climatic changes – Environmental aspects. I. Title.
HD61.B6578 2013
363.34′7–dc23
2012040697

ISBN 978-1-107-03327-6 Hardback
ISBN 978-1-316-60088-7 Paperback

Contents

Figures

Tables

Preface

The importance of the topical theme of climate change and related natural disaster risks in the scientific and policy-making debate has increased considerably since I started my work on this theme in 2006. This debate has been fuelled by images of catastrophes which travelled around the globe and entered the homes of many people via mass media. The immense human suffering caused by Hurricane Katrina in the USA, the severe floods in Pakistan and Thailand, and the earthquake and tsunami-related flooding in Japan readily come to people's minds as recent examples of the potentially destructive force of nature. Even though large natural disasters are low-probability events, such disasters happen almost every year on a global scale, and many smaller extreme weather events that cause significant losses occur regularly. Records by insurers show that natural disaster losses have been trending upwards over the last decades, and it has been projected that these trends will continue in the future as a result of climate change.

Our societies are still learning about how to adequately prepare for disasters and about how to manage the economic risks that extreme weather events pose. Research can speed up this learning process and with this book I hope to make my own contribution to the expanding literature on this topic. While there have been a lot of discussions among insurers and public policy makers about the challenges posed by natural disasters and climate change, progress in the implementation of practical solutions to address these risks has sometimes been cumbersome. Nevertheless, encouraging steps can be observed. For example, in July 2012 President Obama signed the Biggert–Waters Flood Insurance Reform Act of 2012 which reauthorizes and reforms the National Flood Insurance Program (NFIP) in the USA. Among other changes, this reform implies that the NFIP has to move to actuarial insurance rates and improve its financial risk management and policies for flood risk mitigation, which should improve the ability of the Program to cope with future flood disasters. Several academics in the USA who have conducted and published extensive studies on the operation of the Program have provided an important knowledge base for the NFIP reform. Another example comes from the Netherlands where – partly motivated by the research conducted for this book – the introduction of an insurance arrangement for flood damage has been discussed extensively over the last few years. After this book went to press, a Dutch insurer introduced a catastrophe risk insurance policy that covers flood damage. This is an important step forward for the insurability of flood damage in the Netherlands, where insurers have excluded flood coverage from property insurance for the past 60 years. The design of the new insurance product follows some of the proposals made in this book, such as premiums that reflect actual flood risk and combining the insurance policy with incentives for policyholders to limit risk. Homeowners who implement measures that mitigate flood risks are eligible for a discount on their flood insurance premium. However, it remains to be seen whether or not this new catastrophe insurance can achieve a broad nationwide insurance coverage for flood risk, because of its relatively high premium and limited coverage. Exploring a public–private flood insurance scheme, as has been outlined in this book, could be a way forward to arrive at more extensive and affordable coverage for the risk of flooding. These recent developments in the USA and the Netherlands can be seen as an encouragement for bringing together academics and policy makers with the objective of increasing the resilience of our societies to future disasters.

Acknowledgements

This book is the culmination of seven years of research during which I collaborated intensively with Jeroen van den Bergh, Jeroen Aerts and Laurens Bouwer. I am very grateful to Jeroen van den Bergh, who encouraged me to write this book and who has been a source of inspiration for my work. He provided numerous comments and advice on its various chapters. My several pleasant stays with him at Universitat Autònoma de Barcelona (UAB) resulted in a useful exchange of ideas. I am thankful to Jeroen Aerts for initiating several of the projects that provided the basis for this book, as well as for his support of my work and comments on various chapters. I also gained a great deal from collaborating with Laurens Bouwer.

Many other people have supported the work involved in this book. Several colleagues at the Institute for Environmental Studies (IVM) provided feedback on the surveys that I conducted. In particular, I am grateful to Joop de Boer, Roy Brouwer, Sebastiaan Hess, Vincent Linderhof and Marije Schaafsma. I would also like to thank the interviewers and participants in the surveys. Alfred Wagtendonk helped in preparing GIS data. I had helpful discussions with Anna Alberini, Ada Ferrer-i-Carbonell, Thijs Dekker, and Frédéric Reynes regarding econometric analyses. The insurance companies Hagelunie and Interpolis provided data for this book, and I thank Gijs Kloek and Roy Kluitman (Eureko Re) for their assistance in preparing and interpreting the resulting database. I have had many insightful discussions with experts from insurance companies and policy makers over the past few years. In addition, this book has benefited from comments by, and discussions with, a number of other people, notably Frans Berkhout, Marija Bočkarjova, Philip Bubeck, Hans de Moel, John Gowdy, Geert Groen, Chris Harrison, Maarten Heeneman, Dave Huitema, Ekko van Ierland, Heidi Kreibich, Howard Kunreuther, André Lucas, Tim McDaniels, Erwann Michel-Kerjan, Youbaraj Paudel, Boris Porfiriev,

Jennifer Poussin, Piet Rietveld, Isabel Seifert, Oleg Sheremet, Richard Tol, Pieter van Beukering, Peter Wakker and various anonymous referees.

This research was funded by the Dutch national research programmes 'Climate Change Spatial Planning' and 'Knowledge for Climate'. In addition, I received generous financial support through a Veni grant from the Netherlands Organization for Scientific Research (NWO).

Introduction

In recent decades, the economic damage caused by natural disasters has increased considerably worldwide. It has been projected that extreme weather events will become more frequent and harmful in the future as a result of climate change and socio-economic trends. An important question that arises in this context is how to design effective policies that limit the expected increase in natural disaster risks and reduce the impacts of extreme weather events on human societies. The book addresses this question by examining the role that insurance arrangements can play in increasing economic resilience to natural disasters.

The approach followed involves dealing both with the effects of climate change on natural disaster risks and with the implications of these effects for insurers. The contribution that insurance arrangements can make to societies' management of the risks of natural hazards in the context of a changing climate is analysed, paying particular attention to the complex nature of insuring low-probability/high-impact risks. Public–private partnerships are explored as a possible solution for such complexities. The important role of insurance for individuals in preparing for disasters and purchasing natural disaster coverage is recognized, as well as aspects of the 'bounded rationality' of individuals in dealing with infrequent risks. In particular, material is included about individual risk perceptions, risk aversion, willingness to pay for natural disaster insurance, and applications of utility theories to insurance demand, including prospect theory. The book presents empirical case studies from the Netherlands, which are complemented by international perspectives, studies and examples. The book will appeal to readers from a broad range of disciplines that study aspects of natural disaster risk management, while the material is accessible for both academics and practitioners.

W. J. Wouter Botzen is Assistant Professor at the Department of Environmental Economics of the Institute for Environmental Studies,

VU University Amsterdam. He has been involved in several research projects on climate change economics, individual decision making under risk, natural disaster insurance and natural disaster risk management. He has published on these themes in many leading academic journals.

1 *Climate change and natural disaster risk management*

1.1 Background

Recent natural disasters have demonstrated the vulnerability of various countries to extreme weather events such as floods, storms and droughts (Munich Re 2012). The economic losses caused by such disasters can be of such great magnitude that individuals and businesses may be unable to carry them, which could cause major financial distress and bankruptcy unless there is compensation available. In the face of a projected rise in the frequency and severity of natural disasters as a result of socio-economic developments and climate change, the question arises of how to design policies that limit exposure to, and ameliorate the impacts of, natural disasters. The role of financial arrangements for natural disaster risk is important in this respect, since an often raised question is who should pay for the elevated risks faced (Kunreuther et al. 2011). Apart from providing financial relief after a disaster has struck, financial arrangements such as insurance may contribute to adapting societies to rising risk and may enhance economic resilience to disasters by providing incentives for risk reduction (Mills and Lecompte 2006). This is not an easy task for the insurance sector, given the challenges associated with insuring low-probability, high-impact (correlated) disaster risks. As an example of such problems, Kousky (2011) reports that in ten states of the USA, state-run natural disaster insurance arrangements have been established because private insurance at affordable premiums is unavailable for many homeowners.

During the last few decades, a considerable increase in the frequency and economic damage caused by natural disasters has occurred worldwide (Höppe and Pielke 2006; Kunreuther and Michel-Kerjan 2007). Figure 1.1 shows an upward trend in overall and insured losses caused by great natural disasters since 1950. Natural disasters, such as major

Some parts of this chapter have previously been published as Botzen and van den Bergh (2009a). All of these parts have been updated and extended in this chapter.

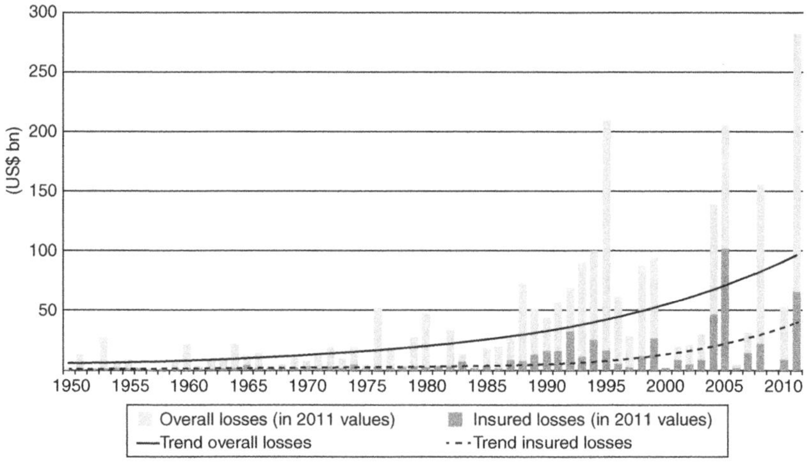

Figure 1.1 Overall and insured losses caused by great natural disasters between 1950 and 2011 (in 2011 values)
Source: Munich Re NatCatService (2012).

storms, floods and earthquakes, have devastating consequences for societies around the world, with especially severe impacts on developing countries, while the consequences for insurers are concentrated in developed countries (Freeman et al. 2003). Specifically, natural disasters may cause many deaths with a single event, result in the spread of diseases in affected areas, and have substantial adverse economic consequences by causing direct damage to property and infrastructure as well as indirect damage, such as business interruption and productivity losses. Some examples of major events in the last decade indicate the diversity of threats posed by nature and their global character. Destructive catastrophes that have exposed the challenges that modern societies face in managing natural disaster risks include the tsunami in Asia in 2004 that caused more than 283,000 deaths; the earthquake in Haiti in 2010 that killed 222,500 people; floods in 2010–11 in Australia with $10 billion of losses, as well as in Pakistan where floods affected 20 million people and put one-fifth of the country under water – resulting in $10 billion of damage; flood damage costing $40 billion in Thailand in 2011 of which only a small portion was insured; and the large earthquake and tsunami in Japan in 2011 which could amount to over $200 billion.

Damaging wildfires occurred in Greece in 2008 and 2009, and major forest fires in Australia in 2009 killed over 200 people and damaged large tracts of land, with devastating effects on wildlife. Summer heat in Russia killed over 55,000 people in 2010. Europe has also experienced severe floods and winter storms in recent years, such as the 2002 floods in central Europe incurring, in total, $15 billion worth of losses, two major floods in the UK in 2007 that caused about $4 billion of damage per event, and the storm Xynthia in south-west Europe in 2010 that caused 63 deaths and €6 billion of damage, as well as destructive flash floods in France (€1.5 billion) and flooding in Germany in 2010 (€1.3 billion).

The USA was hit by a series of very destructive hurricanes in 2005 – hurricanes Katrina, Rita and Wilma killed over 1,500 people and resulted in $180 billion in compensation payments – followed by Hurricane Ike in 2008 with $38 billion of damage (Kunreuther et al. 2011; Munich Re 2009). Hurricane Irene, which hit New York in 2011 as a tropical storm, served as a reminder that huge values of assets are exposed to potentially destructive hurricanes in the USA (Aerts and Botzen 2012a).

The sharp increase in frequency and severity of the recent disasters has made some leading academics suggest that we have entered a new era of natural catastrophes (Kunreuther et al. 2011). Because of their dramatic impacts[1] natural catastrophes are often well publicized in the media and are an important issue for governments, international organizations, such as the World Bank and the United Nations, and the broader research community. An important question that arises as a result of these recent destructive disasters is how to design effective policies that limit the increase in natural disaster losses expected to be caused by climate change and reduce its impacts on human societies. This book provides insights into this subject from an economics perspective and elaborates upon the role that financial arrangements can play in limiting vulnerability and exposure to natural disasters. The insurance sector, which is the world's largest industry in terms of revenue, could be a major partner in managing, spreading and providing incentives for reducing natural disaster risk and, thereby, could promote adaptation to climate change (Mills 2009a). In this respect,

[1] Some authors argue that the effects of catastrophes are minor in macroeconomic terms, even though natural catastrophes have considerable local impacts (Albala-Bertrand 2006). The macroeconomic impacts of disasters may be larger in small countries.

climate change not only poses a threat to insurers, but also entails the development of new business opportunities (Mills 2007).

The remainder of this chapter is structured as follows. Section 1.2 outlines the influence of socio-economic developments and climate change on natural disaster risk. Section 1.3 discusses expert assessment of risk and households' perceptions and behavioural responses to risk. Section 1.4 describes several strategies for managing extreme weather risks. Section 1.5 examines the role that financial arrangements can play in natural disaster risk management. Section 1.6 provides an outline of this book.

1.2 Future natural disaster risks under climate and socio-economic change

1.2.1 Impact of socio-economic developments on natural catastrophe damage

Human actions are a major determinant of the occurrence and consequences of natural disasters. Although natural hazards such as storms, floods, heat waves and storm surges are natural phenomena, the damage caused by these natural hazards is to a great extent influenced by human activities. A natural disaster is usually defined as the impact of a natural event on human societies, for example in terms of loss of life or economic costs (Bočkarjova 2007). Commonly, a certain threshold of economic damage or loss of life needs to be exceeded before an extreme weather event is defined as a natural disaster. Also important in this respect is the ability of societies to prepare for, and manage, the economic disruption caused by a disaster, and this depends on each country's level of economic development (Rose 2003). A major part of the increased damage caused by natural disasters around the globe can be explained by socio-economic developments, such as the increased population and concentration of wealth in areas that are vulnerable to natural hazards (Bouwer 2011; Changnon 2003a; Crompton and McAneney 2008; Miller et al. 2008; Muir-Wood et al. 2006; Pielke et al. 2005). For example, the large rise in hurricane damage in the USA that has been observed in recent decades is mainly due to increased human settlement at coastal locations where hurricanes often make landfall, such as Florida (Pielke et al. 2008; Schmidt et al. 2009b). Also, in a flood-prone country like the Netherlands, the amount of urban area that can be flooded increased six-fold during the twentieth century, and it has been projected that it may double again in the twenty-first century (de Moel et al. 2011). In the future, urbanization in hazard-prone areas, such as coastal agglomerations, is projected to continue, which heightens exposure to extreme weather events, posing challenges for the management and insurability of natural disaster risk (Bouwer et al. 2007).

1.2.2 *Climate change and the frequency and severity of natural disasters*

In addition to socio-economic developments, climate change may increase the intensity and severity of extreme weather events and contribute to an increasing loss burden of natural disasters in the future (IPCC 2007; Mills 2005). Anthropogenic climate change is caused by the emissions of carbon dioxide and other greenhouse gases, such as methane, that have accumulated in Earth's atmosphere since the industrial revolution (mainly since the late nineteenth century), predominantly due to the burning of fossil fuels, deforestation and other land-use changes. These rising concentrations of greenhouse gases result in increases in surface air temperature, because the greenhouse gases in the atmosphere trap heat (see, e.g., Pierrehumbert 2004). The Intergovernmental Panel on Climate Change (IPCC) projects a rise in global average surface temperatures between 1.1 °C and 2.9 °C by 2100 for a low-emission scenario, and between 2.4 °C and 6.4 °C by 2100 under a high-emission scenario (IPCC 2007). The decay rate of greenhouse gases is very slow, and therefore their presence in the atmosphere,[2] with the accompanying negative consequences, will continue in future decades (IPCC 2007; Matthews and Caldeira 2008). Even if current climate policy were to be able to stabilize greenhouse gases to 2000 levels, then a further warming of about 0.2 °C would occur in the next two decades, while warming would be much higher if emissions increase in accordance with the common projections (SRES scenarios) of the IPCC (2007). This implies a clear need for climate change adaptation policy to reduce the vulnerability of human systems

[2] Montenegro et al. (2007) state that 'about 75% of CO_2 emissions have an average perturbation time of 1,800 years, and the remainder have a lifetime much longer than 5,000 years'.

to climate change (Pielke et al. 2007; Stern 2007). Several positive feedback mechanisms of climate change may increase greenhouse gas concentrations in addition to human-induced emissions, and thus result in even more warming than anticipated by some climate models (Stern 2007). Examples are the release of methane due to the melting of permafrost and a reduced uptake of carbon as a result of the weakening of natural carbon sinks, such as the Amazon forest[3] (Heimann and Reichstein 2008; Kennedy et al. 2008).

According to the recent state of knowledge in climate science, which is summarized in the latest IPCC report (2012), it is very likely that climate change will increase the length, frequency and/or severity of warm spells, or heat waves, over most land areas in the twenty-first century. Moreover, climate change is likely to result in an intensified water cycle, which implies that existing regional patterns of scarcity and abundance of water will be amplified, increasing the risk of droughts and floods. In particular, rainfall and floods are likely to increase in high-latitude regions, while southern arid regions are expected to have considerable reductions in rainfall in both hemispheres (Meehl et al. 2007). According to the IPCC (2012), it is likely that the frequency of heavy precipitation will increase in the twenty-first century over many areas of the globe. In other parts of the world, warmer air and oceans could cause more intense storms, such as hurricanes and typhoons. In addition, climate change is expected to cause a rise in mean sea level due to the expansion of warmer oceans and the melting of glaciers and ice caps. The IPCC (2007) projects a global rise in sea levels between 0.2 and 0.6 metres by 2100. An irreversible melting of the Greenland Ice Sheet[4] or the collapse of the West-Antarctic Ice Sheet, which both have a low probability of occurring, could cause a substantial rise in sea level of about 5–12 metres globally, although this is very uncertain and could only occur over the course of several centuries (Alley et al. 2005; Rapley 2006; Wood et al. 2006). The IPCC (2012) estimates that it is very likely that sea level rise will contribute to an upward trend in extremely high coastal water levels in the future. Sea level rise

[3] Deforestation may also increase the vulnerability to disasters caused by climate change, for example because of soil erosion and landslides resulting from more frequent and severe rains and floods.

[4] The surface mass balance of the Greenland Ice Sheet may turn negative at a global average warming larger than 1.9–4.6 °C, which could result in its complete elimination over a very long time period (IPCC 2007).

could result in the inundation of many unprotected low-lying areas, and may increase the likelihood of flooding due to storm surges, which could have considerable consequences for Small Island States and countries with extensively populated deltas and coastal areas, such as the Netherlands, Vietnam and Bangladesh.

The IPCC (2007) states that global temperatures have already increased by approximately 0.76 °C since 1900, while sea levels have risen by about 20 centimetres during that time. There is also evidence that some of the expected effects of climate change on extreme weather events have already materialized. The IPCC (2012) indicates that it is very likely that there has been an overall decrease in the number of cold days and nights and an increase in the number of warm days and nights. There is medium confidence that the length and/or number of warm spells, or heat waves, has increased in many regions. It is likely that anthropogenic climate change has caused these trends. Moreover, there have been significant trends in the number of heavy precipitation events in some regions, and it is likely that more of these regions have experienced increases in these events. There is medium confidence that the increase in heavy precipitation has been caused by anthropogenic (human-induced) climate change. Furthermore, it is likely that an increased incidence of extremely high sea levels has been observed and it is likely that humans contributed to this trend. According to the IPCC (2007) there has been evidence of an increase in the average intensity of tropical cyclones, such as hurricanes and typhoons, in the North Atlantic and some other regions since the 1970s. It is more likely than not that this trend has been influenced by anthropogenic climate change. However, this statement was weakened to 'low confidence' in the latest IPCC report (2012) due to uncertainties concerning tropical cyclone records, incomplete understanding of the physical mechanisms linking tropical cyclone metrics to climate change, and the degree of natural tropical cyclone variability.

There is, however, debate in the scientific community about whether or not the observed upswing in hurricane activity is caused by anthropogenic climate change, meaning that it is likely to persist in the future, or whether it can be attributed to natural climate variability related to the Atlantic Multidecadal Oscillation (Kerr 2006a). Some research suggests that global warming has already resulted in an increased intensity or frequency of hurricanes, and that this may have been caused by higher sea surface temperatures (e.g. Emanuel 2005; Hoyos et al. 2006; Webster et al. 2005). A study by Elsner et al. (2008) shows that an upward trend for wind speeds of strong hurricanes can be observed in each relevant ocean basin. Saunders and Lea (2008) studied the contribution of sea surface temperature to hurricane frequency and activity for the USA and concluded that a 0.5 °C increase in sea temperature is associated with a 40 per cent increase in hurricane frequency and activity. However, it has been argued that current observational databases are insufficiently reliable to analyse trends for hurricane activity, due to subjective measurements and variable procedures over time. Also, the time periods used may be too short to draw definite conclusions about climate change (Landsea et al. 2006; Michaels 2006). This is likely to remain an active and very relevant area of research in the near future, given the high insurance and economic costs that hurricanes can cause (e.g. Höppe and Pielke 2006). Even though it is difficult to attribute past hurricane activity to climate change, several studies have projected that climate change will increase the frequency of severe hurricanes in the future (e.g. Bender et al. 2010). According to the IPCC (2012), it is likely that average tropical cyclone maximum wind speed will increase, although these increases may not occur in all ocean basins.

While most of the future projections for climate change, including those considered in this book, anticipate gradual changes in climate over time, several studies have indicated that abrupt changes in climate, with large impacts on human societies, are possible (Lenton 2011). The possible existence of tipping points in the climate system implies that small perturbations can switch the climate system into a different mode of operation. This means that at one point in time a small change (e.g. more greenhouse gas emissions) could have large long-term consequences. Lenton et al. (2008) made a shortlist of various elements of tipping points,[5] which includes, among other elements:

- large losses of Arctic summer sea ice, which would amplify global warming

[5] Tipping elements are those large-scale components of the Earth system that could be forced past a tipping point and subsequently undergo a transition to a different state.

- rapid melting of the Greenland Ice Sheet and a collapse of the West-Antarctic Ice Sheet, which would cause substantial rises in sea level
- an overturning of the Atlantic thermohaline circulation
- an increase in amplitude of the El Niño–Southern Oscillation
- a destabilization of the Indian summer monsoon.

The uncertainty about these tipping elements is very large but, given their potentially disruptive consequences for humans and ecosystems, it is important to improve our understanding of their underlying physical mechanisms and learn how to manage the risks of abrupt climate change.

Climate change is an externality of economic activity, since individuals and businesses that pollute the atmosphere with greenhouse gas emissions – for example through electricity generation, driving, flying and destruction of forests – do not pay for the costs of the climate change that is caused by increased atmospheric greenhouse gas concentrations.[6] Internalizing these costs for economic agents around the globe via taxes, regulation, or emissions trading systems is complicated by the 'public good' nature of the atmosphere and resulting problems with 'free riding' behaviour. The atmosphere can be defined as a public good because its use by one individual does not reduce its availability to others (non-rivalry) and no-one can effectively be excluded from using it (non-excludability). The latter characteristic may induce free riding behaviour, meaning that economic agents emit more greenhouse gases than is socially desirable. An additional complication is that climate change is global both in its causes as well as in its consequences because the impact of greenhouse gas emissions is independent of where in the world they are emitted (Stern 2007). For these reasons, it is difficult to reach the stringent international agreement on greenhouse gas emissions that is required for stabilizing or reducing atmospheric concentrations of greenhouse gases.

[6] For economic analyses of the cost of climate change and optimal climate policy, see Cline (1992), Tol (1997, 1999), Stern (2007) and Nordhaus (2008). It should be noted that the cost estimates of climate change made by these studies can be debated and hardly take the impacts of climate change on natural disasters into account (Ackerman et al. 2010; Botzen and van den Bergh 2012).

Future greenhouse gas emissions may rise steeply due to the rapid industrialization of Asian economies, with their increasing demands for energy. As an illustration, Figure 1.2 shows the past developments of *cumulative* CO_2 emissions per country between 1900 and 2004 as well as projections until 2080. The figure shows that, currently, the USA has the highest level of *cumulative* CO_2 emissions, followed by Western Europe, China, Japan and India. However, this ranking is projected to change dramatically in the coming decades. In 2031 India will have emitted more CO_2 than Japan. In 2021 China will have larger *cumulative* CO_2 emissions than Western Europe, and by 2052 China will overtake the USA as the largest cumulative emitter. India is expected to have a larger total of cumulative emissions than Western Europe shortly after 2080. Under a business-as-usual scenario, by 2050 the world's largest nations will have a shared responsibility for the climate problem. In the decades to come, they will be responsible for either solving or exacerbating the climate crisis. Effective climate policy will require both the recognition of shared responsibility and an unprecedented degree of cooperation, which will be very difficult in practice, as the recent climate policy negotiations in Copenhagen and Cancun showed. Even in the unlikely case that emissions could be reduced to zero, warming would still continue for several decades because of the lag in response time of the climate system caused, among other reasons, by the past emissions that persist in the atmosphere for a very long time (IPCC 2007). This highlights the necessity to examine the effects of climate change on extreme weather events and the resulting damage, and to design adequate adaptation policies for managing potential changes in these risks (Pielke et al. 2007). Apart from actual reductions in greenhouse gas emissions – defined as greenhouse gas mitigation – adaptation to climate change is an alternative or complementary response strategy that aims at reducing the impacts of climate change and their associated costs. Different definitions of adaptation exist, and this book follows the definition provided by Smit et al. (2000), who define adaptation as 'adjustments in ecological–social–economic systems in response to actual or expected climate stimuli, their effects or impacts'. Throughout this book, the word 'mitigation' is defined in the same way as in the insurance literature, in which it refers to risk mitigation or risk reduction, while the term 'greenhouse gas mitigation' refers to reducing or stabilizing greenhouse gas emissions or

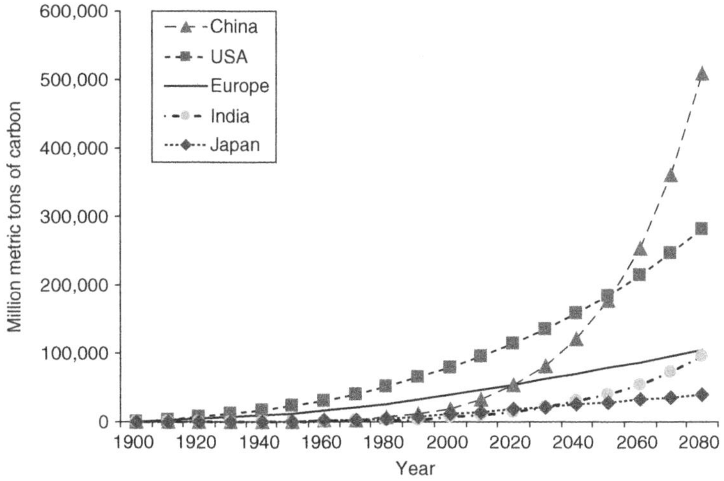

Figure 1.2 Historical and projected cumulative CO_2 emissions from 1900 until 2080 using the EIA (2007) reference scenario
Source: Botzen et al. (2008a).

levels, which is often simply called 'mitigation' in the climate change literature.

1.3 Assessing natural catastrophe risk

1.3.1 Expert modelling of natural disaster risk

Although the scientific basis for climate change has been well established, predictions of regional climate change impacts are still very uncertain (IPCC 2007). Assessments of future natural disaster risk are inherently difficult because of the uncertainties associated with both the impacts of climate change on natural hazards and the consequences of socio-economic developments for exposure and vulnerability to these hazards (IPCC 2007, 2012). Considerable uncertainty and ambiguity are associated with both the frequency of a disaster occurring and the damage that it will cause. Constructing different scenarios of climate and socio-economic change and estimating their influence on risk may be a useful first step in assessing future risk. Risk is defined throughout this book as *probability*damage*. Statistical models can be used to assess how the frequency and severity of natural disasters

Figure 1.3 Main components of catastrophe models
Adapted from Grossi and Kunreuther (2005).

or disaster damage relate to variability in weather and climate (e.g. Saunders and Lea 2008; Schmidt et al. 2009b). Extrapolations of such historical relations under changes in climate conditions may then provide insights into future risks (see Chapter 3). Moreover, catastrophe models are commonly used to assess exposure to natural disaster risks (Grossi and Kunreuther 2005). Such computer-based models estimate the loss potential of catastrophes by overlaying the properties at risk and the potential sources of natural hazards in a specific geographical area using geographical information systems (GIS). Catastrophe models provide important information for insurers because the risk estimates they generate can help insurers to adequately set premiums that reflect risk. Moreover, catastrophe models can assess the overall exposure of insurers to specific natural hazards, which is important information for the financial risk management of insurers, such as the purchase of reinsurance. Governments may use catastrophe models for assessing the geographical exposure to risks, which provides useful information for emergency planning, spatial planning and awareness raising, or for evaluating the effectiveness of existing (and the need for additional) protection measures, such as dikes or flood walls.

Figure 1.3 shows a schematic overview of the main components of catastrophe models (Grossi and Kunreuther 2005). The natural hazard module of a model characterizes the physical characteristics of the hazard, such as the location of a flood, flood depth and flow velocity of water, wind speeds, and frequency of occurrence of the hazard. The portfolio of properties at risk, as a component of the model, can include various characteristics of assets, such as the location, age and type of buildings or land use. The vulnerability component of the model quantifies the impact of the natural hazard on the properties at risk, which may be done with the use of damage functions that describe the relation of physical parameters, e.g. flood depth, with damage to the inventory, such as flood damage to buildings (e.g. Merz et al. 2004). The resulting damage to the portfolio of properties is computed based on these vulnerability measures and may consist of direct losses, indirect losses or both. It should be recognized that all the components of a catastrophe model are associated with a degree of uncertainty, which accumulates in the final estimates of catastrophe damage (see, e.g., de Moel and Aerts 2011).

The output of catastrophe models may be represented as exceedance probability curves that indicate the probability of a certain loss being surpassed, or by geographical maps that show levels of risk (de Moel et al. 2009). The importance of natural disaster risk maps has been recognized by the European Union, which has obliged its Member States to create flood hazard and risk maps by 2013 through the Flood Directive (2007/60/EC). De Moel et al. (2009) provide an overview of existing flood mapping practices in Europe. Their results indicate that about half of these countries have maps that depict flood extents for almost their entire territory, one-third have maps that cover a large part of their territory, while the remainder have hardly, or have not, mapped their flood extents yet. These maps have mostly been created by governmental organizations, but also by insurers who use these maps for determining insurability, for differentiating premiums, and for testing long-term financial solvency. There have been only a few countries that have developed flood risk maps that include information on the potential damage from flooding, which implies that a considerable effort is needed before the Member States comply with the EU Flood Directive (de Moel et al. 2009). Moreover, none of these countries produce 'forward-looking maps' that account for changed flood risk as a result of climate change, which could be useful in guiding climate change adaptation policies.

Catastrophe models need to be updated over time due to socio-economic developments and climate change in order to reflect changes in risk. Insurers can use catastrophe models for updating premium levels if the risk landscape changes. Future projections of natural disaster risk are useful to governments for assessing the benefits of risk reduction measures that often have a long lifetime, such as storm surge barriers. If climate change would increase the frequency or severity of extreme weather, then the 'Natural hazard' component of the model needs to be adjusted to reflect the increased risks. Socio-economic developments, such as increased urbanization in hazard-prone areas, may require changes in the 'Portfolio of properties at risk' component over time.

Only a few studies have assessed the potential consequences of climate and socio-economic change on natural disaster risk (Aerts and Botzen 2011a; Dawson et al. 2009; Feyen et al. 2009; Hall et al. 2005; Mokrech et al. 2008; Pielke 2007; te Linde et al. 2011). An illustration of such a study is given by Bouwer et al. (2010), who examined the influence of land-use change, changes in asset values, and climate change on the risk of river flooding in a Dutch polder area (called 'dike-ring 36') over time. Socio-economic change, which drives land use and property values, 42 flood scenarios, and projections of future flood probabilities were combined in a flood damage model to predict the development of flood risk between 2000 and 2040. The flood risk model calculates flood damage based on land use type and inundation depth. The land use scenarios are derived from a (micro-economic) land use model according to the so-called 'Global Economy' (GE) and 'Regional Community' (RC) socio-economic scenarios that, respectively, reflect high and low economic and population growth (Hilferink and Rietveld 1999). Climate change influenced the return periods of peak river charges and related flood probabilities in the model, according to low (G) and high (W+) climate change scenarios that have been constructed by the Royal Dutch Meteorological Institute (van den Hurk et al. 2006). Figure 1.4 shows the current flood risk and the flood risk projections in 2040 under the RC/G and GE/W+ scenarios apportioned according to the increase in risk due to socio-economic change, the increase due to climate change, and their combined effect. This approach of disentangling future changes in risk due to socio-economic and climate change is very useful for defining effective adaptation strategies. For example, these results indicate the importance of directing adaptation policies to limit both a possible rise in probability and in the damage caused by natural disasters.

1.3.2 Households' assessments of risk and behaviour

Individual risk perception

In evaluating hazards, most people usually rely on intuitive risk judgements, called risk perceptions, which often differ considerably from expert assessments (Slovic 1987, 2000). The understanding of risk perception of individuals is very important in designing adaptation

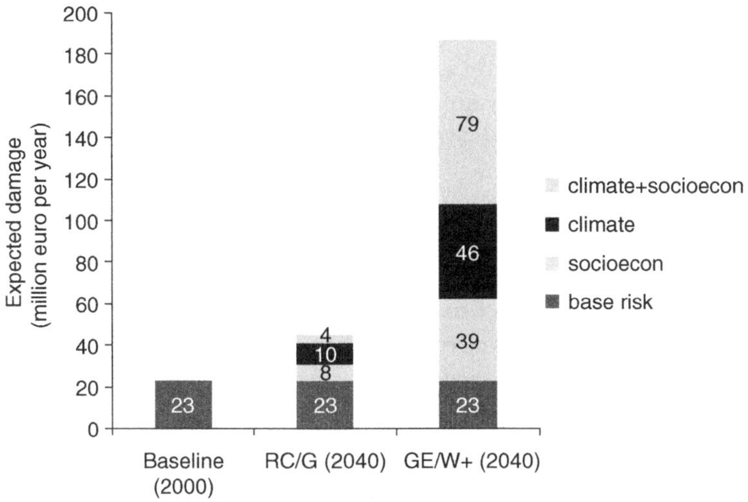

Figure 1.4 Projected annual flood damage in a Dutch polder area in the year 2040 compared with 2000 under scenarios of climate and socio-economic change, assuming no adaptation

Reprinted from Bouwer et al. (2010) with permission from Elsevier.

policies. Household risk judgements can determine the perceived legitimacy of, as well as compliance with, land-use planning and other adaptation policies (Peacock et al. 2005). Moreover, individual perceptions of hazards are important factors behind decision making under risk with respect to buying insurance and undertaking self-protective measures (Burn 1999; Flynn et al. 1999). A low perception of natural disaster risks by individuals may translate into a low demand for natural disaster insurance.

Experts commonly assess risk by making best estimates of the probability and potential damage of a hazard using statistical techniques or catastrophe models. However, expert-generated information on risk, such as probabilistic information, often has little influence on decision making about risk by lay persons (Kunreuther et al. 2001). Individuals often use simple rules when they assess risks, which may be described as heuristics (Kahneman et al. 1982). Individuals may use the 'availability heuristic' in judging natural hazard risk, which implies that they judge an event as risky if it is easy to imagine or recall. For example, individuals who have experienced a disaster may find it easier to imagine that the disaster could happen again in the future and, therefore, feel a higher perceived risk than individuals without this experience. Individuals often rely on affective feelings when they judge the level of risks, which may deviate from pure logical and analytical reasoning (Loewenstein et al. 2001; Slovic et al. 2004). Individuals may have a higher risk perception if natural hazards are associated with negative feelings, which can have been caused or reinforced by experiences of damage caused by natural hazards or evacuation because of a disaster (Finucane et al. 2000; Keller et al. 2006). Such experiences make the hazard more salient to people. Often, natural disasters have a very low likelihood of occurring, so that individuals may have no or few experiences of the hazard. This results in a very low individual risk perception, or individuals may even ignore the risk altogether (see Chapters 9 and 10). Governments can undertake information campaigns if individual risk perceptions deviate considerably from expert risk judgements. Moreover, insurance premiums may be differentiated according to risk, and thereby provide a price signal of the natural disaster risk faced by the policyholder, which can then influence his/her risk awareness.

Individual behaviour under risk and the purchase of natural disaster insurance

Biases in individual risk judgements and the boundedly rational behaviour of individuals may have the effect that people are insufficiently prepared for disasters, and may also result in a low demand for insurance against low-probability natural disaster risk. For example, several studies have shown that many households in the USA do not purchase flood insurance, even though this insurance is provided by the federally operated National Flood Insurance Program (NFIP), which charges much lower premiums than the private market (Michel-Kerjan and Kousky 2010). The NFIP aims to achieve a high market penetration of flood insurance by requiring households who live in the 1/100-year floodplain to purchase flood insurance if they have a mortgage from a federally backed or regulated lender. Nevertheless, Dixon et al. (2006) estimate that only about 49 per cent of single-family homes in the 1/100-year flood zone carry NFIP flood insurance despite this requirement, which is similar to estimates by Kriesel and Landry (2004) for coastal areas in the USA. Market penetration is much lower in flood zones in the USA without the mandatory insurance purchase requirement. Similar findings of low voluntary insurance purchases can be observed in Europe. For instance, most insurance companies in Germany offer an optional insurance coverage for natural hazards – including floods – additional to the standard building or household contents insurance, but few households purchase this coverage. The average market penetration for natural hazard insurance in Germany is only 3–4 per cent in combination with building insurance and 10 per cent in combination with household contents insurance (Seifert et al. 2012).

The low level of demand for natural disaster insurance seen in several countries is in conflict with the expected utility framework that economists often use when analysing individual decision making under risk, such as insurance purchases. Expected utility theory predicts that individuals would buy insurance against a risk if it is offered at a reasonable (actuarially fair) price, and if individuals are risk-averse, which is a commonly made assumption. However, in many cases this framework fails to adequately describe behaviour in practice, especially in the case of low-probability, high-impact risks, such as natural disasters (e.g. Mason et al. 2005). One reason for this is that individuals often deviate from rational behavioural principles when they make decisions under risk (Kahneman 2003). In particular, a common observation is that individuals either overestimate or ignore low-probability risk (Tversky and Kahneman 1992). This processing of risk poses some difficulties when applying the traditional expected utility framework for individual decision making under risk (von Neumann and Morgenstern 1947), which assumes that individuals correctly assess the likelihood of adverse events and that individuals process probabilities linearly. The descriptive failure of expected utility theory in explaining individual behaviour under risk is well documented (Camerer 1998). Alternative theories that allow for the modelling of individual attitudes towards probabilities, or 'probability weighting', may be more suitable for modelling individual behaviour. Important examples are prospect theory and rank-dependent utility theory (Kahneman and Tversky 1979; Quiggin 1982; Schmeidler 1989; Tversky and Kahneman 1992). Allowing for 'bounded rationality', or limitations in individuals' perceptual and cognitive capabilities, is fundamental in correctly anticipating individual responses to risky events, such as the demand for insurance coverage against natural disasters (see Chapter 8).

1.4 Managing natural disaster risks

1.4.1 Economic resilience to natural disasters

An important concept in managing natural disaster risk is the notion of resilience, even though its broad meaning has obscured its use in risk management (Klein et al. 2003). As discussed by Bočkarjova (2007) and Rose (2007), resilience has been defined differently in various disciplines, such as ecology, from where the concept originates, engineering and economics, as well as between different authors. Resilience has two main interpretations; namely, the time necessary for a disturbed system to return to its original state (Pimm 1984) and the amount of disturbance a system can absorb before moving to another state (Holling 1973, 1986). Rose (2004b), who defines resilience from an economics perspective, relates resilience to the time needed for recovery in the aftermath of a disaster, in the sense that a higher level of resilience allows the economy to recover faster at a lower cost. Moreover, Rose (2004a, 2006) regards resilience as a post-disaster characteristic which comprises the inherent and adaptive responses to disasters that result in the avoidance of potential losses, and the ability of a system to maintain its function when shocked. In his definition, resilience encompasses the ability of societies to limit or prevent losses during and after a disaster, and emphasizes the ingenuity and resourcefulness applied. Obviously, risk management policies can be put in place before a disaster occurs in order to increase resilience.

In the context of climate change, Timmerman (1981) defines resilience as the capacity to absorb and recover from the occurrence of a hazardous event. Resilience is related to adaptation, which comprises adjustments prior to the occurrence of a disaster aimed at creating conditions within the human system that enhance the system's resistance to disasters and its capacity to respond to, and cushion the impacts of, a disaster (Bočkarjova 2007; Handmer and Dovers 1996). Bočkarjova (2007) adds to the definition of resilience the ability of the human-induced system to exhibit learning, so as to improve its protective mechanisms (adaptation) in the face of disasters. Resilience may be spatially dependent and can differ between regions within the same country. For example, Porfiriev (2009) argues that megacities may have a higher resilience capacity than small towns, because the latter often lack the economic resources to ameliorate disaster impacts. Climate change increases the need for resilience, since it may lead to more disturbances of the human system due to an increased frequency and severity of weather extremes. A major challenge is that climate change impacts are very uncertain, so that improving resilience requires a coping capacity to deal with uncertain disturbances. Improving resilience (according to the above definitions) and adaptive capacity may thus be seen as a desirable policy instrument to manage natural disaster risks (Tobin 1999). In this respect, it is important that victims are not merely compensated for the damage they have suffered after a disaster has occurred, but that adequate incentives and policies are put in place to foster risk mitigation and encourage rebuilding in a sustainable way so that vulnerability to future disasters is reduced in the long run (Rose 2011).

1.4.2 Risk management strategies

Hazard prevention to reduce the probability of suffering damage and the expected costs of damage

Preventing the hazard from occurring and reducing the probability of suffering damage is an effective strategy for limiting risk associated with certain natural hazards, such as flooding; while it might be more difficult for others, such as wind damage due to storms. Examples of strategies that limit the probability of suffering damage are the creation of dams for flood control, dikes, storm surge barriers, and the relocation of property out of hazard-prone areas. Investments in hazard prevention are usually undertaken by governments, because of the 'public good' characteristics (non-excludability and non-rivalry) of protection infrastructure. If a dike is built to protect a region from flooding, then it is difficult to exclude inhabitants of that region from the protection that it offers (non-excludability), while the value of protection to one inhabitant does not reduce its value for the other inhabitants (non-rivalry). This is an example of a classical market failure which results in an undersupply of flood protection by private markets and provides a rationale for public provision of flood protection. Nevertheless, the public protection from natural hazards is often below economic optimum levels and there seems to be considerable scope to improve cost-effective prevention or damage mitigation strategies worldwide. It has been suggested that larger worldwide investments of $40 billion in disaster preparedness, prevention and mitigation would have reduced global economic losses by $280 billion during the 1990s (IFRC 2001).

Often, public support for large investments in protection infrastructure only arises after a disaster has occurred. For example, strategies to prevent flood damage are well developed in countries around the North Sea and in Japan, where flooding claimed many lives until the middle of the twentieth century. After a catastrophic flood in 1953, the Dutch built their famous Deltaworks: a series of dams, sluices, dikes and storm surge barriers constructed between 1958 and 1997 in the south-west Netherlands (Aerts and Botzen 2012b). This flood protection infrastructure was successful in ensuring high safety standards that in some areas protect against storm surges with a recurrence interval of 1 in 10,000 years. Cost–benefit analysis may guide the determination of safety standards and protection investments, as has been done in the Netherlands in the past (Jonkman et al. 2004; van Dantzig 1956). According to this approach, investments in protection infrastructure are undertaken as long as the total benefits in terms of reduced risk over time outweigh the total costs of protection. Both costs and benefits are measured and discounted over the lifetime of the protection infrastructure. A drawback of hazard prevention with engineering infrastructure is that it may be perceived by households and companies that the risk is eliminated instead of reduced, which can encourage economic development in hazardous areas, thus increasing exposure and potential disaster damage (Vis et al. 2003).

Once in place, a continuous updating of protection infrastructure is needed, especially in areas that are impacted by a rapid increase in the frequency of hazards due to climate change, or by an increase in potential damage that may be caused by socio-economic developments in the protected areas. A proactive or anticipatory approach that reduces vulnerability before climate change results in adverse impacts, such as floods, may be desirable (Klein et al. 2003). The success of adaptation measures in limiting risk will depend on the magnitude and rate of change of the climate; large changes that occur rapidly may be difficult to accommodate. Large regional variations exist in projections of climate change impacts, indicating that a variety of strategies need to be implemented in different areas that may be affected by higher flood, drought or storm risks (IPCC 2007, 2012).

Current prevention measures may be inadequate to deal with climate change. For example, sea level rise in the Netherlands could result in a large increase in flood risk if no additional adaptation measures were to be undertaken. Aerts et al. (2008b) show that the probability that water levels, during a storm surge, will exceed the existing coastal protection near the main cities, Amsterdam and Rotterdam, could increase by a factor of 2 for a 25-centimetre sea level rise, while this probability might increase by a factor of 8–20 under scenarios of, respectively, a 60-cm and an 85-cm sea level rise. At present, the storm surge barriers of the Deltaworks in the Netherlands are insufficiently prepared for (further) rises in sea level and are likely to require adjustments in the future. A cost–benefit analysis performed by Aerts and Botzen (2012b) of the Haringvliet barrier, which is part of the Deltaworks, indicates that adapting the barrier to climate change, instead of replacing it completely, is a good investment. Unfortunately, adjusting the construction of some barriers to sea level rise is not possible. In designing hazard prevention or damage mitigation measures it is therefore advisable to consider flexible infrastructures that allow for adjustments to climate change, especially given the considerable uncertainty that is associated with sea level rise projections (IPCC 2007).

Mitigation of damage at the household level

After the construction of a protection infrastructure there often remains an inherent or residual risk. The undertaking of measures that mitigate damage at the household level may be an effective strategy for reducing residual risks that is complementary to public investment in prevention. Moreover, damage mitigation is sometimes a cheaper or more feasible option for reducing risks than preventing hazards from occurring through large-scale engineering projects, such as storm surge barriers. Such mitigation measures could prevent or limit damage once a natural hazard takes place. Examples are anchoring roofs to withstand strong winds, creating flexible foundations of buildings that prevent collapse during earthquakes, and investing in flood shields that prevent floodwater from entering a building, or 'flood-proofing' of houses through elevation. Several studies suggest that cost savings due to mitigation can be considerable. Kunreuther et al. (2011) modelled hurricane damage in New York, Texas, South Carolina and Florida under scenarios with and without mitigation according to recent building code standards. The results for a 100-year hurricane indicate that mitigation could reduce potential losses by 61 per cent in Florida, 44 per cent in South Carolina, 39 per cent in New York and 34 per cent in

Texas. Savings in Florida alone due to mitigation would result in $51 billion for a 100-year and $83 billion for a 500-year event. Experience of flooding in Europe also indicates that households avoided considerable flood damage due to the implementation of damage mitigation measures. During the 2002 flooding of the River Elbe in Germany, flood damage was limited through adapting the use of buildings to flooding, which means that cellars and storeys susceptible to flooding are not used intensively, and by adapting interior fittings to flooding, which comprises the use of waterproofed building materials and placing easily movable furniture on the lower floors of buildings (Kreibich et al. 2005; Thieken et al. 2005, 2006). In particular, adapting the use of buildings and interior fittings to the possibility of flooding reduced damage to buildings by 46 per cent and 53 per cent, and damage to contents by 48 per cent and 53 per cent, respectively (Kreibich et al. 2005). Given the potential usefulness of mitigation in managing natural disaster risk, further research should focus on identifying cost-effective mitigation measures and how individuals can be encouraged to invest in mitigation, which is likely to depend, among other things, on risk perception. Well-designed insurance arrangements could stimulate damage mitigation, as will be elaborated upon in Section 1.5.

Government compensation for damage

Governments are often under considerable pressure to financially compensate households and businesses after natural disasters (e.g. Downton and Pielke 2001). Compensation for such damage can facilitate the process of economic recovery after a catastrophe. It can speed up the rebuilding of damaged property and prevent bankruptcy of individual households and firms, thereby allowing continuity of business operations and stimulating rebuilding of the capital stock. An efficient financial arrangement for compensating for disaster damage may therefore contribute to economic resilience (Pelling 2003; Rose 2007, 2011).

However, compensation for disaster damage may also provide an incentive to take risks instead of reducing them, if the compensation arrangements are poorly designed. Incentives to settle in safe areas, instead of in hazard-prone areas, are minimal in cases where governments pay compensation for damage, unconditionally, to households who have taken the risk of settling in hazardous areas. In the same arena of a 'moral hazard' effect, incentives for households to limit losses and invest in mitigation measures are minimal if governments generously compensate for the damage caused by natural disasters (e.g. Priest 1996). Furthermore, the uncertainty associated with the ad hoc compensation schemes that exist in some countries, such as in the Netherlands for flood damage, may be undesirable from the perspective of the welfare of risk-averse individuals (see Chapter 4). Often, there is no guarantee that government compensation for disaster damage will be efficiently provided or that it will be distributed to the households who are most in need of assistance. Adequately designed insurance arrangements to compensate those who have suffered natural disaster losses may overcome such complications (see Section 1.5). However, the existence of the possibility of government compensation for disaster damage may reduce individual demand for private natural disaster insurance. This has been called the 'charity hazard', although empirical evidence for this effect is mixed (Raschky et al. 2012).

Diversification of risk management strategies

The combination of climate and socio-economic change that will influence the probability and magnitude of future natural disaster losses results in inherently uncertain changes in risks. A characteristic of a resilient system is that it is diverse; meaning that a number of functionally different components protect the system from various threats in a diversified portfolio (Godschalk 2003). Hence, a resilient natural hazard risk management strategy necessarily involves a package of actions (de Bruijn et al. 2007). De Bruijn and Klijn (2001) and Vis et al. (2003) argue that resilient flood risk management aims both at lowering the probability of the hazard occurring and at reducing its possible impact, i.e. damage. The latter authors suggest that implementing strategies that aim at lowering flood damage in the Netherlands via compartments and green rivers that allow for water storage during peak discharges are a useful and resilient strategy for long-run flood risk management. The objective for policy makers is to find an optimal portfolio of protection measures that prevent and limit damage during and after events. This will be examined in Chapter 5 for investments in flood control in the Netherlands using a portfolio framework that aims for the highest mean and lowest variance in return (avoided damage). Combining investments in dikes, which reduce the probability of inundation, with those in flood-proofing of houses provides for a reduction in risk of an extremely large range of damage, compared with investments in dikes alone. Moreover, efficient financial arrangements for natural disaster damage can contribute to a society's resilience to natural disasters by providing the compensation that is needed for recovery.

1.5 The role of insurance in adaptation to natural disasters

1.5.1 Climate change impacts on the insurance sector

The insurance sector covers a considerable proportion of weather-related risks, especially in developed countries (Hoff et al. 2005).[7] Future insurance claims may increase considerably if climate change projections and socio-economic developments result in an increased frequency and magnitude of natural disaster damage (Dlugolecki 2000, 2008; Mills 2005; Vellinga et al. 2001). From an insurer's perspective, the time pattern of losses due to socio-economic developments may cause fewer problems than the effects of climate change on disaster damage. The reason is that a rise in population and wealth, which increases the monetary value insured, may result in a similar rise in premium revenues, thereby balancing expected insurance payouts and premium revenues. Nevertheless, socio-economic developments that increase the concentration of wealth in high-risk areas may pose challenges for insurers if claims become more interdependent. This implies that insurers are more likely to experience large accumulating losses from a single event, which require large amounts of capital to reimburse claims which are costly to hold as reserves or to acquire through reinsurance.[8]

Climate change can pose considerable challenges for insurers by changing existing risks, creating new risks, and possibly increasing dependencies between risks. In particular, the changes in extreme weather events that are expected to result from climate change can

[7] The insurance sector is here broadly defined as comprising both primary insurers and reinsurance companies. Reinsurance companies, such as Munich Re and Swiss Re, are often the 'last resort' carriers of catastrophe risk. Potential increases in risk as a result of climate change are likely to be passed on from reinsurance companies to primary insurers in the form of higher reinsurance prices or reduced reinsurance coverage.

[8] Reinsurance is insurance coverage that is purchased by a (primary) insurance company from another insurance company (the reinsurer).

be worrisome for insurers. In general, the flexible nature of the insurance business, with short-term contracts and the ability to change premiums and coverage over time, is a desirable characteristic to ensure resilience of the sector to climate change impacts (Vellinga et al. 2001). Uncertainty about climate change effects poses a major challenge for long-term public investments in protection infrastructure, which have a lifetime of several decades, but insurers may more easily adapt to uncertain climate change impacts on risk by gradually adjusting the conditions of annual contracts to actual developments in climate change. Nevertheless, climate change may still pose problems for insurers even if they write mostly annual contracts, while climate change is a long-term process. If climate change increases the probability and/or damage of insured events then premium income will lag behind payouts for claims, unless premiums are adjusted and reflect the changed risk (Mills et al. 2002). Changing premiums to accurately reflect changed climate conditions can be very difficult for an insurer who, in practice, mostly bases premiums on historical losses. Another challenge is that climate change may increase the 'fat tails' of natural disaster loss distributions, which are the extreme risks that are, in general, difficult or expensive to insure (Charpentier 2008). Climate change may, moreover, pose problems for insurers if climate change increases correlations between (extreme) risks. For example, worldwide sea level rise or an increase in hurricane intensity can introduce correlations across insured losses in areas that are geographically distinct (Kousky and Cooke 2009). The best strategy for insurers would be to incorporate changes in the risks of extreme weather events in assessing exposure to, and pricing and management of, natural disaster risk. In practice, this may be difficult, since the low-probability nature of extreme weather events complicates the assessment of climate change impacts on loss trends. Insurers would have to disentangle systematic changes in climate due to global warming and short-term or cyclical changes in weather, which is very difficult in practice (Charpentier 2008; Kunreuther and Michel-Kerjan 2007). Moreover, considerable uncertainty is associated with projected effects of climate change on natural disasters and the resulting damage, which complicates risk assessments (IPCC 2007, 2012). There is a need for insurers not only to better understand the effects of climate change on extreme weather, but also to increase their knowledge about short-run weather variability before their independent effects on risk can be determined (Munich Re 2011). The consequences of climate change on the insurance sector are likely to be more severe if climate change is an abrupt instead of a gradual process, as a study of the effects of tipping points in the climate system on the insurance sector by Lenton et al. (2009) shows.

Even if insurers incorporate changed risk conditions into their risk assessments and pricing of insurance, regulatory obstacles may prevent them from actually charging (higher) risk-based premiums. After the experience of the devastating 2005 hurricane season in the USA, the question arose whether climate change caused an increase in hurricane intensity and frequency, requiring insurance premium adjustments, or whether average hurricane activity had not changed and the 2005 hurricane season reflected natural climate variability (see Section 1.2). One of the main catastrophe-modelling firms, Risk Management Solutions (RMS), projected an increase in hurricane frequency and severity and advised an increase in premiums. Insurance regulators resisted this adjustment in premiums, arguing that rates should be based only on historical losses and should not reflect predictions (Kunreuther et al. 2011). In Florida, insurance regulators have resisted premium increases that were deemed necessary by insurers after the occurrence of a series of hurricanes that revealed an underestimation of the risk. As a consequence, insurers dropped insurance policies and the state initiated a state-run insurance company, which can partly recoup losses from other insurance policies sold in Florida if a major hurricane were to strike (Kousky 2011). In general, one should not constrain the ability of insurers to adjust premiums according to changes in risk, since this could impair the economic viability and solvency of insurers in the face of climate change. Instead, regulators should protect consumers by focusing on whether or not insurers have sufficient capital in place to be able to pay the anticipated claims following a disaster, and thus minimize the risk of insurer insolvency. A challenge for regulators and rating agencies is to accurately assess the insolvency risk of insurers in the face of changing risks as a result of climate change (Dlugolecki 2008).

Regional assessments of insurers' exposure to natural hazards and regional climate change projections of extreme weather risks can be useful information for premium setting and risk management by insurance companies (see Chapter 2). The insurance sector could also play an important role in stimulating and promoting adaptation policies, which limit risks, as will be elaborated upon below. Indeed, climate change is not only a threat for insurers but also presents new profitable business opportunities, such as offering insurance products for greenhouse gas mitigation technologies and projects (Mills 2007).

1.5.2 Demand for financial coverage in a changing climate

Homeowners usually demand financial compensation for damage caused by natural hazards, which can be provided by government relief or insurance arrangements. Consumers' willingness-to-pay (WTP) for financial coverage via insurance schemes is expected to increase if climate change were to result in an increased frequency, or severity, of natural hazards (see Chapters 8, 10 and 11). It is questionable, however, whether consumers would adjust their WTP sufficiently upwards if insurers needed to increase premiums as a result of climate change. If changes in households' WTP under climate change develop in line with changes in expected losses of the insurance, premium changes may have little impact on levels of insurance penetration. Quantitative modelling of insurance demand provides insights into effects of risk and premium changes on the market shares of insurers (see Chapter 11). In modelling insurance demand under climate change, it is important to account for bounded rationality and the commonly observed failure of expected utility theory in the context of low-probability, high-impact risks (see Section 1.3). In addition to changes in demand for existing insurance arrangements, climate change may also raise demand for new financial arrangements. Valuation techniques for consumers' preferences, such as surveys with contingent valuation and choice modelling methods (Mitchell and Carson 1989), can be useful as means of assessing demand for new insurance products. For example, Akter et al. (2009) estimated demand for crop insurance against flood damage in Bangladesh using the contingent valuation method. Their results indicate that crop insurance is (marginally) commercially viable in riverine floodplain areas, since expected WTP values exceed expected payouts of insurance. Such studies of insurance demand provide important information to policy makers and insurers about the feasibility of introducing new financial arrangements for natural hazard risks.

1.5.3 The role of insurers in managing natural disaster risk

It is useful to explore the role that insurance arrangements can play in managing natural disaster risk and promoting adaptation to possible increases in the risk of extreme weather due to climate change (see Chapters 4 and 7). Insurers collect premiums from many individuals in order to be able to pay for damage caused by natural disasters, which is very large for individual households and companies. In this way, insurance arrangements reduce individual loss exposure and thus spread risks. Primary insurers may further pool natural disaster risks with other types of risks that they insure. Moreover, insurers can hedge their risks by buying reinsurance coverage from reinsurance companies that spread risks over large geographical markets or hedge risks on capital markets using weather derivatives, such as catastrophe bonds options and futures (Michel-Kerjan and Morlaye 2008). This risk-spreading function of insurance may be welfare-enhancing for risk-averse individuals, since it improves financial security. Indeed, research on the effects of flood disasters on reported life satisfaction in 16 European countries indicates that the decreased levels of life satisfaction usually observed after flood events are not reported in regions with flood insurance (Luechinger and Raschky 2009). The purchase of natural disaster insurance may be a way for individuals to adapt to increasing natural disaster risk as a result of climate change.

Risk spreading through insurance in itself does not reduce the aggregate risk faced by society. Nevertheless, insurance can promote adaptation to elevated aggregate natural disaster risks by providing incentives for risk reduction. In addition to providing financial security, insurance arrangements may contribute to limiting the damage caused by natural disasters by acting as a price signal of risk, and promoting the undertaking of mitigation measures (Kleindorfer and Kunreuther 1999; Kunreuther 1996; Kunreuther and Pauly 2006). A necessary condition for this is that insurance premiums reflect the risk faced by the insured property (Kunreuther et al. 2011). Risk-based insurance premiums could act as a price signal for settlement in an area and thus stimulate development in less risky areas and restrain development in hazard-prone areas, since premiums would be higher in the latter. Moreover, insurance can provide incentives to homeowners to invest in measures that mitigate damage. This is very relevant, since practical experience shows that although mitigation measures could be useful to manage risks, individuals rarely undertake them (Kunreuther 2006a). Insurance can require the undertaking of mitigation as part of the policy conditions or reward the insured who invest in mitigation measures with premium discounts or increased levels of coverage (see Chapter 7).

Several characteristics of natural disaster risks complicate insurance in (purely) private markets (Freeman and Kunreuther 2003). The low-probability nature of natural hazard risk makes it difficult to assess the frequency of hazards and the resulting damage, and to determine adequate premiums. Catastrophe models may aid rate setting by insurers, but will involve uncertainties too. Moreover, natural hazard risks are correlated and impacts can be very large, so that the insured may face multiple losses when a disaster occurs that could be beyond the capacity of insurers to reimburse. Providing a broad private insurance coverage for extreme risk may not be feasible at affordable prices in some cases. For these reasons, public–private partnerships for covering natural disaster risks could be explored, in which governments cover part of the extreme tail of the loss distribution (Kunreuther 2006b; Michel-Kerjan and de Marcellis-Warin 2006). In designing such partnerships, it is important that the aforementioned desirable characteristics of insurance in managing disaster risk and fostering adaptation are preserved.

1.6 Outline of this book

This book examines the impacts of climate change on the insurance sector, and how insurance can contribute to adaptation to climate change risks by providing financial compensation for natural disaster damage, and by encouraging the mitigation of risk. This book provides the first comprehensive treatment of natural disaster risk management through insurance that specifically focuses on climate change. It aims to present a comprehensive analysis of this subject by examining:

- the impacts of climate change on natural disaster risk and insurance (Chapters 2 and 3)
- problems with insuring low-probability, high-impact natural disaster risk and opportunities for public–private partnerships to overcome such problems (Chapter 4)
- natural disaster risk management and the potential of insurance to stimulate damage mitigation (Chapters 5, 6 and 7)

- individual risk perceptions (Chapter 9)
- the demand for natural disaster insurance in the face of climate change (Chapters 8, 10 and 11).

These themes are examined in this book by combining theory with case studies from the Netherlands, which are complemented with international studies and perspectives. The Netherlands is a unique case study area for research on the issues addressed in this book for several reasons. The Netherlands is very vulnerable to climate change impacts, notably to flooding, due to its low-lying delta in which several main European rivers discharge into the North Sea. Flood risk in the Netherlands is characterized as a low-probability, high-impact risk that is highly uncertain, and this characterization of risk is similar to the extreme risks posed by climate change in many other parts of the world, such as the uncertain projections of more severe hurricanes, floods and droughts. The Netherlands has a very high density of economic values that are exposed to flooding, which is currently the case in several coastal cities around the world, and is likely to reflect the situation in many coastal cities in the future, given the global trend of increased urbanization (Bouwer et al. 2007). Dutch expertise in water management has a long history and is at a very high level because the survival of the country, which for a large part lies below sea level, depends on its flood protection infrastructure. Nevertheless, climate change poses considerable challenges for Dutch water management, and innovative climate change adaptation programmes are needed in order to maintain flood risks at acceptable levels. The financial sector and the insurance sector, in particular, are very well developed in the Netherlands. Research on the potential impacts of climate change on Dutch insurers, and the potential contribution that insurance can make to climate change adaptation in the Netherlands, can provide valuable lessons for other countries. In general, the research methods and results are broadly applicable to other regions. The research results are placed in an international context, and where this is possible the topics addressed in this book are complemented with international examples and case studies from other countries in order to draw more generally applicable conclusions.

1.6.1 Research methods

A broad range of research methods are applied in this book. The potential effects of climate change on the Dutch insurance sector are first examined by reviewing the literature and from consultations with insurers. Statistical analyses are then undertaken in order to identify the relationships between climate conditions and insured hailstorm damage, in order to construct a model that is used for the extrapolation of expected future damage. Historical insurance claims were obtained from the insurance company Interpolis to build and validate these statistical models. The advantages of, and difficulties with, insuring flood risk, and experiences with flood insurance arrangements in other countries are assessed on the basis of a literature review. Modern portfolio theory is used to evaluate investments in damage mitigation and various more traditional water management investments, such as the heightening of dikes. A household survey was conducted to estimate the willingness of homeowners to invest in flood mitigation measures. A binary probit model is employed to model the determinants of this investment decision. Demand for flood insurance is estimated using three main methods: namely by applying theories of decision making under risk with parameters obtained from existing experimental evidence, and by applying contingent valuation and choice modelling. Determinants of demand for flood insurance are estimated with panel Tobit, logit and mixed logit models. Stated preference surveys were conducted to collect data for the latter two methods. In addition, surveys elicited households' perceptions of flood risk. The factors shaping these indicators of risk perception were estimated with binary and ordered probit models, as well as an ordinary least squares (OLS) regression. Data were obtained from geographical information systems (GIS) to model spatial dependencies between households' willingness to invest in damage mitigation, risk perceptions and insurance demand. The empirical case studies are, in every chapter, complemented with a review of the literature that includes international examples, in order to place the results in a broader context and to draw lessons from international practice.

1.6.2 *Chapter outline and structure of this book*

Figure 1.5 provides a schematic overview of the chapters in this book and the main relations between them. The overall picture is as follows. Climate change is expected to increase the frequency or severity of certain extreme weather events in some regions of the world, which increases natural disaster damage and affects the insurance sector, and

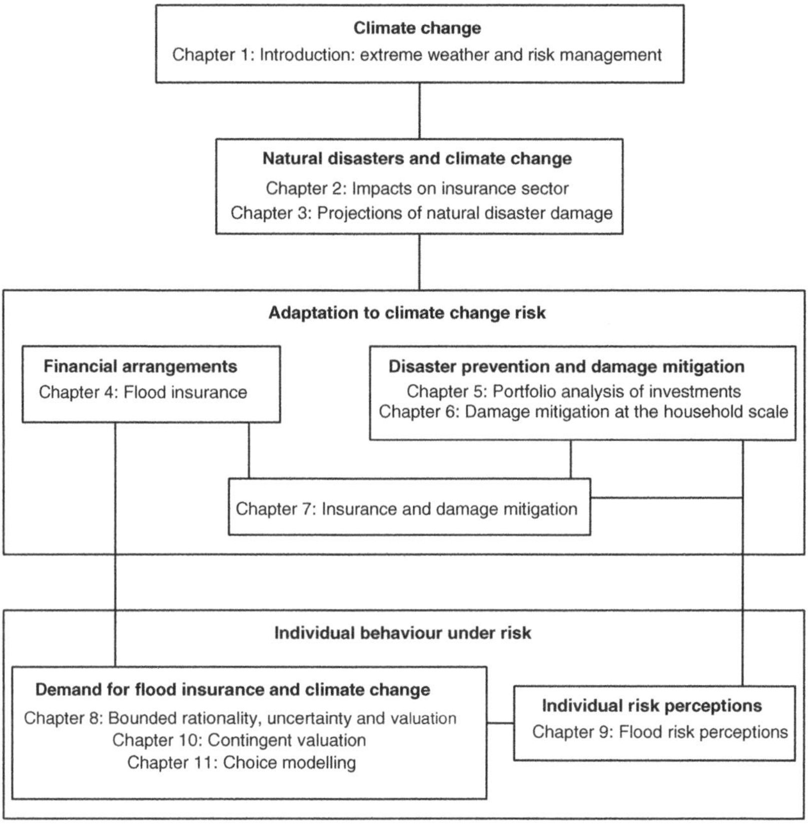

Figure 1.5 Schematic overview of the chapters in the book

this is examined in Chapters 2 and 3. The projected increase in natural disaster losses requires adaptation measures, which can consist of (public–private) insurance arrangements for disaster damage, as discussed in Chapter 4, and investments in prevention and damage mitigation measures, which are dealt with in Chapters 5 and 6. Insurance can be used to stimulate the implementation of the latter at the policyholder level, which is the theme of Chapter 7. If natural disaster risk were to increase as a result of climate change, then insurers will need to revise premiums upwards to reflect the higher risk. An important question is whether or not households are willing to pay these higher premiums, and whether or not demand for insurance is an impediment to the spreading of climate change risks through insurance. This is analysed in Chapter 8 using theories of decision making under risk, and in Chapters 10 and 11, which estimate the demand for flood insurance under scenarios of climate change by using, respectively, the contingent valuation and choice modelling methods. In this respect, individual perceptions of (climate change induced) natural disaster risk, which are addressed in Chapter 9, are an important factor of influence on both the implementation of risk mitigation measures by households and their demand for natural disaster insurance.

2 *Climate change impacts on the insurance sector*

2.1 Introduction

Weather-related catastrophe losses have been significant in the last decade, posing considerable challenges to the insurance industry worldwide (see Chapter 1). For example, between 1996 and 2005 the USA experienced the second most damaging hurricane season of the past century in terms of damage (after being normalized for inflation and wealth). Only the decade 1926–35 suffered higher damage costs due to hurricanes (Pielke et al. 2008). The extent of economic losses suffered due to a major natural disaster such as Hurricane Katrina highlights the vulnerability of modern societies to extreme weather events. Large hurricanes in the USA have shown that major disasters can have considerable impacts on insurance markets, such as operating losses for insurers that cover natural disaster risk, increasing reinsurance and primary insurance premiums, and reduced availability of insurance coverage in high-risk areas.

Problems with natural disaster insurance may become more widespread in the future as a result of climate change. It has been projected by the Intergovernmental Panel on Climate Change (IPCC 2007, 2012) that climate change may increase the frequency and severity of weather extremes, including (extreme) precipitation, the incidence of high sea levels, tropical cyclones, and also heat waves and drought in certain regions. The potential of climate change to increase natural disaster risks is especially relevant for the insurance sector and the economy at large (Vellinga et al. 2001). The insurance industry is the world's largest industry in terms of revenues, and insurers provide coverage against a large proportion of weather-related risks, although the exposure of insurers to natural disasters varies considerably between different insurers and between countries (Mills 2009a).

Parts of this chapter (in particular Sections 2.3 and 2.4) were published as Botzen et al. (2010b). These sections have been updated and extended for this chapter.

Climate change is increasingly recognized as an important issue by the insurance sector, as indicated in the report on climate change of the Geneva Association (2009) and the Insurance Working Group of UNEP (2007), but this concern is not shared by all insurers. For instance, a study of five out of ten of the largest liability insurance companies and three additional insurers[1] in the Netherlands concludes that climate change is not regarded as a worrisome issue (de Rooij and van den Bergh 2005). A review by Mills (2009a) of insurers' responses to climate change worldwide indicates that, even though many insurance companies are taking climate change seriously and react to it (albeit to very different extents), much more needs to be done in this respect. Many insurers argue that adjustments of premiums and levels of coverage are sufficient to accommodate changes in loss patterns. Indeed, the flexible nature of the industry, characterized by mostly annual contracts which allow for fairly quick adjustment of premiums, seems to assure its resilience to climate change. Nevertheless, a lack of concern may turn out to be problematic if it means that increased risks are insufficiently – and with considerable delay – factored into premiums and risk management practices. Adjusting premiums based on experiences of past claims may inadequately reflect changes in the risk of extreme weather events, due to their low-probability nature. Moreover, climate change may increase the tails of the natural disaster damage distribution, which represent extreme damages that are difficult to insure, and can increase the interdependency of natural disaster risk, which impairs the spreading of risk (Kousky and Cooke 2009).

The relevance of weather-related risks for the insurance sector becomes evident by observing past trends in insured and other economic natural disaster losses. Data on past natural disaster losses collected by Munich Re (2011) indicate that global trends of increasing losses can already be observed. The main factor behind this rise in losses has been societal change, in particular an increase in population and wealth in disaster-prone areas. Continued economic and population growth, notably in vulnerable regions (e.g. coastal zones), combined with climate change, is likely to accelerate and magnify the tempo and extent of damage during the decades to come. The influence of climate change on trends in catastrophe damage can be profound, since

[1] These additional companies are the primary insurers Euro Lloyd and SNS, and reinsurance broker Aon (de Rooij and van den Bergh 2005).

a large change in weather extremes can be expected due to a change in mean climate conditions (Dlugolecki 2006). The best strategy for insurers seems to be to take account of changes in the frequency and severity of weather extremes in assessing exposure to, and pricing and management of, natural disaster risk. In addition to commonly used traditional measures to limit risks, such as increasing premiums and limiting coverage, the insurance sector could play an important role in stimulating and promoting climate change mitigation and adaptation policies in collaboration with governments (Mills 2007).

In examining the vulnerability of the insurance sector to climate change, it is useful to assess projected changes in probabilities of extreme weather events using the outputs of regional climate models, and relate these to existing insurance arrangements for natural disasters. Climate change not only poses a threat for insurers but can also present new profitable business opportunities. Insurance demand for currently non-existent insurance products could increase in the future if economic losses from uninsured weather risks increase (see Chapter 8). However, problems with the insurability of extreme weather risks may hamper the development of markets, due to the correlated nature of these risks, the possibility of very large losses, and the uncertainty associated with the likelihood and impact of extreme weather events (Kunreuther and Michel-Kerjan 2007). Because of these difficulties, public–private partnerships have been proposed to insure catastrophe losses in the USA (e.g. Kunreuther 2006b; Kunreuther and Pauly 2006) and flood risks in the Netherlands (see Chapter 4). Such partnerships may be promising solutions for meeting demand to compensate for weather-related damages that are not currently covered by private insurance.

The remainder of this chapter is structured as follows. Section 2.2 highlights the relevance and implications of climate change for the (international) insurance sector. Sections 2.3 and 2.4 examine the potential impacts of climate change on the insurance sector by presenting a case study of the Netherlands. Observed changes in climate and climate change projections for the Netherlands are taken as a starting point in Section 2.3. Section 2.4 examines the current distribution of weather risks between the public and insurance sectors in the Netherlands and explores how these risks may be affected by climate change. The consequences of climate change for the insurance sector are assessed and alternative arrangements are explored. Section 2.5 provides an international perspective by discussing studies that have examined the implications of climate change for the insurance sectors in Canada, the UK, the USA and developing countries. Section 2.6 explores several strategies that the (international) insurance sector can adopt to deal with increasing risks from climate change, and actual responses of insurers to climate change. Section 2.7 concludes the chapter.

2.2 The relevance of climate change for the (international) insurance sector

Some of the future changes in extreme weather events projected by the IPCC can already be observed in reality. Research confirms that climate change is, at least partly, associated with the increased frequency or intensity of certain weather extremes that has been observed during the last few decades, such as heat waves, extreme precipitation, and perhaps tropical cyclones (IPCC 2007, 2012). It is therefore of interest to examine whether or not more extreme weather has already resulted in more natural disaster damage.

Weather-related loss patterns indicate an increased importance of natural disasters for the insurance sector. Global inflation-adjusted insured, and other economic, losses of natural disasters – as well as the frequency of events between 1950 and 2005 – have increased considerably, as Table 2.1 shows. It is relevant for insurance companies that insured losses also increased rapidly, which is partly due to an increase in insurance penetration. This is reflected by the *L/I* ratio, where *L* denotes economic loss and *I* denotes insured loss. This ratio has decreased from 30 in the 1950s to about 3 in the period 1996–2005, because of the development of insurance markets worldwide. The trend in the *L/I* ratio reflects an increased insurance density, notably in developed countries, where a large proportion of the monetary losses from natural disasters occurs. The development of the *L/I* ratio of natural disaster losses signifies the increased importance of extreme weather events for insurers. The proportion of losses caused by weather-related events in total disaster losses is significant. For example, storms and floods accounted for 58 per cent of economic losses and 80 per cent of insured losses between 1950 and 2004 (Munich Re 2005a). Growth trends in losses as a result of non-weather-related events have been rather constant over the past three decades (Swiss Re 2005; Vellinga et al. 2001). This could mean that weather extremes have become more

Table 2.1 ***Natural disaster losses per decade between 1950 and 2005 in billion $ (2005 values)***

Decade	1950–59	1960–69	1970–79	1980–89	1990–99	1996–2005
Number of events	21	27	47	63	91	57
Economic losses (*L*)	48.1	87.5	151.7	247	728.8	575.2
Insured losses (*I*)	1.6	7.1	14.6	29.9	137.7	176
L/*I* ratio	30.06	12.32	10.39	8.26	5.29	3.27

Adapted from Munich Re (2006).

frequent and severe, or that vulnerabilities and exposure to weather and non-weather kinds of hazards have evolved differently over time.

2.2.1 The influence of climate change on natural disaster damage in the past

Several studies have examined whether trends in natural disaster losses can be attributed to climate change (for an overview, see Bouwer 2011). In general, it is very difficult to determine the independent influence of climate change on the increase in losses reported, due to the low-probability nature of extreme events, the relatively short time span of loss records, and the various societal factors that influence disaster loss records. For instance, the effects of improved disaster risk reduction, such as flood protection, building codes and emergency measures, may potentially limit the influence of climatic factors on loss trends. Moreover, natural climate variability, such as an active hurricane season, may increase natural disaster losses. Even though natural climate variability may be consistent with climate change projections, it may not provide evidence that climate change is the causal influence, since it could reflect a short-lived, rather than a persistent, trend in natural disaster damage.

Some studies suggest that climate change may have contributed to the increased losses caused by natural disasters, due to an increased frequency and intensity of extreme weather events (Epstein and McCarthy 2004; Mills 2005; Vellinga et al. 2001). However, caution is warranted because considerable uncertainty remains about the influence of past climate change on the recent growth in losses (Changnon et al. 2000; Pielke et al. 2005). It has been shown that the rise in losses around the world is mainly due to socio-economic developments (Bouwer et al. 2007; Changnon 2003a; Crompton and McAneney 2008; Miller et al. 2008; Muir-Wood et al. 2006). A major part of the trend is attributable to hurricane damage in the USA, which does not exhibit an upward trend after normalization for societal change; in particular, changes in wealth or assets in coastal zones (Pielke 2005; Pielke et al. 2008). Socio-economic factors that have caused increased losses are population growth, growth in number of single-person households, increasing wealth, environmental degradation, industrialization in hazard-prone areas (e.g. floodplains and coastal areas), increased vulnerability of modern societies and technology, and increased concentrations of population and economic values (Berz 1999; Hoff et al. 2003). In addition, increased insurance coverage and changing attitudes towards compensation have resulted in a rise in the reporting of losses (Vellinga et al. 2001).

In order to disentangle the part of the trend in natural disaster damage that may be the result of climate change, the original economic losses caused by natural disasters should be normalized (or corrected) to reflect changes in price levels, exposure and vulnerability that are related to population and economic growth. Bouwer (2011) presents a systematic review of such normalization studies, and examines whether these studies find a trend that remains after this normalization which may be attributed to climate change. The results of Bouwer (2011) are shown in Table 2.2. It appears that most studies do not find a trend in natural disaster damage after correcting for socio-economic developments. For the few studies that do observe a trend, it was found that this trend disappears if a longer time period is considered (Miller et al. 2008; Schmidt et al. 2009b); the results have not been sufficiently corrected for changes in exposure or wealth (Chang et al. 2009; Nordhaus 2010); or there is the possibility that (other) socio-economic factors have increased losses over time which could not be fully accounted for in the normalization (Changnon 2001, 2009a, 2009b; Jiang et al. 2005). This review concludes that, based on the current empirical evidence, it cannot yet be proven that climate change has increased natural disaster damage in the past. Nevertheless, it is likely that climate change will increase natural disaster risks in the future, due to the projected increase in extreme weather events.

Table 2.2 *Review of normalization studies of historical natural disaster damage*

Hazard	Location	Period	Normalized for changes in	Remaining trend	Study
Bushfire	Australia	1925–2009	Dwellings	No	Crompton et al. (2010)
Earthquake	USA	1900–2005	Wealth and population	No	Vranes and Pielke (2009)
Flood	USA	1926–2000	Wealth and population	No	Downton et al. (2005)
Flood	China	1950–2001	GDP	Yes since 1987	Jiang et al. (2005)
Flood	Europe	1970–2006	Wealth and population	No	Barredo (2009)
Flood	Korea	1971–2005	Population	Yes since 1971	Chang et al. (2009)
Flood and landslide	Switzerland	1972–2007	None	No	Hilker et al. (2009)
Hail	USA	1951–2006	Property, insurance market values	Yes since 1992	Changnon (2009a)
Windstorm	USA	1952–2006	Property, insurance market values	Yes since 1952	Changnon (2009b)
Windstorm	Europe	1970–2008	Wealth and population	No	Barredo (2010)
Thunderstorm	USA	1949–1998	Insurance coverage and population	Yes since 1974	Changnon (2001)
Tornado	USA	1890–1999	Wealth	No	Brooks and Doswell (2001)
Tornado	USA	1900–2000	None	No	Boruff et al. (2003)
Tropical storm	Latin America	1944–1999	Wealth and population	No	Pielke et al. (2003)

Tropical storm	India	1977–1998	Income and population	No	Raghavan and Rajesh (2003)
Tropical storm	USA	1900–2005	Wealth and population	No	Pielke et al. (2008)
Tropical storm	USA	1950–2005	Asset values	Yes since 1970 No since 1950	Schmidt et al. (2009b)
Tropical storm	China	1983–2006	GDP	No	Zhang et al. (2009)
Tropical storm	USA	1900–2008	GDP	Yes since 1900	Nordhaus (2010)
Tropical storm	USA	1900–1995 1900–2005	Wealth and dwellings Wealth and population	No	Bouwer and Botzen (2011)
Flood, thunderstorms, hail, bushfires	Australia	1967–2006	Number and value of dwellings	No	Crompton and McAneney (2008)
Hurricanes, floods	USA	1951–1997	Wealth and population	No	Choi and Fisher (2003)
Hail, storm, flood, wildfire	World	1950–2005	GDP and population	Yes since 1970 No since 1950	Miller et al. (2008)

Notes: The selection criteria of Bouwer (2011) are that the study is included only if it is peer-reviewed, if it systematically analyses well-established records of natural disaster damage, and if it covers at least 30 years of monetary damage. The study by Bouwer and Botzen (2011) has been added to the table because it also meets these criteria. Adapted from Bouwer (2011).

2.2.2 Expected consequences of climate change for insurers

In the past, the major causes of increased disaster losses have been socio-economic developments, such as increases in wealth and vulnerability. In the future, a combination of societal change and climate change is likely to result in a considerable increase in insured disaster losses. Bouwer et al. (2007) have indicated that the loss potentials of the ten largest cities of the world are projected to increase between 22 per cent and 88 per cent by 2015 due solely to socio-economic developments. In addition to societal change, insurers are expected to experience changes in the location, frequency and intensity of natural disasters, for example major storms, as well as ordinary (smaller) weather events, such as local precipitation (Mills et al. 2002). The consequences will be an increasing average loss burden and greater yearly fluctuations in losses.

Even though insurance is a short-term business with mostly annual insurance policies where the policy terms can be adjusted over time, the influence on risk of long-term developments such as climate and socio-economic change can pose considerable challenges for insurers. The increasing exposures to natural disasters and concentrations of risks due to socio-economic developments increase the probability of extreme damages and drive up insurers' capital requirements. Moreover, the increasing interconnectedness of the world due to globalization increases correlations of (extreme) risks, which weakens the possibilities for diversification (Coomber 2006; Kleindorfer 2009). Climate change can further increase the probability and amount of natural disaster losses over time in addition to socio-economic change. This can potentially increase the 'fat tail'[2] of the natural disaster damage distribution, which represent extreme damages (US Climate Change Science Program 2008; Weitzman 2009). An increase in the possibility of extreme damage and fat tails can complicate insurability, since it increases the insolvency risk of the insurer (Kousky and Cooke 2009). An increase in extreme risks requires the insurance company to hold larger reserves, to purchase more reinsurance, or to hedge risks on capital markets; all of these strategies for increasing the capacity to sustain losses are costly

[2] 'Fat tail' roughly means that the probability of a loss declines slowly relative to the magnitude of the loss (Kousky 2011).

(Charpentier 2008). Moreover, the effects of climate change, such as sea level rise or increased regional hurricane activity, have implications for several types of natural disaster risks in many regions of the world. This can increase the correlations of natural disaster risks between regions as well as the correlation between risks covered by different insurers' lines of businesses (Kleindorfer 2009; Kousky and Cooke 2009). An increased dependence of risk implies a higher likelihood that large losses are suffered at the same time. Abrupt changes in climate are likely to be more problematic for insurers than gradual change. Lenton et al. (2009) have examined the impacts of tipping points on the insurance sector, and outlined several important vulnerabilities of the sector to abrupt climate change, including the effects of rapid sea level rise on increased hurricane risk, and large-scale drought, which impairs the availability of drought insurance for farmers and increases wildfire risks. Dlugolecki (2008) points out that from an asset management perspective, climate change may also have indirect consequences for an insurance company if climate change adversely affects the profitability of other companies in which the insurer has invested part of its reserves.

Changes in the frequency, severity and dependence of natural disaster risks should be accounted for in financial risk management and premium setting by insurers. Premium income may lag behind the development of claims if the changing risks due to climate and socio-economic change are not properly accounted for in calculating premiums. Merely adjusting premiums based on past experience is unlikely to fully reflect changes in risk in the face of low-probability, high-impact events (Tucker 1997). This can result in declining reserves and profitability of insurers. Moreover, if insurers take inadequate or delayed account of increased risk, then this can imply that their exposure to this risk is too high because too little reinsurance is purchased. Moreover, the capacity of insurers to assess and handle claims may be too low in that case (Dlugolecki 2008).

Insurers should assess how risks have changed, and should account for changes in natural disaster risk as a result of climate change in setting premiums and in their financial risk management, such as reinsurance purchases. In this respect, a distinction should be made between the persistent effects of a changed climate on risk and short-term natural variability in weather which affects natural disaster losses.

However, disentangling the effects of climate change and natural weather variability on risks is challenging, and the uncertainty associated with climate change will complicate the determination of accurate premiums (Charpentier 2008). A better understanding is required both of the processes by which the climate system and extreme weather changes, and of the timing of changes in risks. Knowledge about the timing of changes in risks is important because insurers need to know, for example, whether predicted changes in storm frequency have already partly materialized and need to be reflected in current premiums, or whether these changes will occur in the more distant future and premium adjustments are not yet needed. The model predictions of the influence of climate change on natural hazards and expected costs of insurance are associated with considerable parameter uncertainty, which complicates insurers' risk assessments (Doherty 1997). Moreover, risks are influenced by changes in vulnerability, which can be affected by climate change adaptation policies, such as changed building codes, which have uncertain effects on risk. These uncertainties add to the risk for insurers to provide coverage against natural disasters and may result in premiums that are much higher than the expected value of the loss, because insurers charge an 'ambiguity premium' for uncertainties in risk assessments (Kunreuther and Hogarth 1992). Ultimately, it is questionable whether or not policyholders are willing to pay these higher premiums, and this will be examined in Chapters 8, 10 and 11.

2.3 Observed climate change and projections for the Netherlands

The Intergovernmental Panel on Climate Change (IPCC) states that it is very likely that anthropogenic factors are responsible for the major part of the observed global warming since the mid twentieth century. In particular, it has been acknowledged that greenhouse gases are a significant cause of global warming, even though other factors, such as the natural variability of climate, also play a role (IPCC 2007). This section outlines the observed and projected changes in climate for the Netherlands. The findings and projections discussed are based on studies by the Royal Dutch Meteorological Institute, which used general circulation models (GCM) of climate change to derive a projection of climate change for the Netherlands (van den Hurk et al.

2006; Verbeek 2003).[3] In general, climate change in the Netherlands is in accordance with global climate change. Mean temperature in the Netherlands has risen by 1.2 °C from 1900 onwards, which is slightly higher than the observed rise in global temperature. The increase in temperature has been associated with a strong increase in the number of extremely warm days and a moderate decrease in the number of extremely cold days.[4] All of the ten warmest days since 1900 occurred after 1988 (van den Hurk et al. 2006). Yearly precipitation in the Netherlands increased by 18 per cent from 1906 onwards, which was mainly caused by increased precipitation in the winter, spring and autumn. Furthermore, extreme precipitation increased considerably during winter. The maximum '10-day precipitation sum' per winter has increased by 29 per cent since 1906. An increase in the frequency of storms has not been measured. In contrast, the number of storms gradually decreased from 1962 onwards (van den Hurk et al. 2006).

Climate change projections for the Netherlands in 2050 relative to 1990 are shown in Table 2.3.[5] Projections are made for four scenarios, including a moderate and a warm scenario in which global temperature increases by 1 °C and 2 °C, respectively, and corresponding 'plus' scenarios in which atmospheric circulation changes in addition to the temperature change. Westerly circulation is stronger in winter and easterly circulation is stronger in summer in these plus scenarios. This plus scenario is included since climate in the Netherlands is susceptible to changes in circulation patterns over Europe. Recent scientific insights into the development of global warming suggest that the warm scenarios are more likely to materialize than the moderate scenarios (KNMI 2009).

[3] These scenarios date from 2006 and are based on the latest GCM modelling exercises that have been made for the IPCC's (2007) Fourth Assessment Report (van den Hurk et al. 2006).

[4] A specific day is defined as extremely cold (warm) when the temperature on that day was lower (higher) on only 10% of the days between 1961 and 1990 (Verbeek 2003).

[5] A recent study by the Delta Committee in the Netherlands also presents climate change scenarios for longer time horizons, namely up to 2100 and 2200 (see www.deltacommissie.com). These projections are not discussed here because they provide an upper estimate of extreme climate change, and because the short-term projections up to 2050 are more relevant for the insurance sector, which has a short planning horizon.

Table 2.3 *Projected climate change for the Netherlands in 2050 relative to 1990*

	Scenario			
	Moderate	Moderate +	Warm	Warm +
Global temperature rise	+1 °C	+1 °C	+2 °C	+2 °C
Change in atmospheric circulation	No	Yes	No	Yes
Winter mean temperature	+0.9 °C	+1.1 °C	+1.8 °C	+2.3 °C
Annual coldest day	+1 °C	+1.5 °C	+2.1 °C	+2.9 °C
Mean precipitation (winter)	+3.6%	+7.0%	+7.3%t	+14.2%
Wet day frequency (winter)	+0.1%	+0.9%	+0.2%	+1.9%
10-year return level 10-day precipitation sum (winter)	+4%	+6%	+8%	+12%
Maximum average daily wind speed (per year)	0%	+2%	−1%	+4%
Summer mean temperature	+0.9 °C	+1.4 °C	+1.7°C	+2.8 °C
Annual warmest day	+1.0 °C	+1.9 °C	+2.1 °C	+3.8 °C
Mean precipitation (summer)	+2.8%	−9.5%	+5.5%	−19%
Wet day frequency (summer)	−1.6%	−9.6%	−3.3%	−19.3%
10-year return level 10-day precipitation sum (summer)	+13%	+5%	+27%	+10%
Potential evaporation	+3.4%	+7.6%	+6.8%	+15.2%
Absolute sea level rise	12–25 cm	15–25 cm	20–35 cm	20–35 cm

Source: van den Hurk et al. (2006).

Mean temperature is expected to increase between 0.9 °C in the moderate scenario and 2.3 °C in the warm plus scenario by 2050. This increases the frequency of very warm summers, while the frequency of very cold winters decreases. Average precipitation in wintertime is expected to increase between 3.6 per cent and 14.2 per cent. The expected increase in summer precipitation is smaller in the moderate and warm scenarios with, respectively, 2.8 per cent and 5.5 per cent by 2050. Average summer precipitation is even projected to decrease by 9.5 per cent and 19 per cent in the moderate plus and warm plus scenarios. Extreme precipitation is projected to increase considerably both in the winter and the summer. In contrast, the frequency of wet days decreases during the summer. Evaporation in the summer is also expected to increase significantly, especially when circulation patterns change, which increases the probability of drought. Projected changes in the maximum average daily wind speeds per year are small. The strength of the heaviest storms (which occur less than once a year) could increase, even though van den Hurk et al. (2006) indicate that this is still very uncertain.

Sea level is projected to rise between 15 and 25 centimetres for 1 °C of global warming and between 20 and 35 centimetres for 2 °C of global warming by 2050. This sea level rise is exacerbated by subsidence of land.[6] According to the Royal Dutch Meteorological Institute (van den Hurk et al. 2006), it is very likely that the rate of sea level rise will increase during the twenty-first century. The sea level rise estimates take into account melting of the Greenland and Antarctic Ice Sheets, but exclude the impact of a very rapid melting of Greenland ice or a collapse of the West-Antarctic Ice Sheet, which have a very low probability of occurring but would raise global sea levels considerably (Kerr 2006b). Recent research suggests that these ice sheets may be melting more rapidly than initially anticipated (Dowdeswell 2006; Overpeck et al. 2006). However, considerable uncertainties remain about the melting of these ice sheets. The IPCC (2007) states that the surface mass balance, which is the difference between accumulation and melting of ice, of the Greenland Ice Sheet becomes negative at global average warming in excess of 1.9 °C to 4.6 °C, but that the Antarctic Ice Sheet will remain too cold for widespread surface melting. The Greenland Ice Sheet may melt completely and subsequently contribute to sea level rise by about 7 metres if the negative surface mass balance were sustained for millennia. Obviously, this would have disastrous consequences for flood risks in the Netherlands, but it is highly uncertain.

A potential increase in flood risk is especially important for the Netherlands, which is very vulnerable to sea level rise and flooding due to its low-lying lands. Aerts et al. (2008b) have estimated the independent influence of climate change and socio-economic developments on flood risk, defined as *probability*damage*, in the Netherlands until the year 2100 using a catastrophe model. Two extremes were studied, in order to gain an insight into the effect of urban growth, on the one

[6] Land subsidence in the Netherlands is approximately 10 cm per century.

hand, and climate change on the other.[7] The effects of climate change were modelled using three sea level rise scenarios of 60 cm, 85 cm and 150 cm per 100 years, which would influence the flood probability ('Natural hazard' component in Figure 1.2). Furthermore, changes in urban development were assessed using two scenarios: namely, low economic growth (RC) and high growth (GE). Corresponding changes in the 'Portfolio of properties at risk' module of Figure 1.2 were based on a land use model for the Netherlands (Koomen et al. 2008; Loonen and Koomen 2009). The predicted changes in land use under these scenarios are shown in Figure 2.1; the projected increase in urban areas (shown in black) increases exposure to flooding. The results of the catastrophe model shown in Figure 2.2 indicate that a moderate rise in sea level of 60 centimetres results in a similar increase in flood risk as a high economic growth scenario. Climate change effects only dominate for very large increases in sea level, but flood risks are expected to increase substantially even under the low sea level rise scenario.

2.4 Risk-sharing arrangements for weather risks in the Netherlands

The projected changes in climate shown in Table 2.3 can be expected to have a considerable impact on the Dutch economy and society as well as the insurance industry. In particular, these climate change projections indicate an increase in risk for extreme precipitation, flooding and periods of drought. These projections have been taken seriously by the Dutch Union of Insurers (Verbond van Verzekeraars) who established a Commission on Climate Change in 2008 which has the objectives of examining the consequences of climate change for Dutch insurers, and of promoting adaptation to climate change risk. In particular, this Commission has investigated the impacts of climate change on insured rainfall damage, although a comprehensive in-depth study of climate change risks for insurers is still lacking (Verbond van Verzekeraars 2010a).

This section examines the main impacts of climate change on insured and uninsured weather-related losses in the Netherlands. The purpose is to identify which major weather risks are currently carried by the

[7] In this study, it is assumed that no preventative measures will be undertaken, in order to highlight the relative importance of socio-economic versus climate change (Aerts et al. 2008b).

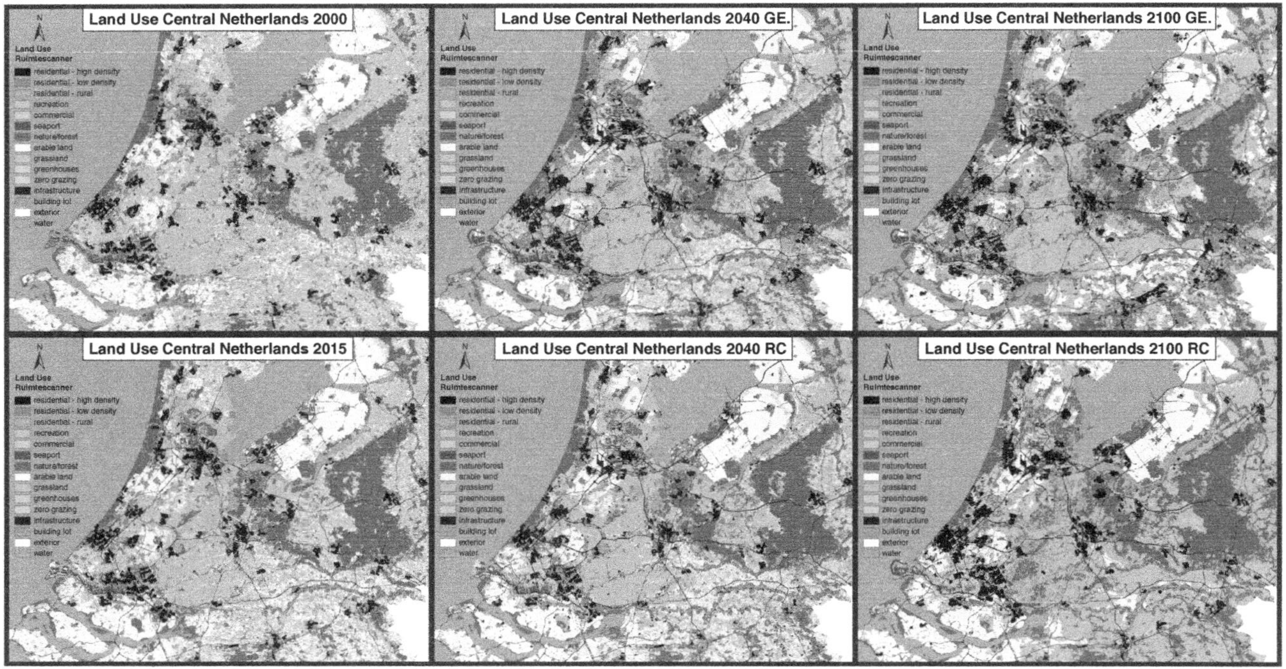

Figure 2.1 Maps that show the projected development of land use in the Netherlands according to a land use model (land use scanner) between 2000 and 2100 under a scenario of high (GE) and low (RC) economic growth

Note: Urban areas are shown in black.

Reprinted from de Moel et al. (2011) with permission from Elsevier.

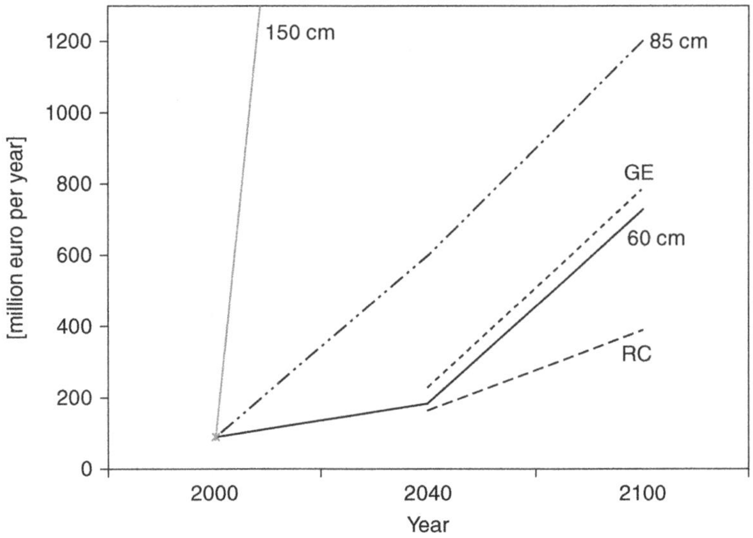

Figure 2.2 Assessment of future flood risk in the Netherlands under a range of climate change (sea level rise 60 cm, 85 cm and 150 cm) and socio-economic (GE, high economic growth; RC, low economic growth) scenarios
Source: Aerts et al. (2008b).

insurance or public sector, as opposed to households and businesses. In addition, it will be shown how existing risk-sharing arrangements may be adapted in response to climate change. This requires analysing the economic consequences of the climate change projections discussed in Section 2.3, which can subsequently be related to specific risk-sharing arrangements that are currently available in the Netherlands or that might be developed in response to changing risk. Climate change impacts will be discussed for extreme precipitation and flooding, windstorms, extreme drought and hailstorms. The main focus is on monetary (property) losses caused by weather events. Although reference will be made to the effects of extreme weather on health and life insurance, the overall impact of climate change on death rates due to extreme weather may be limited in a developed country such as the Netherlands, although health effects may be important in the case of rapid or extreme climate change. Globally, mortality rates due to extreme weather events have decreased considerably over the last century (Goklany 2007; IFRC 2003), which suggests that societies have been able to adapt to increased health risks caused by more frequent natural disasters, for example, by means of improved medical care (Goklany 2007). Of course, human adaptation to climate change may also limit the damage caused by future natural catastrophes. Several adaptation strategies are described that are specific to the weather risk discussed in this section; and general strategies that can be undertaken by the insurance sector are examined in Section 2.5.

2.4.1 Extreme precipitation and flooding

Especially relevant to low-lying countries with large rivers and a coastline, such as the Netherlands, are the economic consequences of increased (extreme) precipitation, peak discharges of rivers, and sea level rise. A distinction is made here between the risks of flooding and extreme precipitation, since different risk-sharing arrangements exist for these risks. Flood damage is defined as damage caused by the failure of primary dikes or dams that protect areas from the flooding of main rivers or lakes, which may be caused by extremely high water discharges and extreme drought, which weakens dikes, or by technical and human failures. The potentially most severe flood damage would be caused by the sea flooding due to failures of coastal defences during a storm surge. Extreme precipitation damage is defined as heavy local rainfall that can cause direct damage to property and crops, or result in an overflowing of secondary waterworks[8] in smaller rivers. Extreme precipitation can also inundate low-lying polders.

Climate change is expected to increase the risk of river flooding in the Netherlands, which can cause significant economic costs, as shown by the river floods of 1993 and 1995. The total damage caused by these floods amounted to approximately €115 and €65 million, respectively (Kok and Barendregt 2004). The Netherlands is very vulnerable to flooding, since about 70 per cent of properties lie below potential sea or river water levels (Kok et al. 2002). Private flood insurance offered by insurance companies has not been generally available since the major storm surge in 1953, implying that flood damage has to be paid for by

[8] Secondary waterworks are dikes or dams located within an area that is surrounded by primary waterworks. Secondary waterworks provide protection against smaller rivers or lakes whose water levels can be regulated by the government via water management practices. In contrast, primary waterworks provide protection against flooding from main rivers, lakes or the sea, whose water levels are difficult or impossible to regulate.

the victims of flood damage and the government. An exception is flood damage to cars, which is covered by motor insurance. Estimated losses for car insurance during the 1993 flood in the Netherlands were less than €0.5 million (van Schoubroeck 1997).

Apart from increased flood risks due to precipitation and higher peak river discharges, sea level rise may increase the intensity of storm surges, which are the main cause of sea floods in the Netherlands. Storm surges occur when water is pushed towards the shore by the force of strong winds. Rising sea levels inflict damage by accelerating shoreline erosion and may increase the likelihood of storm damage and high potential water levels during floods (West et al. 2001). The probability of storm surges may increase if rises in sea level are very extreme. This may render increased coastal protection insufficient. Flood risks are even higher if climate change were also to lead to higher wind speeds on the North Sea, but whether this will occur is very uncertain (Bouwer and Vellinga 2007). The impact of storm surges in the Netherlands can be catastrophic, both in terms of economic losses, which could be in the region of €100 billion (Kok et al. 2002), and in potential loss of life of the order of hundreds to thousands (Jonkman and Cappendijk 2006).

Flood damage can be partly compensated by the government via the Calamities and Compensation Act (WTS: Wet Tegmoetkoming Schade bij Rampen en Zware Ongevallen), which has been in force since 1998. The WTS is an ad hoc compensation arrangement for which no funds have been established, and there are no clearly predefined rules about the criteria for which cases, and by how much, flood damage will be compensated (see Chapter 4 for a more detailed discussion of the WTS). The cabinet needs to give separate approval to compensate damage from saltwater floods, which are explicitly excluded from the WTS (Nielen-Kiezebrink et al. 2005). Flood damage is paid for by the government or by private households and companies in cases where the government decides not to pay compensation.

The ad hoc arrangement via the WTS is inefficient, because eligibility for compensation is uncertain and incentives for households to limit their losses are minimal.[9] Moreover, projected increases in relief

[9] Insurance arrangements may stimulate the undertaking of damage-reducing measures at the policyholder level (Chapter 6). Such measures are effective in reducing flood damage near rivers, but may be unfeasible in very low-lying polders.

Layer 3	Government
Layer 2	Capital markets Reinsurance companies Primary insurance companies
Layer 1	Households and companies

Figure 2.3 A multilayered insurance programme

payments as a result of climate change may exert considerable pressure on public budgets. The advantages and difficulties of introducing insurance against damage from extreme weather events and flood insurance in the Netherlands, in particular, are discussed extensively in Chapter 4. This describes a public–private partnership in the form of a multilayered insurance arrangement for flood damage, as shown in Figure 2.3. The bottom layer of small losses is paid by households and companies. The private insurance sector covers the middle layer and hedges risks on capital markets, while the top layer of large losses is covered by the government in order to overcome problems of insurability of highly correlated flood risks.

Damage resulting from extreme local precipitation can also be considerable, even in the absence of flooding. For example, extreme precipitation events in 1998 caused damage of approximately €408 million, of which 80 per cent was crop damage. The government compensated about €280 million, mainly via the Calamities and Compensation Act (WTS). Only €27 million of the total damage was insured (Kok and Barendregt 2004). A large proportion of the damage was neither covered by private insurance nor compensated by the government, and was therefore carried by households and businesses. In response to these severe precipitation events, the government recommended increasing the insurability of damage resulting from extreme precipitation, in order to reduce future government relief payments (Kok and Barendregt 2004). In 2000, the Dutch Association of Insurers advised its members to include coverage for damage caused by local extreme precipitation in building, home contents and inventory (company assets) insurance.

Currently, insurance against property damage that results from severe precipitation is readily available to households in the Netherlands. About 40 per cent of the insurance companies that offer building and home contents insurance to households cover damage resulting from severe local precipitation. In contrast, only 11 per cent of the insurance companies that offer inventory, assets and building insurance to companies cover this damage (Heerkens 2003).[10] The definition of local extreme precipitation that the Dutch Association of Insurers have suggested should be included in insurance policies is precipitation near the location where damage has been caused by rainfall of at least 40 millimetres in 24 hours, 53 millimetres in 48 hours, or 67 millimetres in 72 hours. The current return period of rainfall greater than 40 millimetres in 24 hours is about 3 years (Wijngaard et al. 2005). The Royal Dutch Meteorological Institute (KNMI) has estimated that climate change may increase this return period to once in 2 or 1.5 years by 2050, depending on the climate change scenario (Groen 2007). The past development of insured extreme rainfall damage and the potential increase in insured rainfall damage according to the KNMI scenarios of climate change have been estimated by the Dutch Union of Insurers (Verbond van Verzekeraars 2010b). The portion of rainfall damage relative to the total claimed damage for home insurance was between 5 per cent and 30 per cent in the period 1987–94 and increased to between 15 per cent and 45 per cent between 1996 and 2010. This increase in precipitation damage relative to total damage cannot be explained by a decrease in non-precipitation damage. Regression analyses have been used to estimate the relations between indicators of insured rainfall damage and precipitation between 1987 and 2008 for different provinces in the Netherlands. The results show that the frequency of claims positively relates to precipitation, while the average amount claimed stays more or less constant even if precipitation is higher. The estimated regression equations have been used to forecast how much insured precipitation damage in two provinces in the west of the Netherlands can be expected to increase by the year 2050 according to the KNMI climate change projections. The results indicate that insured precipitation damage may increase by between 6 per cent and 22 per cent according to, respectively, the moderate and warm climate scenarios. It is assumed that losses increase linearly with

[10] This includes damage that results from the overflowing of waterworks and direct rainfall on buildings caused by severe local precipitation. It also covers damage that results from severe local precipitation in the wider neighbourhood of the building that enters the building due to overpressure of waterworks. Exceptions are damage that results from flooding of primary waterworks, canals and storm surges, as well as damage caused by rising groundwater.

precipitation intensity. This may, however, be a conservative estimate if a rise in frequency and severity of precipitation were to increase damage in a non-linear way. The projected percentage increases in precipitation damage are likely to have a non-negligible impact on the amount of insured losses. As an illustration, in 2008, insurance companies paid out approximately €90 million for claims on home and home contents insurance policies that were directly attributed to precipitation (Verbond van Verzekeraars 2010a), while total insured precipitation damage was about €175 million in 2003 (Kok 2006).

Crop damage caused by severe precipitation can be insured from 2004 onwards by the insurance company Agriver in the form of a public–private partnership. This insurance consists of a mutual fund that is financed by premium contributions by its policyholders. Policyholders can be obliged to increase their contributions to the fund by up to six times their normal premium if yearly payouts exceed collected premiums. The government supports this insurance by doubling premium contributions up to €50 million per year if severe losses arise that are larger than the capacity of the fund. A considerable part of the damage has to be paid for by the agricultural sector, because the deductible is 25 per cent of the total insured damage. The policy conditions for crop insurance against extreme precipitation are more restrictive than for property insurance as described above. Damage will only be paid if rainfall is more than 75 millimetres in less than 24 hours, more than 100 millimetres in less than 48 hours, or more than 150 millimetres in 96 hours. The current return period of rainfall greater than 75 millimetres in 24 hours is once in 70 years. This return period is projected to increase, due to climate change, to between once in 20 years and once in 50 years by the year 2050 (Groen 2007).

The extreme precipitation event in the south-west Netherlands in 1998 illustrates the effects of climate change on crop insurance claims. The government compensated about €140 million of the damage suffered, of which approximately €110 million (80 per cent) was crop damage (Kok and Barendregt 2004). Damage was only compensated in areas where rainfall was more than 100 millimetres per 48 hours, which is more restrictive than the current crop insurance policy. If we suppose that this damage is representative for an event that triggers the crop insurance policy, then under current climate conditions the yearly expected loss of such an event is approximately €1.6 million.[11] Climate change is expected to increase this yearly expected loss from €2.2 million up to €5.5 million in 2050, depending on the climate change scenario. Expected claims will be larger if more extreme precipitation increases damage non-linearly.

Various adaptation measures can be undertaken to partly limit the rise in risks of extreme precipitation. Improved water management could reduce the damage that results from more extreme precipitation. For example, measures that are being considered are to increase capacity to store precipitation water, to increase pumping capacity in low-lying areas, and to improve the possibility of discharging precipitation water via rivers (Aerts et al. 2008b). Such measures only limit damage partially because damage caused by the direct impact of extreme rainfall is not prevented. Agricultural crop damage, in particular, may be limited by improving farm water drainage systems or by shifting production to less vulnerable crop types and cultivation modes. Insurers could play a role in stimulating the undertaking of such damage mitigation measures via requirements in policy regulations or by providing discounts to farmers who invest in better drainage systems. However, it is noted that, in general, economic and societal developments are likely to be more important factors driving crop choice and production modes than climate change (Netherlands Environmental Assessment Agency 2005).

Changes in precipitation may further affect road safety and related insurance claims dealt with by motor vehicle insurers. Koetse and Rietveld (2009) indicate, however, that the impact of changes in precipitation on road safety in the Netherlands is ambiguous. In the summer, the number of wet days is expected to decrease, while extreme precipitation is expected to increase. The combination of these effects has an uncertain impact on accident rates. During winter, both the number of wet days and extreme precipitation are expected to increase, which may increase the risk of accidents. It has been estimated that the number of traffic accidents in the Netherlands

[11] This is computed by dividing the loss of €110 million by the return period of 70 years. This calculation is based on damage due to a single event instead of a yearly loss amount that was used in the previous computation of future claims on home insurance policies. This is done because crop insurance policies only pay out under more extreme rainfall than home insurance policies.

increases between 35 per cent and 182 per cent during precipitation (Verbond van Verzekeraars 2010a). However, traffic is generally slower when it is raining heavily, which may decrease the amount of damage caused by an accident when it occurs. Further research is required to assess the influence of the combination of these effects on accident frequency and severity, before the impact on motor vehicle insurers can be estimated.

2.4.2 *Windstorms*

The Netherlands is susceptible to windstorms, especially during the winter. Past experience with windstorms indicates that storms can inflict considerable damage on properties. For example, the major windstorm Daria in 1990 caused an estimated damage of €1.2 billion, which equalled about 0.5 per cent of the Dutch GDP at that time. Insurance companies covered approximately €0.7 billion, or 60 per cent of the total damage, which represented claims from damage to cars, private houses and businesses, as well as industrial and commercial buildings (Munich Re 1993). The insurance coverage of storm damage to property is almost universal in the Netherlands. Only government properties are not insured. Therefore the effects of climate change on wind speeds and storm frequencies are very relevant for Dutch insurers.

The climate change projections for maximum wind speed shown in Table 2.3 indicate that considerable uncertainty exists about the effects of global warming on the severity of windstorms. Maximum average daily wind speeds per year are expected to remain the same in the 'moderate' and decrease in the 'warm' scenarios, while they are expected to increase by, respectively, 2 per cent and 4 per cent in the 'moderate plus' and 'warm plus' scenarios. This increase in wind speeds in the 'plus' scenarios seems minor. However, a small change in peak wind speeds can have considerable consequences for damage. This is the case because the relation between wind speed and damage is exponential for increases in wind speed between 1 per cent and 10 per cent. In particular, Dorland et al. (1999) have estimated that a 2 per cent increase in daily peak wind speed and scenarios of population and economic growth by the year 2015 may lead to 50 per cent more storm damage. This has been estimated using a geographical and statistical storm model for the Netherlands.[12] Of this 50 per cent additional damage, only 20 per cent is due to population and economic growth, meaning that the change in climate is more important. These results indicate that the effects of changes in wind speeds in the 'plus' scenarios of climate change projections on insured windstorm losses could be substantial if daily peak wind speeds increase as rapidly as the projected maximum average daily wind speeds per year. Further research on changes in atmospheric circulation in the Netherlands is needed before any conclusive statements about the likelihood of the 'plus' scenarios and the corresponding effects on insured windstorm losses can be made.[13]

2.4.3 Extreme drought

Hot summers and periods of extreme drought are likely to become more common as a result of global warming. The shortage of rainfall experienced in the moderately dry summer of 2003 in the Netherlands currently occurs once in every 10 years, whereas it is expected to occur every 6–8 years in 2050 (RIZA et al. 2005). Water shortages can result in increased costs, reduced production and poor product quality. Impacts can be significant, especially for the agricultural sector. Crop failure occurs when summers are extremely warm. In addition, crops will be damaged when farmers do not have sufficient access to water supplies for irrigation during periods of extreme drought. For example, the 2003 heat wave in Europe threatened Dutch agriculture in the low-lying delta due to increased salinity, which makes water unsuitable for irrigation purposes. Problems with salinity could become more severe in the future in the western part of the Netherlands, due to sea level rise and land subsidence (RIZA et al. 2005). Average long-term drought damage to the agricultural sector in the Netherlands is estimated to be €150 million per year and €1.8 billion in the extremely dry years that occur once every 90 years (RIZA et al. 2005).[14]

[12] A simpler model that is based on national average data indicates that damage could even increase by 80% when peak wind speeds increase by 2% (Dorland et al. 1999).

[13] The KNMI climate change scenarios comprise daily mean wind speeds. Further research is needed to obtain insight into prolonged or sustained wind speeds (van den Hurk et al. 2006).

[14] Decreased revenues because of drought are estimated to be on average €675 million per year (using 30 years of data). These losses are calculated by

Until recently, insurance against drought damage for the agriculture sector was not available in the Netherlands, while several studies advocated developing drought insurance in collaboration with the government (van der Meer et al. 2007; van der Meulen et al. 2006). Stakeholder analyses of representatives of the agricultural sector and insurers concluded in 2007 that these parties were positive about introducing a broad weather insurance that covers drought damage bundled with other weather risks, provided that this insurance is financially backed up by the government (van der Meer et al. 2007; van der Meulen et al. 2006). Insuring drought risks is difficult in a purely private market due to a lack of statistical data to estimate loss distributions and the correlated nature of the risk. Moreover, a large deductible and/or limits on insurance coverage will be necessary in order to avoid problems of moral hazard. Moral hazard may arise if insured farmers invest less in irrigating their crops if crop damage caused by droughts is (completely) covered by insurance. A high deductible implies that farmers pay a large share of the drought damage themselves before the insurance starts paying claims, which gives farmers an incentive to limit drought damage, and minimizes the moral hazard problem. From 2010 onwards, crop damage caused by drought can be insured through a 'broad weather insurance policy' (Brede Weerpolis)[15] which can be purchased from the following mutual insurance pools: Onderlinge Weerschade Verzekering UA, Agriver, Onderlinge Fruittelers, Hagelverzekeringsmaatschappij OFH UA and Vereinigte Hagel. The deductible is high and, in the case of drought damage, only 25 per cent of the total insured amount of the damaged crops is covered by the insurance. The government supports this insurance by subsidizing premiums up to 60 per cent of the actual cost. This support is in line with the intention of the government to

subtracting actual revenues from potential revenues obtainable in a situation with optimal circumstances. However, not all losses are perceived as damage. Reductions in revenues during years with average losses are regarded as being normal by the agricultural sector. Therefore, it is better to define damage as the excess reduction in revenues relative to an average situation. Using this definition, long-term average damage amounts to €150 million per year. This damage is adjusted for positive effects of warm summers on crop yields (RIZA et al. 2005).

[15] This insurance policy also covers damage caused by other extreme weather events than drought, such as precipitation, frost, snow, ice storms, windstorms, hailstorms, erosion, and fire caused by lightning.

expand coverage of natural disaster insurance in order to prevent the need for ad hoc public compensation for disaster damage via the WTS regulation.

Insurance coverage against extreme drought damage may be beneficial for economic stability if it enables individual farmers to continue after heat spells, and contributes to the overall resilience of the agricultural sector to climate change by covering potentially catastrophic damage. In addition to insurance of crops against drought, other measures could be undertaken that foster adaptation of the agriculture sector to increased summer drought risk. For example, water storage during wet periods could be improved by creating new freshwater basins so that more irrigation water is available during dry spells. Moreover, farmers near the coast could shift to growing crop types that are less susceptible to salinity of irrigation water, even though market developments are likely to be the main driving forces behind the choice of cultivated crops (Stuyt et al. 2006).

Apart from their effects on agriculture, periods of drought also have a considerable impact on the industrial, energy and shipping sectors (RIZA et al. 2005). Less rain and increased evaporation decreases the supply of water available for cooling and shipping. The decreased availability of cooling water limits the capacity for generating electricity and industrial production. Furthermore, increased salinity of water and higher water temperatures during drought periods decrease the quality of cooling water. Ships on the major transport route Rotterdam–Germany–Basel cannot be fully loaded when discharge from the River Rhine, which is one of the most important rivers in Europe for waterway transport, drops below 1,000–1,200 m^3/s. Transport costs are higher during periods with these low water levels. The costs of the 2003 drought for the inland waterway transport sector have been estimated at €111 million (RIZA et al. 2005), or about €11.1 million expected damage per year with a return period of once in 10 years under the current climate conditions.[16] Climate change is likely to increase the average number of days per year that the Rhine discharge at Lobith is below 1,000 m^3/s from 19 days under present conditions to 26–34 days in 2050 (Middelkoop et al. 2001). As a result, future transportation costs could rise and the profitability

[16] Jonkeren et al. (2007) estimate the welfare loss of the 2003 drought as €91 million for a certain part of the Rhine market.

of shipping companies may decrease if increased costs are not fully passed on through higher prices, or risks are not fully covered by contract design. Costs are mainly passed on through higher prices, which may be undesirable for clients of shipping companies if such cost increases occur frequently in the future, due to climate change. At present, the risks of drought are carried by these sectors themselves and insurance products are lacking. It may be relevant for insurance companies to examine whether profitable insurance products could be designed to hedge these losses in extremely dry periods, especially because other adaptation options are limited.

Another adverse effect of drought is increased fire risk. The probability of fire damage to residences and company buildings, including domestic furniture and machinery, is higher in dry periods, which can increase insurance claims. Moreover, less water is available to extinguish fires during dry periods. The probability of wildfires increases during periods of drought, resulting in more forest and heath fires, as happened in the Dutch national park 'De Veluwe' during an extreme drought in 1976. Such fires potentially damage property located nearby. In recent years, several large heath fires have occurred in the Netherlands, which caused substantial destruction of natural areas. Heat waves can further result in increased daily mortality and morbidity. For example, the 2003 heat wave in the Netherlands was estimated to cause approximately 500 deaths due to heat stress, mostly among elderly people (Kabat et al. 2005). It is complicated to assess the impacts of increased illness on health insurance claims, and mortality on life insurance claims caused by droughts, compared with the effects of climate change on property claims, due to difficulties with measuring the independent effect of drought on health- and mortality-related insurance claims.

2.4.4 Hailstorms

Considerable uncertainty exists about the influence of climate change on the frequency and intensity of small-scale weather phenomena, such as hailstorms, and there are insufficient studies to assess historical changes (IPCC 2007: see p. 315). The possible consequences of climate change on hailstorm damage are, nevertheless, very relevant for insurers in the Netherlands, since such events are very common and hailstorm damage is mostly covered by insurance policies. Hailstorms can inflict damage to, for example, buildings, vehicles, caravans, greenhouses, and especially crops, which are very vulnerable (van Asseldonk et al. 2001). Moreover, severe hailstorm events may disrupt road traffic and increase the probability of car accidents. As a consequence, extreme hailstorms can result in insurance claims for vehicle, property, home contents, crop/agriculture, casualty and life/health insurers. A major hailstorm can cause considerable property damage. For example, a severe hailstorm in 1998 caused damage costing approximately €100 million in the Netherlands.

Some studies suggest that a strong positive correlation exists between hailstorm damage and minimum temperature as an approximate storm predictor in France and Switzerland (Dessens 1995; Willemse 1995). These results suggest that a rise in minimum temperatures due to global warming may increase insured hailstorm damage. The results of Dessens (1995) indicate that hailstorm damage increases by about 40 per cent when the mean minimum temperature increases by 1 °C, using French insured crop damage and temperature data between 1946 and 1992. Chapter 3 estimates the relation between a range of monthly indicators of temperature, as well as precipitation, and agricultural hailstorm damage for the Netherlands in different seasons, using insurance claims between 1990 and 2006. Separate models are estimated for hailstorm damage to outdoor farming and greenhouse horticulture. Results obtained with Tobit models indicate that climate change may increase hailstorm damage considerably, especially during the summer. In particular, estimates made under the KNMI climate change scenarios (Table 2.3) indicate that hailstorm damage to outdoor farming may increase by between 25 per cent and 48 per cent by the year 2050 and hailstorm damage to greenhouse horticulture in summer may increase by between 116 per cent and 219 per cent. In addition to increased hailstorm risk, an increased risk of lightning may be associated with more intense precipitation. However, a detailed study of the effects of climate change on insured lightning losses in the Netherlands is still lacking.

Hailstorm damage is to a large extent influenced by changes in societal factors (Changnon 2001). Losses are, for example, influenced by the amount of crops and number of greenhouses and their vulnerability. Insurers, in cooperation with the agricultural sector, could explore adaptation options that mitigate damage and are effective in terms of costs and benefits. An advantage of the undertaking of such measures is that the reduced risk will be beneficial both to the insurance companies and to the farmers. Examples of mitigation measures that could be effective in limiting hailstorm damage to the farming sector are the placement of stronger glass or plastic in greenhouses and the protection of field crops with hail nets. Farmers may also adapt to larger damage expectations, after consecutive experiences of large hailstorm damage, by switching to crops or agricultural modes that are less vulnerable to hailstorms.

2.4.5 *Summary of current risk-sharing arrangements for weather risks*

The analysis provided in this section is summarized in Table 2.4. It shows that private insurance companies cover a considerable proportion of the main climate risks. In particular, the effects of increased risks of extreme precipitation and flood damage to cars can be substantial. Furthermore, insurers cover a large part of windstorm and hailstorm damage, which may be affected by climate change even though this is still uncertain. Recently, drought insurance has become available for the agricultural sector, but high deductibles and limited insurance coverage imply that a substantial share of the potential increase in drought risk as a result of climate change is still carried by farmers. The major climate change related risk for the Netherlands is damage due to the failure of dikes, caused either by peak river discharges or by storm surge. This is the largest threat in terms of potential damage, although probabilities are relatively low. This risk of river flooding is currently carried by the government via the Calamities and Compensation Act (WTS). However, this arrangement does not provide a right to (complete) compensation, which means that it is likely that at least part of the damage caused by flood events will need to be paid for by the affected households and businesses.

The analysis so far has focused on possible increases in risk due to climate change. Fortunately, climate change can also have positive effects for the Dutch insurance sector. The probability of temperatures below zero is projected to decrease in the Netherlands. This may decrease frost damage to crops. As a result, claims from the agricultural sector can be expected to decrease, since coverage of frost damage during certain time periods is included in some crop insurance policies.

Table 2.4 ***Current risk-sharing arrangements for the main weather risks in the Netherlands***

Weather event	Climate change	Potential damage	Main damage category	Risk-sharing arrangement
River flood	Increase in risk	Large–Very large	Motor vehicles	Private insurance
			Other property	Government (WTS)
Storm surge	Increase in risk	Very large	Property	None
Extreme precipitation	Increase in risk	Large	Property households	Private insurance
			Property businesses	Limited private insurance
			Crop losses	Public–private partnership
Windstorm	Uncertain	Large–Very large	Motor vehicles	Private insurance
			Property households	Private insurance
			Property businesses	Private insurance
Extreme drought	Increase in risk	Medium–Large	Crop losses	Subsidized private insurance
			Waterway transport	None
Hailstorm	Uncertain	Medium–Large	Crop losses	Private insurance
			Motor vehicles	Private insurance
			Property households	Private insurance
			Property businesses	Private insurance

There may be adverse consequences for the insurance sector if the probability of rain at temperatures close to freezing increases, as this will increase the probability of ice storms, which occur when precipitation falls as rain and then freezes upon contact with the earth. Ice storms have adverse economic consequences, for example, because they increase the likelihood of car accidents due to slippery roads. This augments losses for insurance companies that are active in the vehicle insurance market. Moreover, frost and ice storm losses due to business interruption for the transport sector are insured in the Netherlands. Ice storms can also impose significant damage to crops. It is not clear, though, whether this scenario of increased ice storm risk is likely to occur in the Netherlands. The number of days with temperatures below zero is expected to decrease, while winter precipitation is expected to increase, which results in an ambiguous combined effect on the ice storm hazard.

2.5 International case studies of the impacts of climate change on the insurance sector

The insurance industry is a very heterogeneous sector with large regional variations in coverage, exposures to natural hazards, market penetration and country-specific regulations (Mills 2005). The roles of private and public agencies in providing compensation for natural disaster damage can differ considerably between countries as, for instance, Bouwer et al. (2007) show for flood risk in Europe. Also, climate change projections and their related impacts on natural disasters are region-specific. For this reason, the potential effects of climate change on insurance can be best illustrated using country-specific case studies, as in the preceding sections for the Netherlands. This section presents the findings of several representative studies that focus on other countries. A distinction is made between developed countries, where insurance markets have relatively high market penetration rates with broad coverage, and insurance in developing countries where, in general, insurance coverage is lower but is expected to grow considerably in the coming decades.

2.5.1 Canada

An early study of the potential impacts of climate change on the insurance sector is that of White and Etkin (1997) who explored this issue for the Canadian insurance industry by reviewing industry losses caused by extreme weather events within the context of projected changes in these hazards as a result of climate change. The authors start their analysis by providing evidence showing that natural disaster losses have increased in Canada over time, and argue that socio-economic developments have been the major cause of this increase, such as increased concentrations of population and wealth. In the past, large insured losses have been recorded for the following natural hazards: hail, tornadoes, flood, storm and wind, which are listed here in decreasing order of their share in total losses over time in Canada. Several Canadian insurance lines that are sensitive to climate change have been identified, including:

- crop insurance, which covers damages caused by hail, flood and drought
- car insurance, and in particular road accidents, which in one out of three cases are related to extreme weather
- property insurers, who may witness increasing losses from storms, hurricanes and floods
- life and health insurance, which may be affected by a higher number of weather-related diseases and deaths.

Moreover, Canadian insurers write policies outside their country in regions prone to natural hazards, such as in the USA. Part of the projected increase in damage to crops is carried by the general taxpaying population, since federal and provincial governments provide subsidies to crop insurance premiums of 50 per cent on average, and governments usually pay administrative costs and part of the losses of insurance programmes (McLeman and Smit 2006).

2.5.2 United Kingdom

A research study commissioned by the Association of British Insurers (ABI) has examined the implications of climate change for the UK (ABI 2009). In particular, this study combined climate models with risk models to assess how a change in global mean temperature of 2 °C, 4 °C and 6 °C affects the hazards of inland flooding and windstorms in the UK, how these changes in hazards influence insured losses, and how this flows through insurance prices and impacts upon capital requirements. The reason for examining the influence on risk

Table 2.5 ***Implications of climate change for insured losses, insurance premiums and capital requirements for inland flooding in the UK (in 2008 £ values)***

Temperature change	2 °C	4 °C	6 °C
Increase in average insured loss	8% (£47m)	14% (£80m)	25% (£138m)
Increase in insured loss from 1-in-100 year events	18% (£769m)	30% (£1,240m)	56% (£2,353m)
Increase in insured loss from 1-in-200 year events	14% (£832m)	32% (£1,920m)	73% (£4,346m)
Theoretical impact on insurance pricing	16%	27%	47%
Additional minimum capital required for 1-in-200 year flood	£1,065	£2,457	£5,565

Source: ABI (2009).

of scenarios of temperature rise instead of estimating the expected change in risk over time is that this removes the timing of projected changes from the discussion, which is a key uncertainty. The effects on premiums are estimated using a cost-pricing method based on average annual loss, and additional capital requirements are estimated based on the requirements imposed on insurers by the Solvency II regulation. An important implication for the UK of an increase in global mean temperature is more rain and associated inland flooding. All areas in the UK are expected to be affected by increased flood losses as a result of climate change, albeit to varying extents.

Table 2.5 shows the projected increases in flood risk in the UK as a result of climate change and GDP growth of 2.5 per cent annually, and the associated impacts on insurance premiums and capital requirements. GDP growth is used as a proxy for increasing exposure to flooding over time, due to a higher number and greater value of buildings. The results indicate a projected increase in flood insurance premiums of between 16 per cent and 47 per cent, depending on the scenario of global temperature rise. The additional minimum capital required to cover the expected increase in flood risk is substantial and varies between £1 billion and £5.6 billion. If insurers do not hold sufficient capital, then this is likely to result in reduced availability of insurance, according to the ABI. The model estimations did not show a clear relation between global temperature rise and the frequency or severity of windstorms in the UK. However, a change in cyclone tracks can be expected as a result of climate change. A modest change in storm tracks could increase average annual insured windstorm losses by 25 per cent to £827 million. According to the ABI, the projected increases in risk highlight the importance of the implementation of policies that reduce greenhouse gas emissions in combination with adaptation policies that limit the potential impacts of climate change.

2.5.3 United States of America

Mills et al. (2005) examined the potential impacts of climate change on the insurance sector in the USA, where trends of increasing natural disaster losses are of considerable concern. Insured weather-related catastrophe losses have increased faster over time than premiums or population and economic growth, as Figure 2.4 shows. The occurrence of major natural disasters has limited the availability of affordable private natural disaster insurance in some states, which has put pressure on the public sector to act as an insurer of last resort and to establish state-run or state-operated natural disaster insurance programmes (Kousky 2011). Moreover, flood insurance in the USA is, since 1968, mostly provided by the National Flood Insurance Program (NFIP), which is operated by the Federal Emergency Management Agency (FEMA).[17] This implies that part of the projected increase in natural disaster risks is carried by public insurance programmes, although a large share is covered by private insurance.

Several lines of insurance business are likely to be affected (to varying degrees) by climate change because they provide coverage of damage related to floods, windstorms, thunderstorms, hailstorms, ice storms, wildfires, droughts, heat waves, lightning strikes, subsidence damages,

[17] Most insurance policies for households and small businesses in the USA are sold via the NFIP, but commercial flood losses are often privately insured; vehicle losses due to floods are often covered under the comprehensive part of the standard auto policy; business interruption damage caused by floods is not covered by the NFIP and can only be purchased on the private market.

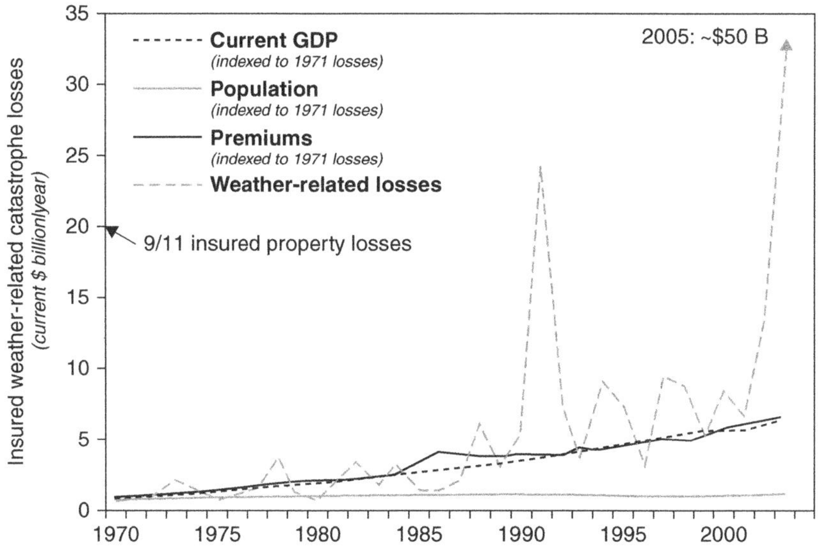

Figure 2.4 Insured weather-related catastrophe losses compared with the development of insurance premiums, GDP and population in the USA
Source: Mills et al. (2005).

coastal erosion, and the implications of all of these on human health (Mills et al. 2005). An increase in these weather-related risks because of climate change may in some cases result in reduced availability and coverage of some types of natural disaster insurance (Mills 2005). This problem may be solved by establishing public–private collaborations that provide affordable coverage for these risks; otherwise reduced coverage for households and businesses might be the consequence. Although the focus in this section is on climate change, it should be appreciated that socio-economic developments, such as increased settlement in disaster-prone areas, have a considerable influence on past and projected future natural disaster risks in the USA (Kunreuther and Michel-Kerjan 2007).

Table 2.6 gives an overview of the potential implications of climate change for the insurance sector in the USA, which is based on Mills et al. (2005). This shows:

- the projected change in specific extreme weather events and their estimated likelihood of materializing, according to the IPCC (2007)
- examples of the consequences of the climate change projections

Table 2.6 *Examples of projected changes in climate and impacts on the insurance sector*

Projected change in extreme climate phenomena and representative examples of impacts of the projected change in climate	IPCC assessed likelihood of the change	Peril or hazard	Insurance sector impacts (+ = increased losses, – = reduced losses)							Implications for insurance consumers	
			Property (buildings, industry)	Property (autos, marine, aircraft)	Liability: business interruption	Crop	Health	Life	Public insurance	More exclusions	Reduced availability
Higher maximum temperatures, more hot days and heat waves	Very likely										
• Hospitalizations, illness, mortality		Heat wave					+	+	+		
• Heat stress livestock and wildlife		Heat wave				+			+		
• Crop damage		Heat wave				+			+	✓	✓
• Soil subsidence		Subsidence	+		+				+	✓	✓
• Decreased ice in northern maritime shipping lanes		Float ice		–					+		
• Road accidents (slower reaction time)		Accidents		+					+		

• Higher energy demand (cooling) and less reliable supply		Power outage			+				+		
Higher minimum temperatures, fewer cold days	Very likely										
• Decreased cold-related human morbidity and mortality		Cold wave					–	–	–		
• Decreased damage to some crops, but more to others		Heat wave				–			–		
• Increased pest and disease vectors		Infestation	+		+	+	+	+	+		
• More avalanches		Avalanche	+		+					✓	
• Melting of permafrost		Subsidence	+		+				+	✓	✓
• More lightning		Lightning	+	+	+	+			+		
More intense precipitation	Very likely										
• Flood, landslide, avalanche and mudslide damage		Previous column	+	+	+				+	✓	✓
• Soil erosion		Rain	+	+		+			+	✓	
• Recharge of some floodplain aquifers		Flood				–			+		
• Road accidents				+					+		

Table 2.6 (*cont.*)

Projected change in extreme climate phenomena and representative examples of impacts of the projected change in climate	IPCC assessed likelihood of the change	Peril or hazard	Insurance sector impacts (+ = increased losses, – = reduced losses)							Implications for insurance consumers	
			Property (buildings, industry)	Property (autos, marine, aircraft)	Liability: business interruption	Crop	Health	Life	Public insurance	More exclusions	Reduced availability
Increased summer drying and drought	Likely										
• Lower crop yields		Drought				+			+	✓	✓
• Ground shrinkage and damage building foundations		Subsidence	+								
• Decreased quantity and quality of water resources		Drought			+	+	+		+		
• Increased risk to wildlife		Wildfire	+	+	+	+	+	+	+	✓	✓
Higher peak wind intensity tropical cyclones, and intensity precipitation	Likely										

• Property damage, business interruption, mortality, disease epidemics		Wind, disease							+	✓	✓
• Coastal erosion, damage to property near the coast		Tidal surge	+						+	✓	✓
• Damage to coastal ecosystems		Tidal surge	+				+	+	+	✓	✓
Intensified droughts and floods associated with El Niño	Likely										
• Decreased agricultural and rangeland productivity		Drought				+			+	✓	✓
• Decreased hydro-power potential		Drought			+				+		
Increased intensity of mid-latitude storms	Unknown										
• Mortality and diseases		Storm					+	+	+		
• Property and infrastructure damage		Storm	+	+	+	+			+	✓	✓
• Damage to coastal ecosystems		Storm	+	+	+				+		

Adapted from Mills et al. (2005).

- the associated expected increase (+) or decrease (–) in losses for property, crop, health, life, liability and public insurance
- implications for insurance consumers in terms of coverage and availability of insurance.

The overview shows that a wide range of insurance lines would be affected by changes in small-scale and large-scale extreme weather events, and that public insurance schemes are assumed to face an increasing exposure to almost all of the weather events considered here. Overall, the expected impacts of climate change are negative and imply increased losses, although several beneficial consequences of climate change can also be expected.

Expected changes in large-scale weather events, such as storms and floods, are obviously important for the insurance sector, given the serious potential damage that they can inflict. As an illustration, a report produced by the Association of British Insurers in collaboration with the catastrophe-modelling firms AIR Worldwide and RMS states that US losses from extreme hurricanes could increase due to climate change by $41 billion and $62 billion above present-day losses of $60–85 billion for, respectively, 1/100- and 1/250-year events, which is a 70–75 per cent increase compared with present risks (ABI 2005a). If climate change increases hurricane risks to this extent, then claims submitted to property insurers that cover wind damage may increase considerably as well as claims for flood damage. Moreover, insured business interruption losses would increase, which typically make up about 20 per cent to 40 per cent of total claims resulting from hurricanes (Mills et al. 2005).

Insured risks related to climate change are not limited to storms and floods. If temperature increases, then droughts and heat waves become more common. This increases the risk of wildfires that damage property, especially in California, and increases the probability of power outages that inflict business interruption losses. Moreover, crop losses are expected to increase due to more droughts and extreme precipitation. Crop damages are covered in the USA by a 'multi-peril' federal crop insurance programme, which includes drought and also insures livestock, while hail insurance is provided by the private market. The correlated nature of drought risk impedes insurability in private markets.[18] The crop insurance programme is a public–private partnership

[18] Farmers can receive disaster assistance for damage to crops that are not covered by crop insurance via the federal Noninsured Crop Disaster Assistance Program.

that is managed by private insurers, and the government pays a proportion of the premium and serves as an insurer of last resort in case private insurers cannot pay out claims. Crop insurance programmes have already faced severe solvency problems, which are expected to worsen in the future (Mills et al. 2005).

Changes in small-scale weather events, such as local precipitation, should receive similar attention as large disasters, since the collective impacts of losses from small-scale weather can be as, or more, significant than major disasters (Mills et al. 2005). In this respect, the automobile insurance sector may be more sensitive than people expect. Automobile damages are, on average, between 10 per cent and 55 per cent of insured losses from catastrophes, according to the Insurance Property Claim Services database, and accident risks may increase during various extreme weather conditions. The life and health insurance segment represents a large share of the insurance business in the USA, and is affected by a broad range of climate change impacts (Table 2.6). However, as has been noted by Mills et al. (2005), apart from single disasters, life insurance losses are not expected to increase considerably in the USA, but life insurance losses may increase more rapidly in emerging markets where US insurers operate.

Another implication of climate change for insurance is the potential increase in (insured) liability claims that is related to climate change. This issue has been explored by Kunreuther and Michel-Kerjan (2007) for US insurers. In the USA, shareholders and several states have initiated lawsuits against some companies for inadequately preparing for financial exposure to climate change, or for emitting too much greenhouse gases. A study by Lloyd's of London (2008) shows that more than a quarter of 183 board members surveyed expect that climate change could trigger a major wave of liability claims. If firms are being sued for the climate change damages they cause, then insurers need to pay out on the claims of firms who have liability insurance. To successfully sue firms for climate change damages is very difficult in practice, because it is almost impossible to prove causal links between a firm's activity (greenhouse gas emissions) and harm (climate change damage).[19] Even if no court verdicts are being reached,

[19] It may still be possible to sue companies if they have contributed to an increased likelihood of an adverse event related to climate change if it is impossible to attribute the specific event to climate change (Allen 2003). This also entails several practical difficulties: scientists may disagree on the extent to

substantial costs may need to be paid to settle court cases if the threat of lawsuits is sufficiently real. Moreover, insurers need to defend those firms with liability coverage for the directors and executive officers who face legal action for failure to report to their shareholders some of the climate-related risks faced by the company (Kunreuther and Michel-Kerjan 2007). Even if such legal charges prove to be invalid, the court cases are still costly for insurers because of the considerable expenses associated with them. Dlugolecki (2008) points out another type of liability case that can be related to climate change, where companies may be held liable for failing to account for climate change in their transactions. For example, an engineering company may be sued for an increase in flood losses that arise due to poorly designed flood defences that did not account for the effects of climate change on flood risk. Such cases can increase the claims on several liability insurance products, such as professional indemnity cover for consulting engineers. Faure (2007) argues that if liability from climate change were to be accepted, then insurers would want to limit this liability.

The projected increase in losses is expected to result in an increase in insurance premiums and, in some cases, exclusions of certain types of coverage and reduced availability of insurance. The results shown in Table 2.6 suggest that coverage and availability is most likely to be impaired for insurance types that are already challenging to insure privately; namely, public–private crop damage insurance; flooding insurance which is provided by a mix of public and private insurance; windstorm insurance; and subsidence and erosion-related risk. Kousky (2011) provides several recent examples of the withdrawal of insurers from regions prone to natural hazards, such as windstorm insurers in Florida.

2.5.4 Natural disaster insurance in developing countries

Although economic and insured natural disaster losses are, in absolute terms, higher in developed countries, fatality rates and economic losses expressed as a proportion of GDP are higher in developing countries

which global warming has contributed to an increased likelihood of the event; it is difficult to determine who is responsible for increasing greenhouse gases; and it is impossible to make all emitters pay for their share of increased risk (Kunreuther and Michel-Kerjan 2007).

(IPCC 2012). The effects of socio-economic developments and climate change on natural disaster damage are unequally distributed around the world (IHDP 2007). Climate change is expected to have greater economic consequences for developing countries (see, e.g., Stern 2007; Tol 2002a, 2002b), mainly because their adaptive capacity is lower and they have fewer resources to expend on an effective adaptation policy. For example, coastal cities such as Mumbai and Rio de Janeiro, in developing countries, are less protected against sea level rise than cities such as Rotterdam in the Netherlands (de Sherbinin et al. 2007). Moreover, the socio-economic effects that augment natural disaster damage, such as increased urbanization, are expected to be more profound in the developing world. As an illustration, 95 per cent of urban growth in the next two decades is expected to occur in developing countries (Sanchez-Rodriguez et al. 2007). Few studies have examined the impacts of socio-economic change on natural disaster risk in developing countries. An exception is the study by Storch and Downes (2011), who show that planned urban development in the period 2010–25 in Ho Chi Minh City is expected to increase flood risk exposure threefold compared with current levels.

Natural disasters can lead to a 'cycle of poverty', when victims who are not financially compensated take out high-interest loans (or default) or sell assets and livestock in order obtain the financial resources to rebuild damaged property and to cope with the aftermath of a disaster. Moreover, some farmers may switch to low-risk and low-yielding types of agriculture in order to reduce their vulnerability to natural hazards. These actions that reduce future income can exacerbate poverty and reduce the capacity to cope with natural hazards in the future.

Insurance and alternative risk transfer instruments could be beneficial in ameliorating the impacts of natural disasters on citizens in developing countries. Adequate compensation arrangements for disaster losses can serve as a safety net for the poor. Often, poor households are more at risk of natural disaster than richer households in developing countries, for example, because they are located in hazard-prone areas where it is cheaper to live (e.g. Brouwer et al. 2007). Poor households in developing countries commonly lack access to insurance markets, so that most of the damage caused by natural hazards is not carried by the insurance sector (Hoff et al. 2005). As an illustration, about 1 per cent of businesses and households in low-income countries, and 3 per cent in middle-income countries, have catastrophe insurance coverage, compared with 30 per cent in high-income countries (Munich Re 2005b). An explanation for this is that insurance is generally not affordable in the developing world without the support of donor organizations. The limited availability of loss data, the lack of an insurance culture in the developing world, and poor contract enforcement are other factors that impede the development of natural disaster insurance. Uninsured individuals have to rely on support from family, friends or the government when a disaster strikes, which may be less reliable than insurance, especially during large-scale disasters that affect large parts of the country, when the ability of people to help others is reduced. Alternatively, victims of natural disasters rely on international aid, which is also uncertain because it is often driven by erratic media coverage of disasters in donor countries. Well-established insurance schemes may provide more certainty of compensation and may expedite the recovery process. There is, especially, a need for crop insurance, since crop yields and losses are very sensitive to weather, while farming is an important source of income and the means to satisfy basic nutritional needs for many households in developing countries.

Innovative insurance schemes such as micro-insurance and index-based insurance could expand insurance coverage against natural disasters in developing countries. Micro-insurance programmes, typically with low premiums or coverage, can be particularly suitable for targeting the poor. Micro-insurance can cover a broad range of risks, but has mostly been used to cover health risks and crop and livestock damage (Warner et al. 2009). Recently, index-based insurance against weather-related losses has been introduced in several developing countries such as Bolivia, Malawi, India, Mongolia, Sudan and Ethiopia (Barnett et al. 2008). Affordability of insurance may be improved by index-based insurance products, which entail fewer asymmetric information problems, lower transaction costs and simpler contract designs than traditional insurance. Index-based insurance pays out if an underlying index either exceeds or falls short of a pre-specified threshold; for example, an area's average crop losses or a weather indicator such as temperature.[20] The index cannot be

[20] Other examples are area average crop yields, area average crop revenues, area livestock mortality, flood levels, cumulative rainfall, wind speeds and Richter-scale measures.

influenced by a single policyholder, so that problems with information asymmetries such as moral hazard are reduced, but the index is related to the damage that the policyholder suffers. Transaction costs are lower with index-based insurance because loss assessments at the policyholder level are not needed. A drawback of index-based insurance is the existence of 'basis risk', which occurs because the index is not perfectly correlated with losses at the policyholder level. This can lead to the situation where the policyholder experiences a loss but is not compensated by the insurance. Basis risk is too high for effective index-based insurance arrangements in heterogeneous regions with many micro-climates. A high basis risk is likely to result in a low take-up of the insurance. In regions with very heterogeneous risks, traditional insurance is more suitable.

The recent growth of insurance schemes in developing countries is promising, but Warner et al. (2009) indicate that there has been insufficient experience to demonstrate that insurance has helped the poor to escape poverty. There is both a need for increased coverage for (poor) households and businesses, and investments in disaster risk reduction that may be coupled with, or complemented by, insurance. Linnerooth-Bayer and Mechler (2006) suggest increasing the coverage of natural disaster insurance in the developing world using a two-tier strategy.[21] In the first tier, the establishment of insurance programmes for weather-related risks is promoted, which could be supported by international donors. Support from developed countries seems to be justified, given the growing recognition that developed countries are, to a large extent, responsible for the consequences of climate change in developing countries. An overall climate insurance programme could support local insurance schemes by providing technical support, reinsurance, or subsidies. Technical support could take the form of risk assessments, data collection, or assistance in setting up new insurance schemes. Subsidies should be carefully applied and should not crowd out existing (private) insurance or reduce investments in risk prevention and mitigation. Reinsurance support and subsidies should be provided in such a way that the capitalization costs of catastrophe insurance schemes are lowered, so that the insurance is affordable for consumers in developing countries (Linnerooth-Bayer et al. 2009). In

[21] A similar kind of two-tiered scheme was later proposed by the Munich Climate Insurance Initiative (Linnerooth-Bayer et al. 2009).

the second tier, disaster relief is provided to countries, dependent on whether or not they have enacted credible risk management strategies. This disaster assistance is important because, even after the implementation of the first tier, there is likely to be a large remaining part of the population in the developing world without natural disaster insurance coverage. This disaster assistance should be targeted towards high-level risks which cause damage that exceeds the ability of the country to recover from on its own (Linnerooth-Bayer et al. 2009).

Climate change may increase the demand for insurance schemes that cover weather-related losses in developing countries but also impact on the solvency of such schemes if losses increase more than reserves and the inflow of premiums can sustain. It is, therefore, important to examine the robustness of (micro-) insurance schemes to climate change. An example of such a study is provided by Hochrainer et al. (2008), who examined the robustness of index-based micro-insurance for crop damage in Malawi by combining catastrophe modelling with climate modelling. Their results show that capitalization of the insurance scheme already needs to be increased, between 2007 and 2017, in order to cope with the projected impacts of climate change. Given the serious consequences of climate change in developing countries, it is especially important to limit the increase in risk by implementing cost-effective preventative measures. There are ample opportunities to couple both of the aforementioned insurance tiers with risk management by providing incentives for the implementation of risk reduction measures, in order to reduce vulnerability to natural disasters (see Chapter 8). At present, risk mitigation is insufficiently embedded in micro-insurance schemes, as has been argued by Warner et al. (2009).

2.6 Possible and actual responses by the international insurance industry to climate change

The previous sections identified extreme weather events that may increase in frequency and severity due to climate change and are covered by insurance policies, such as extreme precipitation, hailstorms and windstorms. Adaptation strategies could limit the increase in risk. However, it is unlikely that adaptation and risk reduction will completely offset the increasing loss trends for natural disasters that have already been observed because of socio-economic developments (see Section 2.2) and are likely to increase more rapidly due to climate change. Mills (2009a) finds that mainstream insurers increasingly view climate change as an important risk to their business. A survey by PricewaterhouseCoopers (2007) of 100 insurance industry representatives in 21 countries showed that climate change is viewed as the fourth most important issue (out of the 33 listed) by insurers, while natural disasters rank second. A 2008 survey by Ernst & Young (2008) of 70 insurance analysts in various parts of the world found that climate change was ranked as the largest risk facing the industry. The major reinsurance firms, Munich Re and Swiss Re, established research departments many years ago that record natural disaster losses and analyse loss trends, and have expressed concerns about climate change.

There are several ways for insurers to respond to increased risks. Insurers can use four main traditional strategies to manage their exposure to natural hazards: they can limit their risk, adjust premiums, control the damage, and transfer the risk (see Dlugolecki 2000). Adaptation of the insurance industry to risks that change gradually will be less difficult and costly than adaptation to rapidly changing risks. At the base of all of these strategies lies a need to understand how risks have developed due to a change in hazard characteristics as a result of climate change, and changes in vulnerability and exposure due to socio-economic influences on risk. This is a challenging task, given the uncertainty associated with all of these factors of influence on risk. It is, therefore, not surprising that a recent study by UNEP-FI (2011) indicates that there is a need for more precise knowledge about the short-term impacts of climate change at the regional level for the financial services sector.

Insurers limit their risk by restricting payouts through upper and lower limits (deductibles) of liability or by imposing restrictions and exceptions in insurance contracts. For example, several restrictions and exceptions are included in insurance contracts that cover damage caused by extreme precipitation in the Netherlands. Insurers can further limit their exposure by not providing insurance against certain risks. For instance, Dutch insurance companies exclude coverage against damage caused by flooding, and some windstorm insurers in the USA have decided not to renew policies that cover wind damage in some hurricane-prone regions. These measures transfer risks to businesses and households. Excluding or limiting coverage may be regarded as undesirable since it decreases the financial security of households, which lowers the welfare of risk-averse individuals. Moreover, a reduced availability of insurance puts pressure on governments to assume a larger share of natural disaster risks by compensating for damage after a disaster or by establishing public insurance schemes.

Increasing premiums is another strategy to deal with increased risks, and has commonly been used after a major catastrophe has occurred and insurers revise their expectations of risk upwards (e.g. Kunreuther et al. 2011). A problem with raising premiums is that this often results in public and regulatory resistance and may result in a decreased market share if competition is fierce. It is, however, fundamental that insurers are allowed to charge premiums that are sufficiently high to cover expected payouts of claims, because otherwise they would sustain shortfalls in reserves and risk insolvency. Insurers' regulators should therefore follow a precautionary approach in not disallowing premium increases if these higher premiums are an accurate reflection of higher risk. Uncertainty associated with risk assessment may complicate agreements between insurers and regulators on what constitutes appropriate premium adjustments in the face of climatic influences on risk.

Strategies that may not impair the availability and affordability of insurance could also be implemented to cope with increased risks. Insurers can diversify risks by operating in various insurance branches or in several geographical areas. Further harmonization of the European financial service market may encourage robustness of the insurance sector to climate change by creating a larger geographically spread market. Also, insurance regulation and accounting standards are crucial factors that can support or hinder adaptation to climate change, but this has not yet been studied in detail. The allowance for insurance to efficiently build up sufficient catastrophe reserves to cover disaster losses is a particular example of regulation which affects the adaptability of the financial sector. Reinsurance – insurance coverage that primary insurance companies buy for their own risk – is commonly used to transfer risk. Large reinsurers are often better able to handle natural hazard risks, as they are better equipped to diversify risks geographically due to operating in large global markets. The limitations of reinsurance are that, in certain cases, prices are considerably above the expected value of losses and that the coverage obtainable is sometimes restricted after disasters have occurred. This is especially the case after very costly catastrophe events, such as major hurricanes in the USA, leading to the available capital in reinsurance markets being limited (Kunreuther 2006b). The high price of reinsurance may be passed on to consumers via higher premiums. Another option for insurance companies is to hedge risks on capital markets using alternative risk transfer (ART) instruments, such as catastrophe bonds and weather derivatives. Such financial instruments can be used to help in acquiring funds at times when an insurance company experiences large losses. The importance of these instruments in hedging disaster risk has grown substantially and is expected to grow further in the near future (Kunreuther et al. 2011; Michel-Kerjan and Morlaye 2008).

Apart from these traditional measures to manage risk, the industry, in collaboration with governments, could promote long-term policies that limit global warming and its consequences. These greenhouse gas mitigation and adaptation policies may help to maintain the insurability of extreme weather risks. This involves undertaking and stimulating greenhouse gas emission reduction policies, investing in risk assessment research, building public awareness for climate change issues and promoting adaptation strategies (Mills 1996, 2007; Mills and Lecomte 2006; White and Etkin 1997). This is likely to be a cost-effective strategy for the insurance sector, given that their exposure to weather risks is significant. The whole insurance sector could benefit from mitigating climate change, implying that there is a rationale for (global) coordinated action. And, indeed, insurance companies have for many years been at the forefront of advocating emissions reductions.

Several studies by Mills (2007, 2009a, 2009b) provide an overview of insurers' responses to climate change around the globe. Given the large diversity of the sector and the varying exposures of different insurance companies to climate change, it is not surprising that the responses of insurers to climate change vary enormously. The insurance community has increasingly accepted the science of climate change, and most agree that reducing vulnerability to weather extremes is a high priority. Nevertheless, some insurers prefer to dismiss the science or view uncertainty as a reason not to take any action, while others view uncertainty as an impetus to react. Many insurers perceive opportunities in responding to climate change. For instance, these opportunities can arise from increased demands for insurance against weather-related damage and risk transfer instruments such as catastrophe bonds and weather derivatives; a demand for insurance of new assets, such as 'green' buildings or renewable energy technology; an increased demand for micro-insurance; and new public–private partnerships which can expand existing insurance markets.

Mills (2009a) obtained information on insurers' responses to climate change from a variety of primary sources, such as corporate social responsibility (CSR) reports, for the period 1999–2008, which are augmented by interviews and a web-based survey of 100 insurance companies. Interestingly, most activities within the insurance sector are at present focused on the mitigation of greenhouse gas emissions, rather than on risk reduction. Most of these activities are done by property–casualty insurance companies, while considerably less action is taken by life–health insurance companies. The most activity has taken place in Europe (47 per cent), followed by the USA (40 per cent). Leadership is demonstrated primarily by larger companies. Even though Mills (2009a) identifies a large number of examples of responses by insurers to climate change, most initiatives are modest and a comprehensive response to climate change is lacking, although many insurers are now moving in that direction. Dlugolecki (2008) provides some examples of how insurers could respond more proactively to climate change. He argues that it is still difficult to obtain insurance coverage for renewable energy projects, which would be desirable for hedging the risk of a lack of the 'renewable' resource in the operational phase (e.g. calm conditions for wind turbines), late construction, crop failure in the biomass industry, and underperformance of new technologies (Dlugolecki 2008).

Another strategy for insurance companies to reduce weather-related losses is to stimulate the adaptation of societies to climate change impacts. This can involve the undertaking of loss-reducing measures at the policyholder level, by rewarding lower risk with lower premiums, higher coverage and lower deductibles (Freeman and Kunreuther 2003). At the least, insurers can give a price signal of natural disaster risk by the premiums they charge, as long as premiums adequately reflect risk. Moreover, providing information about how to reduce the vulnerability of properties and establishing maximum thresholds of acceptable risks may be effective policies to reduce exposure to weather extremes (R.E.T. Ward et al. 2008).

For example, insurers could recommend the use of fire- or flood-resistant building materials (Thieken et al. 2006). Strategies for promoting cost-effective investments by individuals or businesses would be beneficial to both the insurance companies and the insured parties and would promote societal adaptation to climate change broadly. However, what is striking about the findings by Mills (2009a) is that few of the efforts of insurers are targeted at adaptation, although this is beginning to receive more attention. In some cases where insurability is threatened by climate change, insurers could establish public–private partnerships in order to maintain insurability (see Chapter 4).

2.7 Conclusions

The frequency and severity of extreme weather events is likely to increase as a result of climate change, which is expected to increase insured and other economic losses caused by natural disasters. A considerable rise in worldwide natural disaster damage can already be observed during the last few decades. This chapter has shown that this increase in historical insured weather losses has been mainly caused by socio-economic developments such as increases in wealth and economic growth in areas prone to natural hazards. In future, the combination of societal change and climate change is expected to accelerate upward trends in damage caused by extreme weather events. The projected increase in natural disaster risks will have considerable impacts on the insurance sector.

An in-depth understanding of the effects of climate change on the insurance business is important for the sector in order to guarantee the adequate pricing of insurance premiums and good risk management. In particular, insurers should assess the consequences of climate change on their exposure to weather-related risks. This chapter presented several studies of the impacts of climate change on insurance in various countries: namely, the Netherlands, Canada, the UK and the USA. These studies show that a broad range of insurance arrangements can be identified that will be affected by climate change. Moreover, the public sector carries part of the risk posed by climate change through natural disaster insurance or compensation schemes that are run by the government or that are provided jointly with private insurers. Although some effects of climate change on insurance may be positive, the majority of the reported effects are negative and indicate increased risks for insurers. However, climate change can also bring new business opportunities for the insurance sector. In particular, insurers – in collaboration with the public sector – could explore the development of new insurance products for weather-related risks that are currently not covered by insurance. For example, public–private collaborations have increased insurance coverage for extreme precipitation and drought risks for the agricultural sector in the Netherlands. Especially in developing countries, where the market penetration of insurance coverage is very low, there is a need for innovative insurance solutions to provide affordable coverage against weather-related risks.

The increasing weather-related risks of insurers require the implementation of risk management strategies to handle these risks and the undertaking of strategies to reduce insurers' vulnerability. This could involve traditional measures such as limiting risks, raising premiums to reflect changed risks, transferring damage through reinsurance, or using damage control measures. Insurance companies may limit insurance coverage of natural disaster risk if they are unable to raise premiums to a level that reflects higher risk, for example because of regulations or because premiums would need to increase by so much that they would exceed the willingness-to-pay of consumers. Moreover, if climate change increases the frequency of extreme losses, then insurers may want to limit their exposure to such events if they regard their risk of insolvency through offering this cover to be unacceptable. Problems with reduced insurance availability are more likely to arise for risks that are already difficult to insure privately, such as drought risk and flood risk in certain regions. The reduced availability of insurance merely shifts risks to households, businesses and the public sector, which could be regarded as undesirable from a societal viewpoint.

Other strategies that reduce risk are also beneficial for insurance customers in the sense that they limit climate change risks and premiums of associated insurance products and do not impair insurance availability. These strategies involve the promotion of emissions reduction and adaptation measures in cooperation with public agencies. In addition, insurance companies could stimulate the undertaking of cost-effective risk-reducing adaptation measures at the policyholder level. By implementing such strategies, the insurance industry could foster the transition towards climate-resilient societies. This chapter has outlined international responses of insurers to climate change and, indeed, many insurance companies have been very active in this respect. However, the extent to which climate change is taken seriously differs considerably between different insurers. The neglect of climate change and the risks it poses for the insurance business by some could change when the sector recognizes the urgent need to assess their exposure to climate change, and when the projected increases in weather-related damage materialize.

3 *Climate change and future costs of natural disasters*

3.1 Introduction

Climate change may cause an increase in the frequency and intensity of weather extremes, which could have adverse consequences for the insurance industry, as the previous chapter has shown. The Intergovernmental Panel on Climate Change (IPCC) concluded that the historical temperature rise is mainly due to anthropogenic greenhouse gas emissions, and in the last few decades has been responsible for the observed increased frequency and severity of certain weather extremes, such as droughts and extreme precipitation. However, the current state of climate research cannot yet draw firm conclusions concerning the possible influence of global warming on small-scale severe weather phenomena such as hailstorms (Trenberth et al. 2007: p. 300) and hailstorm damage (IPCC 2012; Table 8-1 in Vellinga et al. 2001). The latter is the topic of this chapter, in which the effect of global warming on hailstorm damage in the Netherlands is examined by extrapolating historical relations between insured hailstorm damage to the agricultural sector and indicators of temperature and precipitation. Moreover, this chapter provides an overview of studies that have estimated the potential impacts of climate change on (insured) natural disaster damage for other natural hazards in a variety of countries.

Insight into the potential effects of climate change on damage caused by natural disasters is relevant for the effective design of climate policy. There is considerable uncertainty about the effects of climate change on extreme weather, which complicates the adequate inclusion of their costs in cost–benefit analyses (CBA) of climate policy (Nordhaus and Yang 1996; Stern 2007; Tol 2002a, 2002b). Nevertheless, extreme weather events, such as storms, can have a major economic impact on modern societies. This highlights the importance of considering cost estimates of

Parts of this chapter (in particular Sections 3.2, 3.3, 3.4 and 3.5) have previously been published as Botzen et al. (2010a).

potential changes in the frequency and severity of extreme weather events due to climate change in the assessment of strategies for the reduction of greenhouse gas emissions. Stringent reductions in greenhouse gases may be warranted on the basis of the precautionary principle if research shows that large increases in damage from extreme weather events can be expected as a result of a changing climate (van den Bergh 2004).

Research on the effects of climatic conditions and climate change on insured natural disaster losses is especially relevant for the insurance sector. The latter is particularly vulnerable to weather extremes, such as storms (Mills 2005; Vellinga et al. 2001). Economic losses and insured losses from natural disasters have increased rapidly over recent decades due to socio-economic developments and possibly climate change, as data collected by the reinsurer Munich Re (2009) indicates (Kunreuther and Michel-Kerjan 2007). Insurers, and especially reinsurers, may experience increased losses from climate extremes if these risks are not adequately incorporated into premiums and risk management practices (Schiermeier 2006; also see Chapter 2). The economic sectors that demand natural catastrophe insurance may also be affected by more weather-related impacts. For example, hailstorms inflict considerable damage on agriculture, but not all of these losses are insured. Moreover, if climate change increases hailstorms, then it is likely that insurers will transfer part of the higher risk back to the agricultural sector by providing less coverage or by increasing premiums. This chapter outlines several responses of insurers and farmers in adapting to a potential rise in hailstorm risk.

Data on insured losses are especially useful in assessing possible climate change impacts (Changnon 2003b), as the quality of this data is relatively good. Insurance loss data contain information about both the frequency and the severity of hailstorm damage (McMaster 2001; Vinet 2001), and climate change may affect both. However, measuring the contribution of climate change in historical loss data is complicated due to the low-probability nature of extreme events and problems with recording losses (Brooks 2006). Very few long loss records exist that allow the determination of the impact of climate change on hailstorm damage. Partly due to these problems, great uncertainty remains about the role of climate change in the upward trend of natural disaster losses in the last few decades. It is likely that socio-economic change is the major driver for the observed increase in these losses (Pielke et al. 2005). However, studies that account for socio-economic change (for an overview, see Vellinga et al. 2001) still cannot attribute specific variations in losses to climate variation or climate change (Bouwer et al. 2007). Statistical analyses continue to be necessary to assess the dependence of natural disaster losses on variations in climate variables.

A large part of this chapter examines the potential effect of climate change on future hailstorm damage. Few studies have examined the relations between climate indicators and insured hailstorm damage. For instance, Dessens (1995) and Willemse (1995) have measured simple correlations between temperature and hailstorm damage for France and Switzerland, respectively. Both studies found a positive and significant relationship between temperature and hailstorm damage, which suggests that global warming may increase hailstorm losses in the future. Here the potential change in hailstorm damage is estimated for different climate change scenarios based on historical relations between damage and a range of climate indicators. The analysis covers minimum, average and maximum temperatures, as well as precipitation. It uses data on insured hailstorm damage for the agricultural sector of the Netherlands. These data consist of aggregated monthly payouts to policyholders between January 1990 and June 2006. What are called 'out-of-sample' forecast tests are conducted in order to examine which indicators best serve as predictors of hailstorm damage.

The analysis performed here is richer than in previous studies due to the explicit modelling of time dynamics and seasonal effects. The specifics of loss data are accounted for in this study by applying Tobit estimation methods, which result in more accurate estimations than ordinary least squares (OLS) or simple correlation methods. Furthermore, separate models are estimated using observations for the whole year and only for months with high damage. This makes it possible to assess whether variation in hailstorm damage both throughout the year and in high-damage periods can be largely explained by temperature and precipitation. Moreover, a comparison of results indicates whether relations are similar in these distinct sample periods. Separate models are estimated for hailstorm damage to outdoor farming and to greenhouse horticulture, as changes in temperature and precipitation are expected to have different impacts on these sectors. Some possibilities are presented for adaptation strategies to deal with the impacts of climate change on hailstorm damage.

The remainder of this chapter is structured as follows. Section 3.2 discusses international studies on the potential impacts of climate change on convective weather. Section 3.3 provides information about insurance against hailstorm damage in the Netherlands and explains the data used in the empirical analysis. Section 3.4 analyses the estimation results for a range of models that relate hailstorm damage to climate indicators. The forecast performance of various indicators of hailstorms is tested, and the results are compared with other studies. Section 3.5 extrapolates hailstorm damage to the future, in accordance with a range of climate change scenarios. Section 3.6 discusses the economic implications of the projected increase in future hailstorm risk, as well as adaptation strategies for insurance and agricultural sectors. Section 3.7 discusses several international studies of the potential impacts of climate change on future storm and flood damage. Section 3.8 concludes the chapter.

3.2 International studies on the potential impacts of climate change on the risks of convective weather

It has been shown that temporal fluctuations in natural disasters coincide with fluctuations in weather conditions, such as temperature and precipitation. In particular, high minimum temperatures are related to increased hailstorm damage (Dessens 1995). The daily minimum temperature is usually recorded in the early morning and can be regarded as a rough estimate for the 'wet bulb potential temperature' during the following day, which reflects the atmospheric moisture content. It has been suggested that it is a useful indicator of hailstorm activity (Dessens 1995). In particular, hailstorm activity and subsequent economic losses could be greater during periods when minimum temperatures are higher. Furthermore, hailstorm activity and hailstorm damage are likely to increase in the future if average temperatures or extremes in temperatures rise due to anthropogenic climate change.

A few studies have examined the relationship between hailstorm damage and climate indicators. Simple relations between temperature and insured hailstorm damage have been estimated for countries such as France and Switzerland (Dessens 1995; Willemse 1995). Studies for Australia have used temperatures and dewpoint temperatures at different levels in the atmosphere to project potential changes in hailstorm damage with the use of climate models (McMaster 1999; Niall and Walsh 2005). Kunz et al. (2009) used a climate model to examine relations between temperature and water vapour and hail damage days in Germany. Lightning-related insurance claims have been related to temperature in the USA (Changnon and Changnon 1992; Mills 2005; Mills et al. 2002). The results of these studies will be discussed briefly below.

The relation between hailstorm damage and temperature in France has been investigated by Dessens (1995) for the period 1946–92. He estimated year-to-year correlations between insured hailstorm damage and both minimum and maximum temperatures. The insured damage was normalized with a (constructed) dimensionless loss-to-risk ratio. His analysis suggests that a 1 °C rise in summer mean minimum temperature increases hailstorm damage by approximately 40 per cent. Willemse (1995) also found that hailstorm damage in Switzerland is positively related to minimum temperatures. The meteorological study of Willemse is based on the frequency of her constructed hail-clusters – which include geographical information about areas hit by hailstorms – and the frequency of crop insurance claims per hail season and is, therefore, less suitable for assessing the possible impacts of changes in temperature for the insurance sector. The reason is that the data used do not include direct information on the amount of damage suffered, but only on the frequency of claims and the geographical characteristics of the areas hit by hailstorms.

Similar results have been found for temperature and the number of lightning-related insurance claims, indicating that high temperatures do indeed give rise to increased severity of convective weather (Mills 2005). Mills et al. (2002), using simple correlations and monthly data between 1990 and 1995, showed that lightning-related insurance losses are strongly related to average temperature in the USA. Moreover, Changnon and Changnon (1992), using simple correlations, indicated that the frequency of disastrous thunderstorms in the USA is related to mean temperatures. Trapp et al. (2009) examined the effects of anthropogenic climate change on future thunderstorm activity in the USA using a climate model, and projected an increase in the frequency of thunderstorms. Their results show, moreover, that the key factors influencing increased thunderstorm activity are low-level atmospheric water vapour and temperature. Results of correlation analyses by Kunz et al. (2009) show a positive influence on past hail damage days of water vapour and surface temperature in Germany. In contrast, studies by McMaster (1999) and Niall and Walsh (2005) used projected temperatures and dewpoint temperatures at different levels in the atmosphere from climate models, and found no significant changes in hailstorm damage for Australia.

In addition to temperature, relations between hailstorm damage and other meteorological parameters could also be established. For example, Piani et al. (2005) applied a climate model to examine hailstorm frequency and intensity in Tuscany and central Italy and projected an increase in future hailstorm frequency mainly caused by an increase in precipitable water. The relation between temperature and hailstorm damage examined in previous studies is a simplified approach, as temperature is correlated with season, and hailstorm damage is caused by intense precipitation events, which are correlated with the warm seasons of late spring, summer and early autumn. Precipitation intensity may be a suitable related parameter. The relation between precipitation intensity and hailstorm damage would be a more direct approach. Damaging hailstorms result from severe convective weather and may therefore be associated with the occurrence of intense precipitation events. It has been shown that precipitation events in the Netherlands are related to temperature (Buishand and Klein Tank 1996; Klein Tank and Können 1994). Furthermore, an anthropogenic signal has been detected in historic precipitation changes around the globe (Zhang et al. 2007). Historical and projected changes in intense precipitation events can more easily be related to temperature (van den Hurk et al. 2006) than hailstorm losses can, and intense precipitation events can possibly be related to hailstorm intensity and subsequent damage.

For several reasons there is a need to further study the relationship between hailstorm damage and climate indicators. It is of interest to estimate how hailstorm damage relates to precipitation in addition to temperature indicators, and to examine which set of indicators predict damage best. Moreover, it is uncertain whether the relations between hailstorm damage and temperature observed in the aforementioned studies are also applicable to other regions. Studies in several regions are needed before any firm conclusions can be drawn on the relations between damage and climate indicators and on the possible effects of climate change. It is also relevant to make a distinction between hailstorm damage to different farming sectors with varying susceptibility to hailstorms, such as outdoor farming and greenhouse horticulture. Finally, it is likely that a better insight into the relationship between damage and climate indicators, as well as into the temporal dynamics of damage, is obtained when more sophisticated statistical methods are used rather than simple correlation measures.

3.3 Hailstorm damage insurance and data used in the analysis

3.3.1 Hailstorm damage insurance for the agricultural sector in the Netherlands

Insurance coverage for losses caused by hailstorms is generally available for the agricultural sector in the Netherlands. Hailstorm damage insurance faces fewer supply-side failures than other types of catastrophe insurance and, as a consequence, hailstorm insurance schemes have operated successfully in a number of countries (van Asseldonk et al. 2001; van der Meulen et al. 2006). This is because correlated risks and consequent high loss accumulation are smaller for hailstorm damage than for hazards such as floods and droughts because of the greater geographical spread and variability of hailstorms and the widespread use of insurance policies. These features have facilitated the development of a private insurance market for hailstorm damage at affordable premiums. The market is privately organized without government support. In contrast, floods and droughts are more systemic risks and therefore are more difficult to insure in a private market (see Chapter 4). Moral hazard problems are minor in hailstorm insurance markets because no action can be taken to prevent hailstorms from occurring. Adverse selection problems may prevail in insurance markets if people with a high risk of loss seek out more insurance than people with an average or low risk. In the Netherlands, some areas are more prone to hailstorm damage than others. Nevertheless, problems with adverse selection in hailstorm damage insurance markets can be overcome with risk-based premium differentiation, so that adverse selection is not an impediment for the adequate functioning of the market (van Asseldonk et al. 2001). Premium differentiation is applied in the Dutch hailstorm insurance market for outdoor crops, but not for greenhouse insurance schemes. Insurance policies that cover hailstorm damage can be bought for periods of several years for both these sectors, while premiums are determined on a yearly basis.

In the Netherlands, greenhouse insurance generally provides comprehensive coverage for material damage to structures, glass, crops, and other farm buildings apart from greenhouses caused by weather events such as hail and windstorms, and non-weather events such as fire and failure of equipment. Crops grown in Dutch greenhouses include flowers, vegetables, herbs and ornamental plants. Hailstorms can inflict catastrophic damage on greenhouse-based horticulture companies. Examples of direct damage include damage to the greenhouses, which mainly consists of broken glass or plastic, but also damage to other buildings, machinery and cultivated crops. Indirect losses can be considerable as well; these comprise idle production factors, such as lost revenues because of the inability to use damaged machinery or greenhouses. Crop damage can be categorized as destruction of (part of) the crop, mechanical damage to plants (e.g. defoliation, breakage or bruising of stems), or a reduced product quality that results in lower selling prices. Greenhouse horticulture companies can buy three complementary insurance products: namely, 'greenhouse insurance', the 'standard building and machinery contract', and 'crop insurance'. Farmers can receive discounts if multiple insurance products are purchased from a single insurance company. Detailed information about the type of damage and the events causing the damage needs to be provided by farmers when they submit a claim. Damage is assessed by specialized personnel, or by mutual agreement between the farmer and insurer. Structures are insured based on their reconstruction value, and equipment on its replacement value. The payout of the insurance company is based on the percentage of the property damaged, and the indemnity is computed by multiplying this percentage by the insured value. Payouts of the greenhouse crop insurance equal reduced revenues as a result of both a lower quantity and lower quality of crops. Coverage is provided from the time of seeding to marketing. Payouts are computed based on the expected gross returns of cultivated crops. Market penetration of greenhouse insurance is high: about 99 per cent of commercial greenhouses are insured (van Asseldonk et al. 2001).

Insurance against hailstorm damage to outdoor crops is generally available in the Netherlands and can be purchased on a voluntary basis. Most crops that are susceptible to windstorm damage can be insured against storm damage in addition to hailstorm damage, and also against non-weather events such as fire and vandalism. Payouts of crop insurance equal the reduction in revenues due to both a lower quantity and lower quality of crops caused through hailstorms. Damage to plants that can be grown for more than one year is reimbursed on the basis of their replacement value. Hailstorm damage to outdoor agriculture consists, for example, of damage to arable crops, trees, ornamental plants, fruit, flower bulbs and vegetables. Hailstorms can inflict considerable damage on these crops, which can be catastrophic for an individual farmer and is, moreover, difficult to prevent. This explains why many farmers decide to purchase insurance coverage against hailstorm damage. In particular, the survey results (n = 333) of van der Meer et al. (2007) indicate that 46 per cent of arable farmers have purchased insurance against hailstorm damage for at least one of their crops.

3.3.2 Hailstorm damage data used in the analysis

The Dutch insurance company Hagelunie, which is part of the insurer Interpolis, provided the data on insured damage. Hagelunie currently serves about 76 per cent of the insurance market in the Dutch greenhouse sector, and about 56 per cent of the outdoor agricultural sector. Most of the insured losses are actual payouts to policyholders, but some are estimates of damage that are not yet paid. Three hailstorm damage series were constructed from the data: namely, hailstorm damage to greenhouse horticulture; damage to outdoor farming; and the total of these two series, which approximates hailstorm damage to the total agricultural sector.[1] Total hailstorm damage to agriculture represents payouts between January 1990 and June 2006, while the hailstorm damage to greenhouse horticulture and outdoor farming represents claims until September 2005, since after that date no information was available for the subcategorization. As a consequence, small differences exist in the number of observations used in estimating the models of total agricultural hailstorm damage and the models that allow for the subcategorization. This time period is too short to draw any conclusions about long-term trends in hailstorm damage, but the analysis of these 198 observations provides useful insights into the determinants of hailstorm damage in relation to meteorological variations.

[1] Hailstorm damage to greenhouse horticulture includes damage to greenhouses as well as damage to crops grown inside them, such as flowers, vegetables, herbs and ornamental plants. Hailstorm damage to outdoor agriculture consists of damage to arable crops, trees, ornamental plants and vegetables that are not grown in greenhouses.

The data consist of 20,242 individual claims, which have been aggregated into monthly totals. Total agricultural hailstorm damage until 2004 can be divided into three subcategories: namely, damage caused to greenhouse horticulture (18.5 per cent of total damage); to tulip bulb farming (16.25 per cent of total damage); and to other agricultural crop farming (65.25 per cent of total damage). About 8 per cent of all clients of Hagelunie are foreign. These clients could be excluded from claims made by greenhouse farmers, but not from claims made by tulip bulb farmers. Claims made by other agricultural crop farmers exclude almost all foreign clients; only about 2 per cent of total clients who are actually foreign cannot be excluded, since this information is not available. The hailstorm damage data can be regarded as being representative for the Netherlands because the share of claims by foreign farmers in total damage is very low.[2]

Over time, insured losses generally change as a result of factors unrelated to weather, such as inflation, increases in wealth, and the number of insurance policies sold. Therefore, it is common to normalize historical losses for socio-economic developments before climate conditions and change impacts are analysed (see Brooks 2006; Changnon 2007; Pielke and Landsea 1998). Equation 3.1 shows how our monthly hailstorm damage variables are normalized. Hailstorm damage was normalized using an index that has been constructed to represent changes in the total insured value of hailstorm insurance policies relative to January 1990. The purpose of this normalization is to de-trend the hailstorm damage series for changes in the value of the insured portfolio, which can be caused by changes in the price of the insured crops and greenhouses, as well as by changes in the insured value per policy and the number of current insurance policies. Changes in hailstorm damage over time reflect changes in weather after this normalization.

$$Normalized\ damage_t = \frac{Hailstorm\ damage_t}{Insured\ value\ index_t} * 100 \qquad (3.1)$$

[2] Hailstorm damage may be underestimated in our data when small farmers or farmers in regions with fewer hailstorms do not take out insurance because of transaction costs. Nevertheless, we do not expect that this has a large effect on the relations between damage and the indicators of temperature and precipitation estimated in Section 3.4, because there is no a priori reason to expect that correlations between the climate indicators and damage to insured and to uninsured farmers would be very different.

Hagelunie provided the yearly insured amounts that have been used to create the insured value index for the normalization (denominator of Eqn 3.1). The insured value index at time *t* is computed by dividing the insured amount at time *t* by the insured amount in January 1990 and multiplying by 100. In this way, the index equals 100 for January 1990, and subsequent index values represent changes compared with this base date. Setting the insured value index equal to 100 in the first month of our sample provides an intuitive interpretation of the normalized damage series, since damage in all subsequent months is normalized with respect to the insured value at the beginning of the sample.[3] Changes in damage from January 1990 onwards would then indicate changed meteorological conditions compared with that date, since changes in insured values are already accounted for.

A few observations of the insured value were missing and needed to be generated using interpolation and extrapolation. Subsequently, a monthly index was generated from the yearly data. These procedures were carried out as follows. The original insured amounts between 2004 and 2006 exclude outdoor agricultural crops. These were extrapolated using a growth of 2 per cent per year, which has been the trend growth in insured amounts of outdoor agricultural crops from 1981 onwards. The insured amounts from 2001 to 2003 are not available and were estimated by assuming a constant linear growth between 2000 and 2004. Finally, monthly series were estimated by assuming a linear growth from month to month. The latter is consistent with the insurance company that provided the data (Interpolis), indicating that no abrupt changes in the value of insured amounts occurred during the period under investigation.

The natural logarithm of normalized hailstorm damage is taken for two reasons.[4] First, this reduces estimation problems with large monthly variations in hailstorm damage. Second, it allows for a semi-elasticity or elasticity interpretation of the regression coefficients (Section 3.4). The time series plot of the logarithm of total normalized hailstorm damage to agriculture between 1990 and 2006 is shown in Figure 3.1.

[3] For example, the insured value is approximately twice as high at the end of the sample than in January 1990, i.e. the insured value index increased to about 200. Hailstorm damage at the end of the sample is then divided by a factor 2 to normalize damage for the observed increase in insured value.

[4] The logarithm of damage has been set equal to zero in months without hailstorm damage.

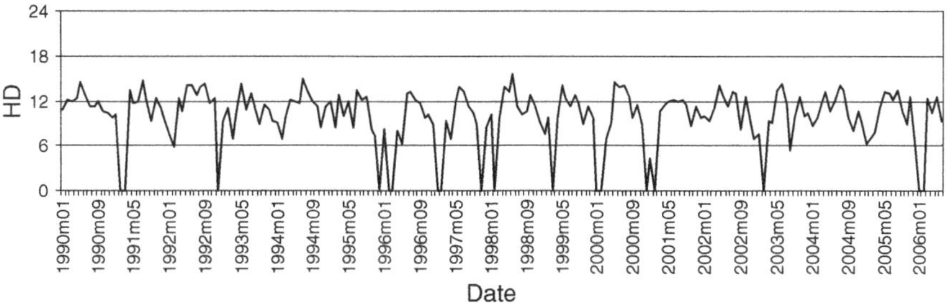

Figure 3.1 Logarithm of the normalized hailstorm damage to agriculture between January 1990 and June 2006

No clear trend is apparent from it. Several severe hailstorms occurred in the period under investigation. A hailstorm with normalized insured agricultural losses of more than €1 million occurred 12 times during this period. The largest monthly losses were suffered in July 1991, July 1994 and June 1998, with insured losses of about €2.5 million, €3.2 million and €5.9 million, respectively. Months with zero hailstorm damage occurred 18 times, as indicated by the troughs in the figure.

Total agricultural hailstorm damage can be separated into hailstorm damage to greenhouse horticulture and to outdoor farming. Equation 3.1 was used to construct hailstorm damage series for these separate sectors.[5] Figure 3.2 shows the logarithm of normalized insured hailstorm damage to greenhouse horticulture, to outdoor farming and to the total of these two types of farming per month between January 1990 and September 2005. For example, the damage amounts in January represent the sum of damage in every month of January to greenhouse horticulture (Inside), to outdoor farming (Outside) and to the total of these two (Total). From the figure it is apparent that the total annual damage to outdoor farming was larger than damage to greenhouse horticulture. Damage to greenhouse horticulture is concentrated in the months of May, June, July and August (Figure 3.2) because of the greater severity of hail in these months. Only severe hailstorms with large hailstones can break the glass of greenhouses and inflict damage on greenhouse horticulture operations. Peaks in hailstorm damage to outdoor farming can also be observed between

[5] The same index was used to normalize insured losses to greenhouse horticulture and outdoor farming, assuming that changes in the total insured values were the same for both insurance types.

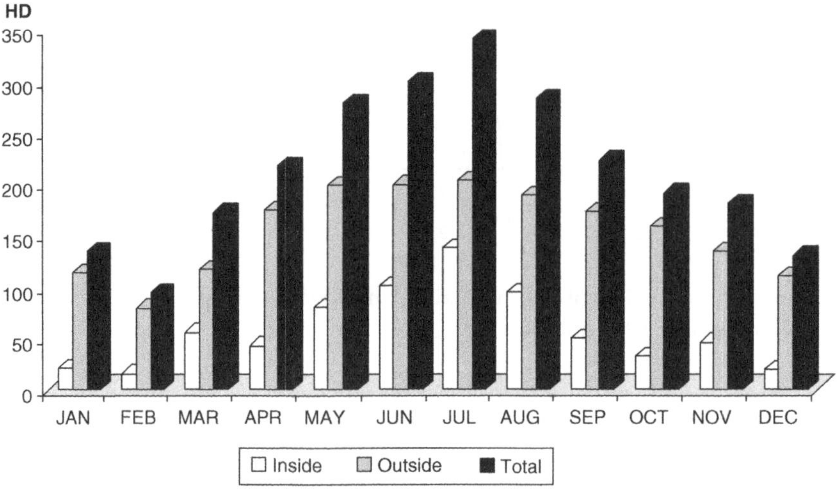

Figure 3.2 Logarithm of the normalized hailstorm damage to greenhouse horticulture (Inside) and to outdoor farming (Outside), summed per month for the period January 1990 to September 2005

May and August, indicating that patterns are rather similar, but considerable damage also occurs in April and September. The seasonal pattern can be explained by two factors. First, in the Netherlands, most heavy hailstorms with large hailstones occur in the summer months. Second, the potential damage is also larger in the spring and summer months because most crops are grown during that time.

Three dependent variables of hailstorm damage will be used in the subsequent analyses: namely, hailstorm damage to greenhouse horticulture, to outdoor farming and to the sum of these.[6] The variables are defined as follows:

- HD_{inside} logarithm of monthly normalized hailstorm damage to greenhouse horticulture
- $HD_{outside}$ logarithm of monthly normalized hailstorm damage to outdoor farming

[6] Augmented Dicky–Fuller tests were applied to the temperature and hail damage variables in order to test for non-stationarity. No clear trend is apparent in these data series. Lags are included when significant. The null hypothesis that these variables are integrated of order 1 can be rejected at the 1% level for all variables. This implies that no problems of spurious regression are to be expected when these variables (levels) are included in the analyses.

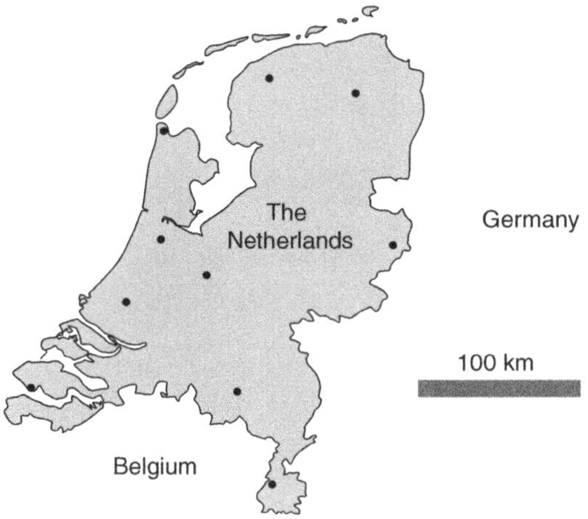

Figure 3.3 Map of the Netherlands with the locations of the ten meteorological stations

- HD_{total} logarithm of monthly normalized hailstorm damage to both greenhouse horticulture and outdoor farming.

3.3.3 *Temperature and precipitation data used in this study*

Long time series of recorded temperature and precipitation are available from the Royal Netherlands Meteorological Institute (KNMI) for the ten main meteorological stations in the Netherlands.[7] The locations of the weather stations are shown in Figure 3.3. Monthly means of minimum, average and maximum daily temperatures, as well as monthly total precipitation, which are representative for the Netherlands as a whole, were estimated by taking the averages of data obtained from these ten stations. Descriptive statistics for these temperature and precipitation measures are provided in Appendix 3A (Table 3.A1). The four climate variables used in the analysis are then defined as follows:

- T_{min} monthly mean minimum temperature
- $T_{average}$ monthly mean average temperature
- T_{max} monthly mean maximum temperature
- *Prec* monthly total precipitation.

[7] Data are available from http://www.knmi.nl/klimatologie/maandgegevens/index.html.

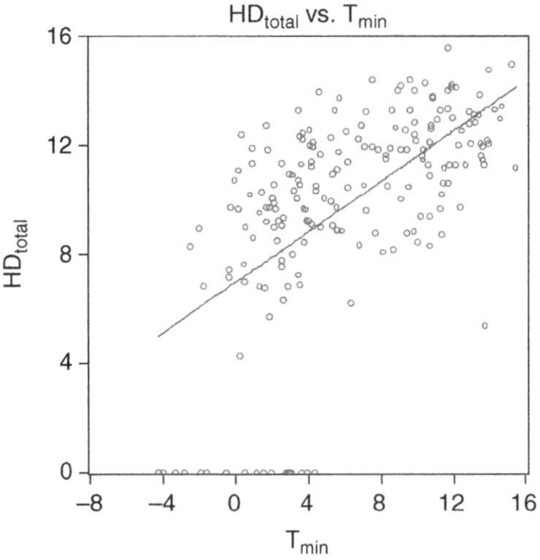

Figure 3.4 Scatter plot of total agricultural hailstorm damage, HD_{total}, versus minimum temperature

Note: The line represents an OLS regression with a constant, a slope parameter of 0.46 (significant at the 1% level) and an R^2 of 0.34.

3.3.4 Main relations between hailstorm damage and temperature

A first indication of the relation between temperature and hailstorm damage can be obtained by examining the scatter plots of the damage variables and temperature. For this purpose, scatter plots of minimum temperature and the HD_{inside}, $HD_{outside}$ and HD_{total} variables are shown in Figures 3.4 and 3.5. Minimum temperatures are used in constructing these plots, because other studies have found that they are a good indicator of hailstorms (Dessens 1995; Willemse 1995). Nevertheless, correlations with hailstorm damage were found to be similar for the $T_{average}$ and T_{max} variables (see Appendix 3A for the simple correlation statistics). The scatter plot of the logarithm of total agricultural hailstorm damage (HD_{total}) shown in Figure 3.4 indicates a strong positive relation between hailstorm damage and minimum temperature, as is apparent from the large positive coefficient and the R^2 value. Minimum temperatures can explain about 34 per cent of the variation in damage.

The scatter plot of hailstorm damage to greenhouse horticulture (HD_{inside}) shown in Figure 3.5 (left) indicates a positive relation between hailstorm damage and minimum temperature. The explanatory power

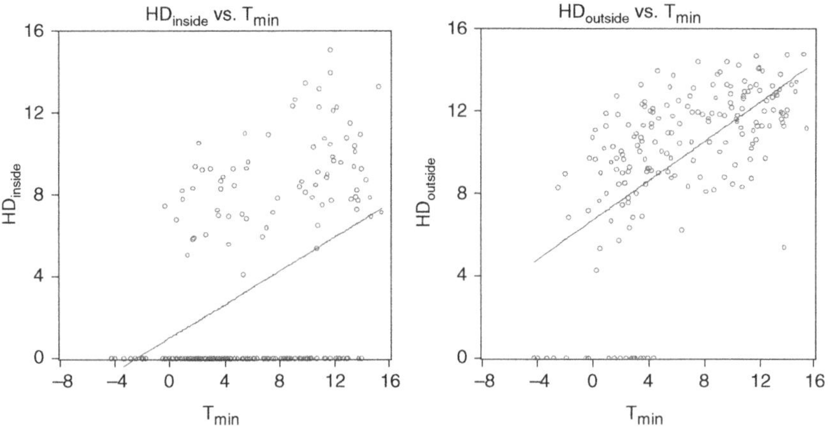

Figure 3.5 Scatter plots of hailstorm damage to greenhouse horticulture, HD_{inside} (left), and to outdoor farming, $HD_{outside}$ (right), versus minimum temperature

Note: The lines represent OLS regressions with a constant, a slope parameter of 0.41 (significant at the 1% level) and an R^2 of 0.18 for HD_{inside}, and a slope parameter of 0.48 (significant at the 1% level) and an R^2 of 0.35 for $HD_{outside}$.

of this model is lower than of the model with total agricultural hailstorm damage, as reflected by a decrease in the adjusted R^2 to 0.16, which may be caused by the larger number of observations with zero damage. The slope coefficient reduces slightly to 0.41 in the regression with HD_{inside} compared with 0.46 in the regression with HD_{total}. The scatter plot of the logarithm of hailstorm damage to outdoor farming ($HD_{outside}$) versus minimum temperature is also shown in Figure 3.5 (right). The plot indicates a strong positive relation between hailstorm damage and minimum temperature, which is reflected by a slope coefficient of 0.48 and an R^2 value of 0.35.

3.4 Estimation results

3.4.1 Tobit estimation results of hailstorm damage models

Statistical model and estimation method

The simple analysis in the previous section suggests that a positive relation exists between minimum temperature and hailstorm damage.

This is further examined by estimating alternative models for the different damage and temperature variables, which also include precipitation. Time dynamics in hailstorm damage are explicitly modelled in order to account for autocorrelations that exist in the data and for seasonal dependence of damage. Autocorrelations may be present in hailstorm damage if the damage is larger (smaller) in several subsequent months because the value of crops growing in fields is higher (lower). Figure 3.2 indicates that hailstorm damage is larger in the spring and summer seasons. This implies that it is relevant to model the seasonal dependence of damage. For this purpose, three dummy variables *SPRING*, *SUMMER* and *FALL* are included in the analysis. The *SPRING* dummy takes the value 1 in the months March, April and May, and 0 otherwise. The *SUMMER* dummy takes the value 1 in the months June, July and August, and 0 otherwise. The *FALL* dummy takes the value 1 in the months September, October and November, and 0 otherwise. Estimations are performed for two sample periods. First, several models are estimated using observations for the whole year. Then these models are estimated for the seasons in which hailstorm damage is concentrated.

The insured hailstorm damage is analysed using the following model:

$$\begin{aligned} HD_t &= \beta_0 + \beta_1 * SPRING_t + \beta_2 * SUMMER_t + \beta_3 * FALL_t \\ &\quad + \beta_4 * HD_{t-1} + \beta_5 * Temperature_t + \beta_6 * LN(Prec_t) + \varepsilon_t \end{aligned} \tag{3.2}$$

where HD_t is the hailstorm damage variable HD_{total}, $HD_{outside}$ or HD_{inside}; $Temperature_t$ is the temperature variable T_{min}, $T_{average}$ or T_{max}, depending on the specification; the natural logarithm of monthly precipitation, $LN(Prec_t)$, is included; and ε_t is the error term. The seasonal dummies capture the variation in hailstorm damage across seasons, and the temperature and precipitation variables capture their influence on monthly variations in hailstorm damage, independent of the seasonal correction of the dummies. The model includes one lag of the dependent variable. Higher lags are not included because they proved to be insignificant. Hailstorm damage is in logarithmic form so that the coefficients on temperature have a semi-elasticity interpretation, and the coefficient on the logarithm of precipitation has an elasticity interpretation.

Hailstorm damage in a particular month is either zero or positive, so that the dependent variable in Eqn 3.2 is censored. An ordinary least squares (OLS) regression yields biased coefficient estimates if the data is censored, since $E[y_t] \neq x_t'\beta$. This problem can be overcome with the Tobit model (Tobin 1958). A formal representation of the Tobit model is given in Appendix 3B. An advantage of applying the Tobit model to the insurance data compared with OLS estimation is that unbiased coefficients of the relation between hailstorm damage and the indicators of temperature and precipitation can be obtained.[8] Moreover, the effect of temperature and precipitation variability on expected hailstorm damage can be separated with the Tobit model into changes in the frequency of observing hailstorm damage and changes in the size of losses in months where hailstorm damage occurs. In this way, the Tobit model allows for a breakdown of the total effect of a change in the climate variables into a damage frequency and a damage size effect.

The distribution of the error terms in the Tobit model (see Appendix 3B) is usually chosen in accordance with the normal distribution (Heij et al. 2004). Huber–White robust standard errors are estimated using quasi-maximum likelihood to allow for possible misspecification of the likelihood function (Huber 1967; White 1982). The marginal effects in the Tobit model are not equal to the coefficient values. The marginal effects are $\frac{\partial E[y_t]}{\partial x_t} = \Phi(x_t'\beta / \sigma)\beta$, where Φ is the cumulative normal density function. The mean marginal effect (MME) of the jth explanatory variable is computed as:

$$\beta_j \frac{1}{n} \sum_{t=1}^{n} \Phi(x_t'\beta / \sigma). \tag{3.3}$$

The formula below shows that the effect of a temperature change on hailstorm damage can be broken down into a change in the probability of observing damage and a change in the amount of damage suffered (McDonald and Moffitt 1980):

$$\frac{\partial E[y_t]}{\partial x_t} = \frac{\partial P[y_t > 0]}{\partial x_t} E[y_t \mid y_t > 0] + P[y_t > 0] \frac{\partial E[y_t \mid y_t > 0]}{\partial x_t}. \tag{3.4}$$

The fraction of the MME due to a response above the limit, i.e. due to a change in damage in months with generally positive hail-

[8] All models are estimated with ordinary least squares (OLS) regressions and robust standard errors. Overall, the results and conclusions are similar to those of the Tobit models and are therefore not reported here.

storm damage under current conditions, is calculated using the formula:

$$\left[\begin{array}{l} 1-\left(\frac{1}{n}\sum_{t=1}^{n}\frac{x_t'\beta}{\sigma}\right)\frac{1}{n}\sum_{t=1}^{n}f\left(\frac{x_t'\beta}{\sigma}\right)/\frac{1}{n}\sum_{t=1}^{n}F\left(\frac{x_t'\beta}{\sigma}\right)-\left(\frac{1}{n}\sum_{t=1}^{n}f\left(\frac{x_t'\beta}{\sigma}\right)\right)^2 \\ /\left(\frac{1}{n}\sum_{t=1}^{n}F\left(\frac{x_t'\beta}{\sigma}\right)\right)^2 \end{array}\right] \qquad (3.5)$$

where f and F are, respectively, the normal and cumulative normal densities.

Forecast performance of hailstorm damage indicators

The ability of the different climate indicators to predict hailstorm damage is examined by performing out-of-sample forecast tests in order to find the best model specification. This is a commonly used method to examine the ability of regression models to predict values of the dependent variable. It works as follows. First, 1,000 random subsamples are generated containing 80 per cent of the total number of observations. Second, Tobit models of total agricultural hailstorm damage (HD_{total}), outdoor farming ($HD_{outside}$) and greenhouse horticulture (HD_{inside}) are estimated on these subsamples, including the seasonal effects, a lagged damage term and a range of climate indicators. Separate models are estimated with only one of the following climate indicators: namely, precipitation, minimum temperature, average temperature and maximum temperature. In addition, models are estimated that include both precipitation and one of the temperature variables (Eqn 3.2). Third, the estimated models are used to forecast outcomes $\overline{y_t}$ for the 20 per cent of total observations that were excluded from the subsamples. For each subsample the root mean squared error (RMSE) is calculated to test forecast performance:

$$\text{RMSE} = \sqrt{\frac{\sum_t^n (\overline{y_t} - y_t)^2}{n}}. \qquad (3.6)$$

Finally, the average RMSE values of the 1,000 random subsamples are computed for each model specification. Table 3.1 shows the results of the forecast tests for the different hailstorm damage models. The three columns provide the average RMSE estimates for models with the dependent variables HD_{total} (left column), $HD_{outside}$ (middle column) and HD_{inside} (right column). The rows in the table represent the various climate indicators included as explanatory variables.

Table 3.1 ***Root mean squared error (RMSE) for models of total agricultural hailstorm damage, HD_{total}, hailstorm damage to outdoor farming, $HD_{outside}$, and to greenhouse horticulture, HD_{inside}, for a range of climate indicators***

Model with climate indicators	RMSE forecast of HD_{total}	RMSE forecast of $HD_{outside}$	RMSE forecast of HD_{inside}
T_{min}	2.9505	3.0041	4.2366
$T_{average}$	2.9567	3.0071	4.2440
T_{max}	2.9650	3.0160	4.2480
LN (Precipitation)	2.7567	2.8631	4.1854
T_{min} and LN (Precipitation)	2.7545	2.8492	4.2057
$T_{average}$ and LN (Precipitation)	2.7408	2.8302	4.1983
T_{max} and LN (Precipitation)	2.7342	2.8221	4.1926

The RMSE estimates sketch a picture that is consistent for the models of total agricultural hailstorm damage, damage to outdoor farming and damage to greenhouse horticulture. Minimum temperatures are a better indicator for hailstorm damage than average and maximum temperatures when the only climate indicator included in the model is one of the three temperature variables. This is reflected by a lower RMSE for models with T_{min} than for the models with $T_{average}$ and T_{max} (comparison of the first three rows in Table 3.1). Models that include only precipitation as a climate indicator have lower forecast errors than models that include only temperature. This implies that precipitation is a slightly better predictor of hailstorm damage than temperature. Forecast errors of damage are lowest when both temperature and precipitation are included in the model. An exception is the model for damage to greenhouse horticulture, where the mean RMSE slightly increases when a temperature variable is added to a model with only precipitation; however, this difference is statistically insignificant.[9]

[9] We conducted pair-wise Welch tests to examine whether the mean RMSE values of the models are statistically different from one another. These tests use the mean RMSE as well as the standard deviation of the 1,000 model runs. The results indicate that the mean RMSE of the models with only precipitation is significantly lower than the mean RMSE of the models with only one of the temperature variables. The models with maximum temperatures and

Interestingly, the models with maximum temperatures and precipitation have the lowest RMSE overall and are, therefore, the best for predicting hailstorm damage. Thus, while minimum temperatures are better indicators of hailstorm damage in models without precipitation, maximum temperatures are better indicators in models with precipitation. An explanation for this result may be the higher correlation between precipitation and minimum temperatures than between precipitation and maximum temperatures (see Table 3.A3 in Appendix 3A). In other words, maximum temperatures provide more complementary information to precipitation when explaining and predicting hailstorm damage.

Estimation results of models using the observations of the whole year

The hailstorm damage models were analysed for the two subcategories of hailstorm damage: damage to greenhouse horticulture and to outdoor farming. The estimation results of Eqn 3.2 with minimum temperatures (traditional model) and maximum temperatures and precipitation as climate indicators (best prediction model) for these categories are shown in Table 3.2. The first two columns of the results relate to the models with $HD_{outside}$ as the dependent variable and represent outdoor farming. The last two columns show the results of the models with HD_{inside}, which represent greenhouse horticulture. The main results for total agricultural damage are similar to the results of models of hailstorm damage to outdoor farming. *F*-tests indicate that the coefficients of the models of greenhouse horticulture (HD_{inside}) and outdoor farming ($HD_{outside}$) are significantly different (at the 1% level), which means that the specification of the pooled model of total agricultural damage is not appropriate. Therefore, only the results of the separate models are reported here, and the results of the pooled model (HD_{total}) are shown in Appendix 3D.

The first and second columns of the estimation results in Table 3.2 relate to the models estimated with minimum temperatures (the traditional model) and with maximum temperatures and precipitation (the best prediction model) as climate indicators. The results are robust to

precipitation result in significant lower RMSE values than the models with only precipitation for total agricultural hailstorm damage and damage to outdoor farming, while differences are insignificant for damage to greenhouse horticulture.

various specifications. Appendix 3D contains the results of the models with minimum temperatures and precipitation, and with average temperatures and precipitation. The number of censored observations, that is, observations with zero hailstorm damage, is given in the third last row of Table 3.2. The overall fit of the best prediction models is good, as is reflected by the R^2 of 0.7 for the model of outdoor farming, and 0.5 for the model of greenhouse horticulture (Appendix 3B gives the formula used to compute R^2).

Next, the results of the models of hailstorm damage to outdoor farming are discussed. The regression results indicate that hailstorm damage is significantly higher in the spring months. The first lag of damage is positive and statistically significant at the 1% level. This implies that hailstorm damage in a specific month is larger when a lot of damage was suffered in the previous month.[10] An explanation for this is that months with high exposure to hailstorm damage tend to follow each other. If many crops are grown in several consecutive months, for example because of a high demand for agricultural products or favourable growing conditions, then damage caused by hailstorms may be relatively large during such a period. In other words, if an above-average value of crops is grown in a certain period then exposure to hailstorms is relatively high as well, which would be reflected by higher damage in consecutive months. This effect is not fully captured by the seasonal effects, resulting in a positive and significant coefficient of the lagged damage variable. The coefficient of temperature is significant at the 1% level in both the traditional and the best prediction models. Precipitation is significant at the 1% level in the best prediction model.

The results for the models with hailstorm damage to greenhouse horticulture differ from the previous findings. The first lag of damage is statistically insignificant, which suggests that no relation exists

[10] Similar results are obtained with models that include monthly dummies instead of seasonal dummies. Moreover, models are estimated that include lagged weather variables, as was suggested by a reviewer. It may be that bad weather in previous periods weakened crops so that they were more likely to be damaged if subsequent hailstorms occurred. Lagged weather variables may capture this effect. However, the results of models with 1-month lags, as well as with 1- and 2-month lags of the weather variables, indicate that the lagged weather variables are statistically insignificant. The lagged dependent variable remains positive and significant (at the 1% level) and the size of the coefficient is about the same as in the models without lagged weather variables.

Table 3.2 ***Tobit results for hailstorm damage to outdoor farming, $HD_{outside}$ (left columns) and hailstorm damage to greenhouse horticulture, HD_{inside} (right columns) of the traditional model and the best prediction model***

	Dependent variable: $HD_{outside}$		Dependent variable: HD_{inside}	
Variables	Traditional model	Best prediction model	Traditional model	Best prediction model
Constant (β_0)	2.8646***	-5.8377**	-8.1820***	-27.1353***
	[1.0911]	[2.8891]	[2.0240]	[6.5070]
SPRING (β_1)	2.8318***	3.1553***	6.0560**	7.2618***
	[0.9800]	[1.0203]	[2.6549]	[2.8024]
SUMMER (β_2)	1.0248	1.3911	9.1075**	9.0109**
	[1.3661]	[1.2057]	[4.2157]	[4.2271]
FALL (β_3)	0.9444	0.8748	3.4763	2.9250
	[1.0911]	[0.9722]	[3.2055]	[3.0219]
$HD_{outside}$ $(t-1)$ (β_4)	0.3584***	0.3363***	n.a.	n.a.
	[0.1006]	[0.0971]		
HD_{inside} $(t-1)$ (β_4)	n.a.	n.a.	0.0988	0.1077
			[0.1594]	[0.1596]
T_{min} (β_5)	0.3205***	0.2041***	0.3165	0.2337
	[0.1134]	[0.0740]	[0.3164]	[0.2448]
LN (*Prec*) (β_6)	n.a.	1.9594***	n.a.	4.2921***
		[0.5467]		[1.4259]
Observations				
Total	188	188	188	188
Censored	19	19	111	111
R^2	0.58	0.72	0.38	0.49
Log likelihood	-460	-450	-335	-331

Notes: Significance levels are indicated as follows: *** indicates significance at the 1% level, ** indicates significance at the 5% level, and * indicates significance at the 10% level. n.a. stands for not applicable. Coefficients are stated without brackets and Huber–White standard errors are given in brackets below the coefficients.

between hailstorm damage suffered in a specific month and the previous month. This may be explained by the larger concentration of hailstorm damage to greenhouse horticulture in the spring and summer months, which is reflected by the seasonal dummy variables.

Table 3.3 ***Mean marginal effects for the Tobit model for hailstorm damage to outdoor farming,*** $HD_{outside}$ ***(left columns) and hailstorm damage to greenhouse horticulture,*** HD_{inside} ***(right columns) for the traditional and the best prediction models***

	Dependent variable: $HD_{outside}$		Dependent variable: HD_{inside}	
Mean marginal effect (MME)	Traditional model	Best prediction model	Traditional model	Best prediction model
Temperature	0.32***	0.20***	0.21	0.15
LN (Precipitation)	n.a.	1.92***	n.a.	1.89***
Fraction of MME due to response above limit	0.93	0.94	0.48	0.48

Notes: Significance levels are indicated as follows: *** indicates significance at the 1% level, ** indicates significance at the 5% level, and * indicates significance at the 10% level. n.a. stands for not applicable.

Moreover, the financial exposure to hailstorms is rather constant from one month to another, since there is little monthly variation in the stock of greenhouses. In contrast, for outdoor farming, months with a large exposure, i.e. when there are many crops grown on fields, tend to follow each other. This may explain why the first lag of hailstorm damage is significant in the model for damage to outdoor farming and not significant in the model for damage to greenhouse horticulture. The relation between temperature and hailstorm damage to greenhouse horticulture is positive but statistically insignificant. A positive and significant relation can, however, be observed between hailstorm damage to greenhouse horticulture and precipitation in the best prediction model.

Table 3.3 shows the mean marginal effects (MME) of the temperature and precipitation variables for the models of damage to outdoor farming ($HD_{outside}$) and to greenhouse horticulture (HD_{inside}). The mean marginal effects of the model for hailstorm damage to outdoor farming are close to the coefficient values, since the scale factor is 0.99 for both the traditional and the best prediction models. The results indicate that hailstorm damage to outdoor farming is expected to increase by approximately 20 per cent if temperatures are 1 °C higher for the traditional model, while this increase is 30 per cent according to the best prediction model. The MME indicates that hailstorm damage increases by about 1.9 per cent if precipitation is 1 per cent higher. Computation of the fraction of the MME due to a response above the limit (Eqn 3.5) indicates that 94 per cent of an increase in hailstorm damage due to a rise in temperature and precipitation is caused by a greater amount of damage in the months with positive damage. This implies that higher temperatures and more precipitation mostly result in larger total agricultural hailstorm damage in months where damage was already expected to occur, while increases in the probability of observing hailstorm damage in months with no damage under current conditions will be negligible.

The mean marginal effects (MME) for the models of hailstorm damage to greenhouse horticulture (HD_{inside}) are considerably smaller than the coefficient values, due to a scaling factor of 0.65. The MME indicates that hailstorm damage increases by approximately 15 per cent if the temperature is 1 °C higher according to the best prediction model, but this effect is statistically insignificant. The MME indicates that hailstorm damage increases by about 2.79 per cent if precipitation is 1 per cent higher. The fraction of MME due to the response above the limit is 48 per cent for the model of hailstorm damage to greenhouse horticulture. This implies that the effect of higher temperatures and more precipitation will be caused almost equally by a higher probability of observing hailstorm damage in months with no damage under current conditions and by higher hailstorm damage in months where hailstorm damage was already expected to occur.

Estimation results of models using the high-damage months April to September

Relations between hailstorm damage and indicators of temperature and precipitation are also analysed for the months of April to September only, when most of the damage occurs. A similar time period has been adopted by Kunz et al. (2009) for analysing hail losses in Germany. The purpose of this analysis is to examine whether variations in hailstorm damage in these months with relatively large amounts of damage can also be explained by temperature and precipitation. The model estimated is similar to Eqn 3.2, except that the lagged damage term is excluded because the dependent variable is no longer continuous over time. Obviously, the *FALL* dummy is excluded, since the winter months and the last two autumn months are not included in this subsample.

The estimation results for models of the two categories of hailstorm damage to greenhouse horticulture (HD_{inside}) and to outdoor farming ($HD_{outside}$) are shown in Table 3.C1 in Appendix 3C. The results for outdoor farming indicate that both temperature and precipitation have a positive and significant effect on hailstorm damage when only the high-damage months are considered, confirming the robustness of our earlier findings. The marginal effects (Eqn 3.3) are approximately equal to the coefficient values. The marginal effects of temperature of the traditional and the best prediction models, respectively, indicate that hailstorm damage is 31 per cent and 24 per cent higher if temperature rises by 1 °C. The marginal effect of the best prediction model is about 4 per cent higher than that of the model that uses observations for the whole year. The marginal effect of precipitation shows that hailstorm damage is expected to increase by 1.35 per cent if precipitation increases by 1 per cent, which is slightly less (0.57 per cent) than given by the model with observations for the whole year (1.92 per cent). Similar results are obtained with the model of total agricultural hailstorm damage (see Appendix 3D).

The results of the models for HD_{inside} indicate a considerably stronger and statistically significant response of hailstorm damage to temperature in the sample with only the high-damage months compared with the estimation results based on the whole sample. The scale factor of the marginal effects (Eqn 3.3) is 0.73 for the traditional and the best prediction models. The marginal effects indicate that a rise in temperature of 1 °C is associated with a rise in hailstorm damage to greenhouse horticulture of 109 per cent and 83 per cent according to the traditional and the best prediction models, respectively. The marginal effect on precipitation increases slightly to 3.39 per cent.

In summary, rather similar relations are found between hailstorm damage to outdoor farming and precipitation, and between such damage and temperature, when the sample is restricted to the high-damage months. This indicates the robustness of the earlier results. In addition, the model of hailstorm damage to greenhouse horticulture using the restricted sample provides relevant new insights. In particular, the results show that a strong positive relation exists between hailstorm damage to greenhouse horticulture and temperature during the high-damage months, while this relation is considerably smaller when observations for the whole year are used. These findings are robust to various specifications with different temperature indicators, as shown in Appendix 3D.

3.4.2 Comparison of results with international studies

There are few studies that are directly comparable with our results. Overall, our results confirm findings from other countries that temperature is a statistically significant indicator of hailstorm or thunderstorm activity (as discussed in Section 3.2). The only statistical study that we are aware of with correlation estimates of insured hailstorm damage and temperature is that of Dessens (1995). He found that hailstorm damage in France is more strongly related to minimum than to maximum temperatures, while the correlation statistic with minimum temperature indicates a 40 per cent increase in hailstorm damage for a 1 °C temperature rise. From these results, Dessens (1995) concluded that minimum temperature is a better storm predictor than maximum temperature, and that hailstorm damage is expected to rise if climate change causes higher minimum temperatures. The conclusion that hailstorms are better correlated with minimum temperatures is consistent with the findings of Willemse (1995) for Switzerland.

Our correlation analysis presented in Section 3.3.4 is similar to the method used by Dessens (1995) and indicates a 46 per cent increase in total agricultural hailstorm damage for a 1 °C temperature rise. This is very close to the 40 per cent increase in total hailstorm damage predicted by Dessens (1995). Moreover, the forecast results in Section 3.4.1 show that minimum temperature is a better predictor of hailstorm damage than maximum temperature, as long as precipitation is not included as an explanatory variable. This confirms the findings of Dessens (1995), who did not include precipitation as an additional variable. Our richer model, which incorporates time dynamics and precipitation as explanatory variables, provides new insights in addition to the findings of previous studies. In particular, precipitation is a useful predictor of hailstorm damage in conjunction with temperature. The superiority of minimum temperature as a predictor of hailstorm damage over maximum temperature is not observed when precipitation is an additional explanatory variable; the combination of maximum temperature and precipitation actually results in the best forecast performance. Moreover, our results indicate that relations between temperature and hailstorm damage are different in months with high damage compared with the rest of the year, and vary between outdoor and greenhouse farming. As a consequence, climate change impacts vary between seasons and affect these sectors differently, as the next section shows.

3.5 Extrapolations of hailstorm damage using climate change scenarios

The estimation results in the previous section indicated that historical damage from hailstorms is positively related to temperature and precipitation. Over the past 50 years, average monthly mean, minimum and maximum temperatures increased by, respectively, 1.2 °C, 1.0 °C and 1.3 °C in the Netherlands.[11] Climate change projections indicate that temperature and precipitation levels could increase further in the future. As a consequence, hailstorm damage may change according to the relations found in the statistical analysis in Section 3.4, provided that other things remain the same.

The extrapolations of damage presented in this section are based on the specification of the regression models (Eqn 3.2) and on the assumption that the relations between hailstorm damage and indicators of temperature and precipitation would remain the same in the future. The extrapolations are made with the traditional model that includes minimum temperature as a climate indicator, and with the best prediction model with maximum temperature and precipitation as climate indicators. It is noted that relations between hailstorm damage and temperature and precipitation constitute a simplification of the meteorological cause of hail severity. Other parameters, such as atmospheric moisture content, vertical instability and the depth of hail-producing events, as well as atmospheric circulation patterns, also determine the occurrence of hailstorms (McMaster 1999; Niall and Walsh 2005). Nevertheless, we have confidence in our results, given the reasonable explanatory power of our models in Section 3.4.

Climate change projections of temperature and precipitation for the Netherlands provided by the Royal Netherlands Meteorological Institute (KNMI) for the year 2050 relative to 1990 are shown in

[11] According to our analysis of changes in 15-year averages between 1951–65 and 1992–2006, based on data from http://www.knmi.nl/klimatologie/maandgegevens/index.html.

Table 3.4 *Climate change projections for mean precipitation and temperature, according to KNMI scenarios for 2050*

KNMI scenario	Moderate	Warm
Global temperature rise	+1 °C	+2 °C
Yearly changes		
Mean precipitation	+3.4%	+6.5%
Temperature	+0.9 °C	+1.8 °C
Summertime changes		
Mean precipitation	+2.8%	+5.5%
Temperature	+0.9 °C	+1.7 °C

Source: van den Hurk et al. (2006) and http://www.knmi.nl.

Table 3.4. Projections are shown for two scenarios: namely, a moderate and a warm scenario, in which global temperature increases by 1 °C and 2 °C, respectively.[12] The KNMI constructed climate change scenarios for annual average changes, as well as for summer and winter seasonal changes (van den Hurk et al. 2006). Scenarios of yearly changes in climate are provided for several weather stations as additional information at http://www.knmi.nl./index_en.html. Averages of these weather stations were taken in order to arrive at annual changes for the Netherlands as a whole. Hailstorm damage is extrapolated here according to the summer scenarios in addition to the annual changes, since most hailstorm damage occurs in the warmer seasons. Projected annual average temperature change for the Netherlands by the year 2050 varies from 0.9 °C to 1.8 °C under scenarios of yearly changes, and from 0.9 °C to 1.7 °C in the summer season. Average precipitation is expected to increase under the moderate and warm scenarios by,

[12] It is noted that the KNMI also produced corresponding 'plus scenarios', in which atmospheric circulation changes, in addition to the temperature change (van den Hurk et al. 2006). Average precipitation is projected to decrease in these 'plus scenarios', while extreme precipitation is projected to increase. It could be that part of the correlation observed between hailstorm damage and precipitation in Section 3.4 is due to a relation between hailstorm severity and extreme precipitation. Extrapolating hailstorm damage based on the projected decrease in average rainfall in these 'plus scenarios' may result in biased predictions, since the increase in extreme rainfall is not accounted for. For this reason, extrapolations of hailstorm damage are made using the moderate and warm scenarios presented in Table 3.4 and not their corresponding 'plus scenarios'.

respectively, 3.4 per cent and 6.5 per cent throughout the year, and by 2.8 per cent to 5.5 per cent in summer.

Climate change projections for average temperature will be used for minimum and maximum temperatures, since they are expected to change by the same amount (van den Hurk et al. 2006).[13] Extrapolations of annual changes are based on the Tobit results (Eqn 3.2) of the traditional and best prediction models that use observations for the whole year (Table 3.2) and the yearly climate change scenarios. Extrapolations of summertime changes in hailstorm damage are based on the estimation results of Tobit models that use observations for the high-damage months April–September (Appendix 3C) and the summer climate change scenarios. Estimation results of models with the high-damage months are used, since severe damage – most relevant to the agricultural sector and the insurance industry – mostly occurs in summer; the winter season is excluded from this subsample.

Table 3.5 shows the expected changes in hailstorm damage (as a percentage) caused by the projected changes in temperature and precipitation under the different climate change scenarios (Table 3.4). Point predictions and 95% interval predictions are estimated with the statistical software package Stata, as described by Xu and Long (2005). Differences between the traditional and the best prediction models are small, and differences in point estimates are not statistically significant. The best prediction model results in slightly more conservative estimates in most scenarios, apart from yearly estimated changes in hailstorm damage to greenhouse horticulture. The latter estimates of increased hailstorm damage are, furthermore, highly uncertain, as indicated by the prediction intervals that include negative values, which is due to the insignificance of the temperature variable in the yearly model of greenhouse horticulture. The prediction intervals of all the other predictions include only positive values, indicating projected increases in damage. The prediction intervals of the change in hailstorm damage obtained with the best prediction models are smaller than the intervals of the traditional models for all estimates.

[13] This expectation is based on the fact that the diurnal temperature range has not changed in the past (IPCC 2007: p. 9), and is not expected to change in the future. At the same time, regional trends in the diurnal temperature range are highly variable (IPCC 2007: p. 9), i.e. some degree of uncertainty remains. This uncertainty is accounted for in the KNMI 2006 climate scenarios.

Table 3.5 ***Estimated change in hailstorm damage in per cent to outdoor farming (left columns) and to greenhouse horticulture (right columns) under climate change scenarios per year and for the summer season (95% prediction intervals are shown between brackets)***

KNMI	Dependent variable: $HD_{outside}$		Dependent variable: HD_{inside}	
Scenario	Traditional model	Best prediction model	Traditional model	Best prediction model
Yearly				
Moderate	+29 [+11, +47]	+25 [+12, +38]	+28 [−25, +82]	+36 [−6, +77]
Warm	+58 [+21, +94]	+49 [+23, +76]	+57 [−50, +163]	+70 [−13, +153]
Summer				
Moderate	+28 [+10, +46]	+25 [+11, +39]	+134 [+45, +224]	+116 [+42, +190]
Warm	+53 [+19, +87]	+48 [+21, +75]	+269 [+90, +448]	+219 [+89, +360]

This implies that the estimated increase in hailstorm damage using the model specification with maximum temperature and precipitation is most reliable.

The results concerning annual changes indicate that hailstorm damage to outdoor farming is expected to increase by 29 per cent and 25 per cent in the moderate scenario and up to 58 per cent and 49 per cent in the warm climate change scenario according to the traditional and the best prediction models, respectively. Hailstorm damage is projected to increase more under the warm climate change scenarios than under the moderate scenarios, because of the greater temperature rise under the warm scenario. Expected changes in hailstorm damage to outdoor farming are similar for both summer and the whole year. In contrast, the expected increase in hailstorm damage in the summer is considerably larger for greenhouse farming. In particular, hailstorm damage to greenhouse horticulture in summer is expected to increase by 134 per cent and 116 per cent under the moderate climate change scenario and by 269 per cent and 219 per cent under the warm climate change scenario according to the traditional and the best prediction models, respectively.

3.6 Economic implications: Adaptation by agricultural and insurance sectors to the projected increase in hailstorm damage

Projected or actual increases in hailstorm damage because of climate change are likely to trigger responses by the main affected economic agents – namely, farmers and insurance companies that cover agricultural hailstorm damage.[14] The adaptation measures that these agents undertake in response to higher hailstorm risk may affect the actual damage that will materialize in the future. It is well known that hailstorm damage, as well as other weather-related losses, is to a large extent modulated by changes in societal factors (Changnon 2001). Damage is influenced by changes in exposure (amount of crops and number of greenhouses) and the vulnerability of property and crops. Since these socio-economic developments may affect future damage, it is relevant to examine the adaptation options available to the insurance and agricultural sectors for coping with the projected increase in hailstorm risk. Insurance companies that offer hailstorm insurance will initially face increased risk if climate change results in more severe hailstorms, since most hailstorm damage is covered by insurance policies in the Netherlands. For this reason, possible responses by insurers to higher hailstorm risk are outlined first. Adaptation options for farmers who are the purchasers of hailstorm insurance are then examined.

There are four traditional strategies that insurers can undertake in response to a rise in natural disaster risk, such as hailstorms (e.g. Dlugolecki 2000). Insurers can limit their risk, adjust premiums, transfer the risk, and control or mitigate the damage. Payouts of insurance coverage of hailstorm damage can be limited by lowering upper bounds of coverage or by raising deductibles. This reduces the damage reimbursements that the insurance company needs to pay. Moreover, the restrictions and exceptions of insurance policies can be expanded so that farmers are less often eligible for insurance coverage of hailstorm damage. These measures that limit risk for insurance companies in fact transfer the risk to the agricultural sector. The transfer of risk to farmers may be regarded as undesirable from a social welfare

[14] In addition, hailstorms affect other lines of business, in particular building (windows and roofs) and automobile insurance policies, which are not discussed here.

perspective, if farmers are risk-averse, and therefore place a positive utility value on insurance coverage.

Raising premiums is another strategy to cope with higher risks. The revenue of insurers could be boosted if premiums were higher, which would provide the financial resources to pay for increasing claims for damage reimbursement. Premiums in hailstorm insurance markets are partly risk-based, so that a rise in expected damage would be reflected by a higher insurance premium (van Asseldonk et al. 2001). However, it is of interest for insurers to look for other options to limit their risk than raising premiums or transferring risk to farmers, because of the competitive nature of insurance markets in the Netherlands, where many companies offer insurance products. Raising premiums or transferring risk to farmers can result in a decline in market share, which may be undesirable for the insurer. The purchase of reinsurance coverage may be an alternative strategy for primary hailstorm insurance companies to transfer risk. Reinsurance is insurance coverage that primary insurance companies commonly purchase from reinsurance companies to hedge their risk (Kunreuther 2006b). From the perspective of the primary insurer, purchasing reinsurance may be an effective adaptation strategy to limit increasing risks, since it shifts part of these to the reinsurance companies (for an insurance premium). However, it should be noted that the projected rise in risk due to climate change is not reduced for the aggregate insurance sector, since the risk is merely transferred from primary insurers to reinsurers. A disadvantage of reinsurance is that prices are sometimes considerably above the expected value of losses, which would imply that the associated costs have to be passed on to the customers of insurance via higher premiums. Moreover, an increase in risk caused by climate change would increase reinsurance prices. Insurers can further use alternative risk transfer (ART) instruments to transfer risk, such as catastrophe bonds, futures or options (Michel-Kerjan and Morlaye 2008). Such financial instruments could be used to acquire funds from capital markets at times when insurance companies experience loss of revenue due to severe hailstorm damage.

A promising strategy to limit hailstorm damage may be to examine opportunities to undertake measures that reduce the vulnerability of crops and greenhouses to hailstorms and mitigate damage. Insurers, in cooperation with the agricultural sector, could explore strategies to mitigate damage that are effective in terms of costs and benefits. Such mitigation measures are cost-effective if the time-discounted value of the expected value of reduced damage exceeds the upfront investment costs. An advantage of undertaking such mitigation measures is that the reduced risk will be beneficial to both the insurance companies and the farmers, as long as premiums reflect the risk reduction induced by mitigation (Kunreuther et al. 2011). Insurers could encourage farmers to undertake investments that mitigate damage by providing premium discounts or increased coverage. Examples of mitigation measures that could be effective in limiting hailstorm damage to the farming sector are the installation of stronger glass or plastic in greenhouses and the protection of field crops with hail nets.

Even though mitigation measures may be able to limit part of the damage that could be caused by climate change, it seems unlikely that mitigation can fully prevent the projected rise in damage reported in Section 3.5. As a consequence, it is likely that farmers will experience a higher premium for hailstorm damage insurance in the future, or face a partial transfer of hailstorm risk from the insurance sector because of lower coverage. It should be noted that a transfer of risk from primary insurers to reinsurers or capital markets entails costs as well, which may also be partly reflected in higher premiums or lower coverage. A rise in premiums does not necessarily imply that total demand for hailstorm insurance will decrease. Willingness to pay for insurance may rise, at least proportionally with the increased expected value of damage for risk-averse agents in the face of a higher frequency or severity of damage (see Chapter 8). Nevertheless, self-insurance of farmers becomes more attractive if insurance premiums were to rise considerably more than the expected value of the loss. Climate change could result in a higher increase in premiums than the expected value of the loss if insurers demand compensation for the uncertainty posed by pricing (increasing) weather risk (Kunreuther et al. 1995). Other adaptation options than self-insurance and mitigation measures are available to farmers. The choice to grow specific crops or the use of particular agricultural methods may depend on the expectations of damage caused by extreme weather events, such as hailstorms, in addition to expectations of market developments, such as costs of inputs and selling prices. Farmers may adapt to larger damage expectations after consecutive experiences of large hailstorm damage by switching to crops or agricultural methods that are less vulnerable to hailstorms.

3.7 International studies on the potential impacts of climate and socio-economic change on natural disaster damage

Chapters 1 and 2 examined how a combination of climate change and socio-economic developments is expected to increase (insured) damage from natural disasters in the future. While the previous sections of this chapter presented a study of the potential impacts of climate change on insured hailstorm damage in the Netherlands and other countries, this section takes a broader perspective and provides an overview of several representative studies that have estimated how climate and socio-economic changes influence future natural disaster losses for other natural hazards in a wide range of countries. An overview of such studies that have projected future economic losses is presented in Table 3.6. This table is constructed on the basis of a detailed review by Bouwer (2012) in addition to studies that have been published more recently.

The impact on the risk of socio-economic change, such as changes in wealth and population, is modelled as a change in exposure by most studies, while vulnerability is usually assumed to be constant. The studies shown in Table 3.6 have taken different approaches to estimating the change in hazard under a scenario of climate change. Several studies have employed simulations of a general circulation model (GCM), or a regional variant (RCM) which uses scenarios of future greenhouse gases as an input to project a range of changes in the frequency of occurrence of weather conditions in a specific geographical region. Most studies have applied a factor of the expected change in a hazard's probability due to climate change, which is a simpler approach than applying a GCM or RCM. Moreover, a few studies have examined how socio-economic scenarios influence future risk as a result of changes in exposure and vulnerability to a natural hazard. Four different approaches to model changes in risk can be distinguished.

1. Risk approaches, which take into account spatial characteristics of a hazard and exposure, have most often been used.
2. Hybrid approaches, which use a similar but less elaborate approach than the risk approach, were used by 4 out of the 27 studies.
3. A few studies employed integrated assessment models (IAM) that integrate physical and economic models.
4. Other approaches can be identified that do not use a detailed model but make simple calculations of hazard and exposure without considering spatial differentiation of risk.

All the studies included in Table 3.6 make projections of future natural disaster damage under climate change, while only eight also make projections under scenarios of socio-economic change. All projections of damage are for the year 2040.

Projected impacts of climate change on storm damage

Studies of windstorm damage are divided into tropical storms that include hurricanes and typhoons and extratropical storms that occur at mid and high latitudes. Most estimates of storm damage also include flood damage caused by storm surge. Global tropical storm damage has been estimated by Narita et al. (2009) to increase on average by 46 per cent. Mean estimates of future tropical storm damage in the USA range between 9 per cent and 47 per cent, and most studies expect an increase in damage between 20 per cent and 50 per cent. The highest mean estimate of 417 per cent for the Atlantic was made by Pielke (2007), who estimated an upper range of potential climate change impacts. Bender et al. (2010) found a potential decline in tropical storm damage due to a decline in the frequency of category 1–3 tropical storms that some GCMs predict. The Association of British Insurers (ABI 2005a, 2005b, 2009) anticipate a mean increase in tropical storm damage by 30 per cent in Japan and 14 per cent in China.

Narita et al. (2010) expect a mean increase in extratropical storm damage in the high latitudes of 22 per cent. Mean damage projections for extratropical storms for Europe, the UK and Germany range between 11 per cent and 16 per cent. Dorland et al. (1999) expect a much larger increase for the Netherlands; namely 120 per cent. In general, uncertainties in the expected change in future storm damage are substantial, as the differences between minimum and maximum estimates indicate.

Projected impacts of climate change on flood damage

A distinction has been made in Table 3.6 between damage projections for coastal floods, river floods and local floods caused by excessive rainfall. Dawson et al. (2011) estimate that climate change is expected to increase damage by coastal flooding between 2 per cent and 25 per cent in the Thames Estuary. Gaslikova et al. (2011) project that the impacts of climate change on coastal flood damage in the Netherlands can be considerably higher (up to 432 per cent). The mean projected increase in damage caused by river floods is between 7 per cent and 135 per cent, of which the smallest increase is expected in the UK and the

Table 3.6 *Results of studies of natural disaster damage in 2040 under scenarios of climate and socio-economic change for a variety of countries*

	Hazard	Region	Approach	Hazard change	Climate scenarios /GCMs	Exposure scenarios	Estimated change in damage in 2040 due to climate change in per cent and, if applicable due to exposure change, in parenthesis			Study
							Min	Max	Mean	
1	Tropical storm	Atlantic	Other	Factor	2	2	58 [164]	1365 [545]	417 [355]	Pielke (2007)
2	Tropical storm	USA	Other	Factor	1	n.a.	12	92	47	Nordhaus (2010)
3	Tropical storm	USA	Other	Factor	1	n.a.	7[a]	76	40	Bouwer and Botzen (2011)
4	Tropical storm	Global	IAM	Factor	4	n.a.	23	130	46	Narita et al. (2009)
5	Tropical storm	USA	Hybrid	Factor	1	n.a.	n.a.	n.a.	22	Hallegatte (2007)
6	Tropical storm	USA, Caribbean	Risk	Factor	3	n.a.	19	46	32	ABI (2005a, b)
7	Tropical storm	Japan	Risk	Factor	3	n.a.	20	45	30	ABI (2005a, b)
8	Tropical storm	China	Risk	Factor	3	n.a.	9	19	14	ABI (2009)
9	Tropical storm	USA	Hybrid	Factor	1	1	n.a.	n.a.	9 [240]	Schmidt et al. (2009a)
10	Tropical storm	USA	Hybrid	GCM	4	n.a.	–27	36	14	Bender et al. (2010)
11	Extratropical storm	High latitude	IAM	Factor	2	n.a.	–11	62	22	Narita et al. (2010)
12	Extratropical storm	Europe	Risk	GCM	14	n.a.	–11	55	8	Donat et al. (2011)

13	Extratropical storm	Europe	Risk	GCM	3	n.a.	6	25	16	Schwierz et al. (2010)
14	Extratropical storm	Europe	Risk	GCM	2	n.a.	−2	44	12	Pinto et al. (2007)
15	Extratropical storm	UK, Germany	Risk	GCM	4	n.a.	−6	32	11	Leckebusch et al. (2007)
16	Extratropical storm	UK	Risk	Factor	1	n.a.	n.a.	n.a.	14	ABI (2005a, b)
17	Extratropical storm	UK	Risk	RCM	3	n.a.	−33	67	15	ABI (2009)
18	Extratropical storm	Netherlands	Hybrid	Factor	2	4	80 [12]	160 [93]	120 [50]	Dorland et al. (1999)
19	Coastal flood	North Sea region	Risk	GCM	2	n.a.	54	432	177	Gaslikova et al. (2011)
20	Coastal flood	Thames Estuary	Risk	Factor	4	4	2 [8]	25 [36]	12 [25]	Dawson et al. (2011)
21	River flood	Netherlands	Risk	Factor	2	2	46 [35]	201 [172]	124 [104]	Bouwer et al. (2010)
22	River flood	Germany, Netherlands	Risk		2	2	57 [10]	213 [36]	135 [23]	te Linde et al. (2011)
23	River flood	Europe	Risk	RCM	1	n.a.	n.a.	n.a.	83	Feyen et al. (2009)
24	River flood	UK	Risk	RCM	3	n.a.	3	11	7	ABI (2009)
25	River flood	Spain	Risk	RCM	1	1	n.a.	n.a.	36 [349]	Feyen et al. (2009)
26	Local flood	Australia	Risk	GCM	2	n.a.	67	514	361	Schreider et al. (2000)
27	Local flood	Netherlands	Risk	Factor	1	0/4	16 [−4]	70 [72]	47 [29]	Hoes (2007)

Notes: Projections of future damage are given as a percentage change in the year 2040 relative to the year 2000. Estimates of studies that have projected changes in damage for other periods have been scaled to reflect the percentage change between 2000 and 2040 by assuming linear incremental changes per year (Bouwer 2012).

Source: Adapted from Bouwer (2012).

largest in Germany and the Netherlands. Damage by river flooding in Europe is projected to increase, on average, by 83 per cent, according to Feyen et al. (2009). Flood damage caused by local floods is expected to increase on average by 47 per cent in the Netherlands (Hoes 2007) and 361 per cent in Australia (Schreider et al. 2000). Large uncertainty is associated with this estimate for Australia. On average, flood damage is projected to increase more rapidly than storm damage, according to the studies shown in Table 3.6.

Impacts on future storm and flood damage as a result of socio-economic change

Several studies (8 out of 27) in Table 3.6 also estimate the influence of socio-economic change on future natural disaster risks. Pielke (2007) estimates that socio-economic change is expected to increase mean extratropical storm damage in the Atlantic by 355 per cent, which is less than the projected increase due to climate change alone of 417 per cent. In contrast, Schmidt et al. (2009a) expects that mean tropical storm damage in the USA will have increased by 240 per cent in 2040 due to changed exposure, which is much larger than their projection of a 9 per cent increase in tropical storm damage due to climate change. Dorland et al. (1999) estimate that the mean damage caused by extratropical storms in the Netherlands would increase less as a result of socio-economic change (50 per cent) rather than as a result of climate change (120 per cent).

For flooding, Dawson et al. (2011) found that future flood risks in the Thames Estuary are mostly driven by socio-economic change. Feyen et al. (2009) project a substantial increase in river flood damage in Spain (349 per cent) due to socio-economic change, which is much larger than the influence of climate change on future flood damage (36 per cent). In contrast, projected increases in flood damage in the Netherlands are lower as a result of socio-economic change rather than climate change, according to Bouwer et al. (2010), Hoes (2007) and te Linde et al. (2011) who, in addition, have estimated future flood damage in Germany.

3.8 Conclusions

Knowledge about the potential impacts of climate change on natural disaster risks is important for insurers who want to assess their exposure to climate change, as well as for policy makers who aim to design strategies to limit the projected increase in risk. Several studies have estimated the potential impacts of climate change on future flood and storm damage, but relatively few studies have examined the influence of climate change on small-scale weather events such as hailstorms. An overview of studies of the determinants of hailstorm damage in a variety of countries showed that, in some regions, hailstorm risk may increase as a result of climate change, and that increased risk may be related to higher temperature and more precipitation.

This chapter also presented an in-depth study of the potential impacts of climate change on hailstorm damage in the Netherlands. In particular, relations between normalized agricultural hailstorm damage and a range of indicators of temperature and precipitation have been estimated. Temporal dynamics are explicitly modelled and Tobit models account for the censoring of the dependent variable. The historical relations between hailstorm damage and temperature and precipitation, as identified in this chapter, imply that projected climate change may lead to increased damage caused by hailstorms in the future. Extrapolations of hailstorm damage under a range of climate change scenarios indicate that hailstorm damage could increase considerably. In particular, the increase in damage is expected to be large, according to the climate change scenarios where global temperatures are expected to increase by up to 2 °C by 2050. According to the best prediction model, annual hailstorm damage to outdoor farming may rise by 25 per cent and 48 per cent under, respectively, the moderate and the warm climate change scenarios. Damage projections further indicate that the increase in hailstorm damage is expected to be considerably larger during the summer season, where damage is already concentrated under current climate conditions. According to the best prediction model, hailstorm damage to greenhouse horticulture in summer is expected to increase by 116 per cent under the moderate climate change scenario, and by 219 per cent under the warm climate change scenario.

The last section of this chapter provided an overview of 27 representative studies of the potential impacts of climate and socio-economic change on future damage from storms and floods in a variety of countries. These studies show that both climate change and socio-economic change are expected to have a non-negligible impact on future natural disaster damage for most of the hazards and countries considered here. It is difficult to draw conclusions on whether the influences of climate or of socio-economic change on risk are likely to dominate, because their relative effects on risk differ considerably per hazard and per region. The large differences between some projected minimum and maximum changes in damage indicate that there is great uncertainty associated with these projections of future damage. These uncertainties appear to be especially large for projections of tropical storm damage and flood damage, and seem to be a bit smaller for extratropical storm damage.

The projected rise in natural disaster damage indicates the need for insurers and policy makers to investigate possible adaptation strategies. Some of these strategies have been explored in this chapter for hailstorm damage. The extent and type of adaptation undertaken is likely to influence the damage that will actually materialize in the future. Insurers may reduce exposure by transferring risk to the consumers of insurance, or by adjusting risk management practices by purchasing reinsurance or hedging risk on financial markets. In addition, insurers could raise premiums to reflect changes in risk when they actually materialize. A combination of these adaptation options is likely to be effective to cope with increased weather risk. Another strategy for insurers that does not impact on insurance availability or affordability is to stimulate the undertaking of cost-effective investments that reduce, or limit, vulnerability to natural disaster damage. Future research on evaluating such measures in terms of costs and benefits could be promising, since damage mitigation may be beneficial for both insurers and insured, even in the absence of climate change. Adaptation measures that entail mitigation of damage may increase the resilience of our society to the impacts of natural disasters and may limit the future costs of climate change.

Appendix 3A Descriptive statistics

Table 3.A1 *Descriptive statistics for temperature and precipitation indicators*

	Minimum temperature	Average temperature	Maximum temperature	Precipitation
Mean	6.32	10.16	13.82	66.94
Standard deviation	4.74	5.47	6.28	32.44

Table 3.A2 *Correlations between the temperature and precipitation indicators and hailstorm damage*

	Minimum temperature	Average temperature	Maximum temperature	LN (Precipitation)
HD_{total}	0.58	0.59	0.59	0.33
$HD_{outside}$	0.59	0.59	0.59	0.31
HD_{inside}	0.42	0.43	0.43	0.18

Table 3.A3 *Correlations between the temperature indicators and precipitation*

	LN (Precipitation)
Minimum temperature	0.24
Average temperature	0.16
Maximum temperature	0.11

Appendix 3B Statistical methods

Formally, the Tobit model relates the observed outcomes of $y_t \geq 0$ to an index function:

$$y_t^* = x_t'\beta + \sigma\varepsilon_t. \quad (3.3B.1)$$

By means of:

$$y_t = y_t^* = x_t'\beta + \sigma\varepsilon_t \quad \text{if } y_t^* \geq 0. \quad (3.3B.2)$$

$$y_t = 0 \quad \text{if } y_t^* \leq 0. \tag{3.3B.3}$$

Here σ is a scalar parameter; x'_t are the explanatory variables with parameters β; and the error terms ε_t have a known symmetric distribution function f with cumulative distribution function F and the characteristic that $E[\varepsilon_t] = 0$. The R^2 in the Tobit model is computed as the correlation coefficient between y_t and $\hat{y}_t$, where $\hat{y}_t = F(x'_t\beta/\sigma)x'_t\beta + \sigma f(x'_t\beta/\sigma)$ and is the estimate of $E[y|x = x_t]$ (Wooldridge 2002).

Appendix 3C Estimation results for high-damage months

Table 3.C1 ***Tobit results for the traditional and best prediction models of hailstorm damage to outdoor farming,*** $HD_{outside}$ ***(left columns), and hailstorm damage to greenhouse horticulture,*** HD_{inside} ***(right columns), using observations from April to September***

Variable	Dependent variable: $HD_{outside}$		Dependent variable: HD_{inside}	
	Models with temperature variables		Models with temperature variables	
	Traditional model	Best prediction model	Traditional model	Best prediction model
Constant	7.5088***	0.4070	−16.6649***	−42.4307***
	[1.1508]	[2.0702]	[5.7935]	[12.5310]
SPRING	2.1777***	2.3466***	7.2935**	6.9591**
	[0.5922]	[0.5556]	[3.4819]	[3.4448]
SUMMER	0.9643**	1.0695**	3.9521	4.0606
	[0.4565]	[0.4573]	[2.6677]	[2.7625]
Temperature	0.3119***	0.2408***	1.4920***	1.1411***
	[0.1019]	[0.0779]	[0.4795]	[0.4000]
LN (*Prec*)	n.a.	1.3542***	n.a.	4.6503***
		[0.3610]		[1.8509]
Observations				
Total	96	96	96	96
Censored	0	0	43	43
R^2	0.41	0.51	0.41	0.44
Log likelihood	−185	−180	−213	−212

Notes: Significance levels are indicated as follows: *** indicates significance at the 1% level, ** indicates significance at the 5% level, and * indicates significance at the 10% level. Coefficients are stated without brackets and Huber–White standard errors are given in brackets below the coefficients.

Appendix 3D Estimation results for various specifications

Table 3.D1 ***Tobit results for total agricultural hailstorm damage,*** HD_{total}, ***for three models with the temperature variables*** T_{min}, $T_{average}$ ***and*** T_{max}

	Model with temperature variables		
Variable	(1) T_{min}	(2) $T_{average}$	(3) T_{max}
Constant (β_0)	-4.3202	-5.0155*	-5.6227**
	[2.6625]	[2.6505]	[2.7069]
SPRING (β_1)	4.2904***	3.9752***	3.8136***
	[0.9218]	[0.9560]	[0.9776]
SUMMER (β_2)	3.0465***	2.6006**	2.4847**
	[1.1154]	[1.1479]	[1.1452]
FALL (β_3)	1.9258**	1.7356**	1.6834**
	[0.8656]	[0.8716]	[0.8705]
HD_{total} (t–1) (β_4)	0.2513***	0.2432***	0.2425***
	[0.0946]	[0.0939]	[0.0932]
Temperature (β_5)	0.1727*	0.1817**	0.1646**
	[0.0915]	[0.0814]	[0.0700]
LN (*Prec*) (β_6)	2.0184***	2.0822***	2.1480***
	[0.5130]	[0.5102]	[0.5153]
Observations			
Total	197	197	197
Censored	18	18	18
R^2	0.72	0.72	0.72
Log likelihood	-468	-467	-467

Notes: Significance levels are indicated as follows: *** indicates significance at the 1% level, ** indicates significance at the 5% level, and * indicates significance at the 10% level. Coefficients are stated without brackets and Huber–White standard errors are given in brackets below the coefficients.

Table 3.D2 ***Tobit results for hailstorm damage to outdoor farming, $HD_{outside}$ (left columns), and hailstorm damage to greenhouse horticulture, HD_{inside} (right columns), for two models with the temperature variables T_{min} and $T_{average}$***

Variable	Dependent variable: $HD_{outside}$		Dependent variable: HD_{inside}	
	Models with temperature variables		Models with temperature variables	
	(1) T_{min}	(2) $T_{average}$	(1) T_{min}	(2) $T_{average}$
Constant (β_0)	-4.1379	-5.0775*	-25.1320***	-26.1002***
	[2.8007]	[2.8131]	[6.0856]	[6.2422]
SPRING (β1)	3.6806***	3.3262***	8.2297***	7.7124***
	[0.9837]	[1.0082]	[2.5747]	[2.7067]
SUMMER (β2)	1.9476	1.4824	10.6462***	9.6817**
	[1.2123]	[1.2221]	[4.1213]	[4.2400]
FALL (β3)	1.1030	0.9132	3.7100	3.2571
	[0.9884]	[0.9814]	[2.9988]	[3.0215]
$HD_{outside}$ (t–1) (β4)	0.3405***	0.3346***	n.a.	n.a.
	[0.0983]	[0.0975]		
HD_{inside} (t–1) (β4)	n.a.	n.a.	0.1196	0.1134
			[0.1614]	[0.1602]
Temperature (β5)	0.2285***	0.2299***	0.1661	0.2168
	[0.0984]	[0.0864]	[0.3245]	[0.2843]
LN (*Prec*) (β6)	1.7898***	1.8776***	4.1206***	4.1974***
	[0.5417]	[0.5402]	[1.4168]	[1.4147]
Observations				
Total	188	188	111	111
Censored	19	19	77	77
R^2	0.71	0.71	0.48	0.49
Log likelihood	-451	-450	-331	-331

Notes: Significance levels are indicated as follows: *** indicates significance at the 1% level, ** indicates significance at the 5% level, and * indicates significance at the 10% level. Coefficients are stated without brackets and Huber–White standard errors are given in brackets below the coefficients.

Table 3.D3 ***Mean marginal effects (MME) for the Tobit model for hailstorm damage to outdoor farming,*** $HD_{outside}$ ***(left columns), and hailstorm damage to greenhouse horticulture,*** HD_{inside} ***(right columns), for two models with the temperature variables*** T_{min} ***and*** $T_{average}$

	Dependent variable: $HD_{outside}$		Dependent variable: HD_{inside}	
	Models with temperature variables		Models with temperature variables	
Mean marginal effect (MME)	(1) T_{min}	(2) $T_{average}$	(1) T_{min}	(2) $T_{average}$
Temperature	0.22***	0.23***	0.11	0.14
LN (Precipitation)	1.75***	1.84***	2.68***	2.73***
Fraction of MME due to response above limit	0.93	0.93	0.48	0.48

Notes: Significance levels are indicated as follows: *** indicates significance at the 1% level, ** indicates significance at the 5% level, and * indicates significance at the 10% level.

Table 3.D4 ***Tobit results for total agricultural hailstorm damage,*** HD_{total}***, for three models with the temperature variables*** T_{min}***,*** $T_{average}$ ***and*** T_{max}***, using observations from April to September***

	Model with temperature variables		
Variable	(1) T_{min}	(2) $T_{average}$	(3) T_{max}
Constant	2.9674*	1.0046	-0.0358
	[1.6474]	[1.8352]	[2.0207]
SPRING	2.6091***	2.5301***	2.3989***
	[0.5944]	[0.5681]	[0.5584]
SUMMER	1.3115***	1.1678**	1.1708**
	[0.4677]	[0.4741]	[0.4710]
Temperature	0.2768***	0.2716***	0.2392***
	[0.1014]	[0.0900]	[0.0782]
LN (*Prec*)	1.1373***	1.3473***	1.4674***
	[0.3968]	[0.3673]	[0.3585]
Observations			
Total	99	99	99
Censored	0	0	0

Table 3.D4 (*cont.*)

	Model with temperature variables		
Variable	(1) T_{min}	(2) $T_{average}$	(3) T_{max}
R^2	0.51	0.52	0.52
Log likelihood	-189	-188	-188

Notes: Significance levels are indicated as follows: *** indicates significance at the 1% level, ** indicates significance at the 5% level, and * indicates significance at the 10% level. Coefficients are stated without brackets and Huber–White standard errors are given in brackets below the coefficients.

Table 3.D5 ***Tobit results for hailstorm damage to outdoor farming, $HD_{outside}$ (left columns), and hailstorm damage to greenhouse horticulture, HD_{inside} (right columns), for two models with the temperature variables T_{min} and $T_{average}$, using observations from April to September***

	Dependent variable: $HD_{outside}$		Dependent variable: HD_{inside}	
	Models with temperature variables		Models with temperature variables	
Variable	(1) T_{min}	(2) $T_{average}$	(1) T_{min}	(2) $T_{average}$
Constant	3.5406**	1.4920	-28.6646***	-37.9735***
	[1.6832]	[1.8819]	[9.4210]	[11.2797]
SPRING	2.5111***	2.4715***	8.4212**	7.7270**
	[0.5941]	[0.5668]	[3.6172]	[3.5130]
SUMMER	1.2323***	1.0725**	4.679*	4.0144
	[0.4556]	[0.4607]	[2.6950]	[2.7658]
Temperature	0.2653***	0.2707***	1.3933***	1.3192***
	[0.1001]	[0.0892]	[0.4971]	[0.4523]
LN (*Prec*)	1.0288***	1.2337***	3.0046*	4.1093**
	[0.3950]	[0.3678]	[1.8300]	[1.8239]
Observations				
Total	96	96	96	96
Censored	0	0	43	43
R^2	0.49	0.51	0.45	0.45
Log likelihood	-181	-180	-212	-211

Notes: Significance levels are indicated as follows: *** indicates significance at the 1% level, ** indicates significance at the 5% level, and * indicates significance at the 10% level. Coefficients are stated without brackets and Huber–White standard errors are given in brackets below the coefficients.

4 *Climate change adaptation through insurance against flooding*

4.1 Introduction

The previous chapters have shown that a combination of socio-economic change and climate change may result in a non-negligible increase in damage caused by natural hazards such as storms and flooding in several regions of the world. These projected changes in risk can be limited and the impacts of natural disasters on societies can be ameliorated by undertaking adaptation measures. Several adaptation measures to climate change that focus on engineering have been proposed, such as improved flood protection infrastructure, in order to limit the economic impacts of climate change and reduce risks (Kabat et al. 2005). This chapter focuses on the potential of 'softer' adaptation measures to limit losses and spread risks, such as insurance arrangements against flood damage, which have received much less attention by policy makers. Climate change and the insurance sector affect each other in two main ways, as discussed in detail in Chapter 2. First, the insurance industry can contribute to the climate-proofing of societies via efficient risk-sharing mechanisms. Second, the insurance industry itself is likely to experience heightened risks of extreme weather events, which requires adequate incorporation of climate risks in setting premiums and risk management (Mills et al. 2005). While the previous chapter examined the latter issue, this chapter discusses the first issue, with a special focus on risk sharing arrangements for flooding in the Netherlands, given that a large part of the Netherlands is a delta region that is especially vulnerable to flood-related weather extremes.[1] Nevertheless, many of the concepts discussed in this chapter are also relevant for flood insurance

Parts of this chapter have previously been published as Botzen and van den Bergh (2008).

[1] About 70% of properties in the Netherlands lie below sea level or below river water level (Kok et al. 2002).

markets in other countries, and the final part of this chapter provides an international comparison of flood insurance arrangements.

It is expected that economic losses caused by extreme weather events, such as extreme precipitation and flooding, will increase in the Netherlands. Requests for public compensation can be expected to rise when damage is not covered by insurance arrangements, as is the case for flood damage. In general, insurance companies in the Netherlands do not cover flood damage because the risks associated with them have been regarded as uninsurable by the insurance sector since the catastrophic North Sea flood that occurred in 1953 (Heerkens 2003). That flood took more than 1,800 lives and inflicted damage of about €30–40 billion at current price levels. After 1953 the Dutch constructed their famous 'Deltaworks' which is an extensive system of storm surge barriers, dikes and dams that protect their low-lying lands from flooding and ensure high safety standards. Despite several decades of dike reinforcements, several small flood events have occurred. High river discharges in 1993 and 1995 and extreme precipitation events in 1998 inflicted considerable damage (Kok and Barendregt 2004). The absence of private flood insurance resulted in pressure on the government to provide compensation as an insurer of last resort. In this respect, the Netherlands is not alone and there are many other countries where flood damage is at least partly compensated by governments in an ad hoc way. Although there are legal arrangements for compensation for disaster losses by the government (de Vries 1998), the Dutch government has become increasingly reluctant to provide this ad hoc compensation, and instead encourages the development of new private insurance arrangements for damage caused by extreme weather in order to alleviate budget pressures (Kok 2005). Because of this, the insurability of crop damage and property damage against severe local precipitation has increased over the last few years (see Chapter 2).

Apart from public budget considerations, it is questionable whether the absence of private flood insurance is desirable from an economic efficiency perspective. A larger role for private insurance companies in spreading the risks of climate change will provide better incentives to reduce economic losses. It is widely recognized that existing adaptations in the Dutch water sector are insufficient to cope with climate change. Therefore, insurance can play an important role in alleviating the negative impacts of climate change by spreading risks among individuals (Bouwer and Vellinga 2005). It will be argued in this chapter that economic efficiency will be improved when private insurance coverage is extended to flood damage. However, a continuing role for the government in insurance arrangements is warranted to overcome problems of the insurability of weather-related disasters that exist in the private market. In addition, substantial public investments in risk-reducing measures, such as dike infrastructure, are necessary to cope with increasing climate risks. Most insurance contracts are provided on an annual basis, which implies that insurance premiums reflect current risk and that insurance incentives for undertaking risk reduction are targeted at mitigating current risks. Although reducing vulnerability to current natural disaster risk and achieving gradual adaptation to changing risk over time through insurance may increase resilience to climate change impacts, additional risk prevention measures and building code policies may need to be implemented to reduce vulnerability to future natural disaster risk (see Chapter 5). Alternatively, several leading experts in the field of catastrophe insurance have proposed introducing long-term insurance (LTI) contracts that have a duration of several years, in order to provide better incentives for insurers and policyholders to limit risks that are faced over a longer period of time, as discussed in Chapter 7.

Dlugolecki (2008) points out that worldwide there is an unfulfilled need for natural disaster insurance. To fill that gap about $250 billion of additional capital is needed for extreme events, and perhaps $100 billion for slow-onset events such as drought. These markets would need to grow between threefold and fourfold in real terms by 2030. It is questionable whether this gap can be filled by private insurance markets alone, given the challenges with insuring natural disaster risks. Collaboration with the public sector may be needed to solve these issues. Recently, several lead authors in the field of catastrophe insurance have proposed forms of public–private partnerships to insure natural disaster risks, in particular for the USA (see, e.g., Kunreuther 2006b; Kunreuther and Pauly 2006; Kunreuther et al. 2011; Litan 2006; Michel-Kerjan and de Marcellis-Warin 2006). The objective of this chapter is to identify possible future directions for efficient and robust risk-sharing arrangements for flood risks that can cope with climate change. For this purpose, the possible advantages of a larger role of private insurance are outlined and solutions for overcoming difficulties with offering flood insurance are provided. Existing private or public arrangements for compensating flood losses in the Netherlands, the UK, Germany, France and the USA will be compared in order to identify how the possible benefits and problems of private insurance are addressed in each of these countries.

The remainder of this chapter is structured as follows. Section 4.2 gives an overview of existing risk-sharing arrangements for flood damage in the Netherlands. Section 4.3 examines the role that private insurance arrangements can play in covering flood damage. Section 4.4 identifies several problems that exist with offering private flood insurance, and suggestions to overcome these problems are given. Section 4.5 discusses international experiences with private flood insurance arrangements. Section 4.6 concludes the chapter.

4.2 Current risk-sharing arrangements for flood damage in the Netherlands

From 1998 onwards, government compensation for disaster losses including flood damage in the Netherlands has been arranged through the Calamities and Compensation Act (WTS). This law was partly developed in response to the floods in 1993 and 1995, when the government provided (almost complete) damage relief on an ad hoc basis. The total damage caused by the 1993 and 1995 floods amounted to about €115 million and €63.5 million, respectively (Kok and Barendregt 2004). Initially, the Dutch government aimed to extend private insurance coverage to freshwater floods. However, this proposal was rejected by the Dutch insurance sector (de Vries 1998). The WTS only provides compensation if a flood results in a considerable disruption to public safety and requires a coordinated effort by organizations and civil services. Moreover, damage caused by storm surges is excluded from the WTS, because the financial costs of such a flood can be considerable and difficult to estimate beforehand (Kok et al. 2002). The WTS operates as an *ex post* damage compensation arrangement for which no funds have been established, and no premiums or other kinds of contributions are charged before a flood occurs. Currently, private insurance coverage against flood damage is not generally available.[2] Obviously, the existence of public compensation crowds out private market alternatives. Therefore, most of the risks of flooding are

[2] An exception to this is flood damage to cars, which is covered by motor insurance (van Schoubroeck 1997).

carried by the public sector or by households and businesses if the public sector decides not to allow compensation.

The rationale behind compensation by the government instead of private insurance companies might be that the government is regarded as liable for flood damage because of its responsibility for dike maintenance. After the floods in 1953, 1993 and 1995, a widely held public viewpoint was that the government was partly liable for the incurred damage, due to insufficient investment in coastal protection and dikes (de Vries 1998). Furthermore, feelings of solidarity can be a reason to provide damage compensation paid out of tax revenues. A disadvantage of the current system with the WTS is that it is not clear in which cases flood damage will be compensated. The decision on whether WTS compensation will be provided, as well as the determination of the extent of the compensation paid, lies with the government that is in office when the disaster takes place. Therefore, these decisions are influenced by political decision making and public pressure, which can be regarded as arbitrary and subjective. Decisions concerning compensation are likely to be driven by equity and political motives rather than by rational economic grounds, as research on flood damage compensation by the US federal government indicates (Downton and Pielke 2001; Garrett and Sobel 2003). Uncertainty for individuals concerning compensation for damage is less when private insurance is available, since purchasing insurance provides the buyer with a contractual right for compensation according to the terms of the insurance policy.

A major drawback of a public compensation scheme is that incentives to limit or reduce losses for individuals are suboptimal (Harrington 1999; Kaplow 1991). These loss-reducing incentives are minimal when individuals expect that the government will provide compensation regardless of individual risk characteristics or the prevention measures undertaken. Why would individuals in flood-prone areas invest in elevating newly built homes or installing water-resistant floors if they expect the government to compensate them for the damage if a flood occurs? Prevention measures are costly for individuals, and benefits are minimal when the government unconditionally compensates potential damage. Although this is partly accounted for by the WTS in compensating damage only partially, loss-reducing incentives are suboptimal due to a lack of transparency and the absence of any reward for *ex ante* damage-reducing behaviour, for example via lower premiums. In addition, damage compensation is not adjusted for the higher risks taken by individuals who settle in flood-prone areas. The WTS is financed out of general taxation. This implies that individuals who settle in areas with a higher flood risk do not contribute more towards compensation of flood damage than individuals who live in areas with a low flood risk or who face no flood risk at all. In this way, the WTS government compensation results in a governmentally subsidized incentive to take on risk. Incentives to limit losses are likely to be better with insurance arrangements, as elaborated upon in Section 4.3. Another disadvantage of the current public compensation scheme is that disaster compensation by the government may hamper economic development more widely if disaster relief is financed through additional taxes or reduced public investment in other areas.

4.3 Suggestions for managing risk with private flood insurance

Several characteristics of insurance can be identified that make it a potentially useful instrument to stimulate adaptation by firms and households to increasing climate change losses. Indeed, well-designed arrangements are able to limit total economic losses and catastrophic impacts of disasters for individuals. The ability of insurance arrangements to spread and segregate risks, reduce damage by providing loss-reducing incentives, and monitor and control policyholders can be a rationale for extending private flood insurance coverage with risk-based premiums.

4.3.1 Risk spreading and segregation

The economic consequences of flooding events or other natural disasters for individual businesses and households can be catastrophic. Damage can be beyond the ability of the affected parties to pay, which could even result in bankruptcy if compensation is not provided. Insurance spreads the economic costs of such events across many policyholders. In this way, the potential catastrophic consequences for individual households and businesses are shifted to insurance companies. In general, insurance companies are better able to carry these risks because they collect premiums from many individuals in order to pay for damage when adverse events take place. In this way, insurance arrangements reduce individual loss exposures by spreading risks. Moreover, insurance contracts reduce the uncertainty faced by individual parties, which directly increases the welfare of risk-averse policyholders and is beneficial for stimulating business activity and economic stability. The government could, in principle, also fulfil this role of risk spreading by providing ad hoc compensation for flood damage, but ultimately the costs of doing so would be borne by all taxpayers. Insurers are able to spread risks outside the borders of a country by hedging part of the risk on international reinsurance and capital markets, which implies that a country can attract inflows of international capital after it has been hit by a major natural disaster.

Insurance arrangements can be used to segregate risks. Risk segregation implies that insurance companies discriminate between different policyholders, using individual risk characteristics, classes of business, or groups with different risk exposure (Freeman and Kunreuther 2003). Segregating risks means that insurance companies set different premiums and policies for different risk categories from a risk pool. For example, insurance companies can charge higher health insurance premiums for older policyholders, and provide different policies for commercial versus recreational vehicles. With regard to flood insurance, insurance providers could offer different policies with higher premiums for households and businesses that are settled in flood-prone areas, such as areas that are not protected by primary river dikes. Varying premiums across risk classes could inhibit development in floodplains, which may reduce flood risks indirectly (Clark 1998; Pearce and Smale 2005). For instance, if insurance premiums reflect actual flood risk then insurance costs will be lower in areas with a relatively low flood risk, which could stimulate development in safer areas and restrain development in high-risk areas. Differentiated flood insurance premiums also provide incentives for individuals to limit their risk exposure so that they become eligible for lower premiums. For example, policyholders who build or reconstruct their house in such a way that vulnerability to flood damage is reduced should be charged a lower premium if premiums adequately reflect risk. Another rationale behind premium differentiation for flood insurance could be to encourage transparency and awareness of the costs that settlement in risky areas entail. Furthermore, it may be regarded as unfair that all policyholders pay for larger risks taken by a few if premiums are not differentiated. It should be noted that premiums for annual insurance contracts reflect the risk during the duration of the contract, which provides incentives to reduce the risks that prevail during that time period. Nevertheless, if climate change results in a gradual increase in natural disaster risks, then risk-based natural disaster insurance premiums will be gradually revised upwards over time, providing incentives to adapt to these gradually increasing levels of risk.

4.3.2 Limiting disaster damage by loss-reducing incentives and monitoring

Insurance companies adopt techniques to modify the behaviour of potentially insured parties in order to create uniform risk categories (Freeman and Kunreuther 2003). Insurance arrangements can provide loss-reducing incentives with the use of deductibles, premium discounts and policy regulations. By rewarding behaviour that reduces risks and potential damage, insurance arrangements are able to reduce total costs and moderate the economic impacts of adverse events (Board on Natural Disasters 1999). In this way, extending private coverage to flood insurance has the potential to limit total flood damage. For example, vehicle insurance is less expensive for drivers with a good driving record, which provides an incentive to drive carefully. This valuable characteristic of insurance arrangements has reduced risks and their associated costs for society throughout history. Often, insurance companies have been the driving force behind new safety procedures and protective measures (Brainbridge 1952).

The ability of insurance companies to provide incentives to mitigate damage is very relevant in the context of adaptation to climate change. In this respect it should be noted that the efforts of insurers to encourage mitigation will be targeted at current levels of risk and not future risk, because most insurers write annual policies. Nevertheless, if insurers are focused on reducing the risk during the time frame in which they provide insurance coverage, then the resulting reduction in vulnerability and exposure to natural hazards may be effective in limiting climate change impacts as long as the influence of climate change on risks follows a gradual process. Effectively designed insurance arrangements can limit societal costs of future disaster losses by providing incentives to implement damage-reducing adaptation measures. Market discipline is likely to be more efficient in reducing damage than public insurance or compensation arrangements (Priest 1996). Insurance arrangements can limit damage by rewarding well-designed buildings with lower premiums, lower deductibles and higher coverage limits on insurance policies (Kunreuther 1996). Flood damage can also be limited when flood insurance arrangements result in strict building codes in flood-prone areas. Financial institutions could help enforce these regulations by requiring guarantees that structures meet building code standards before mortgages are granted. In addition, the undertaking of cost-effective loss-reducing measures could be encouraged by bank loans that are tied to structures. In practice, investments in loss-reducing measures are often limited in the absence of these bank loans because (yearly) premium reductions are relatively small compared with the (often one-time) costs of mitigation measures. Bank loans can overcome investment problems for individuals with short time horizons and budget constraints (Kleindorfer and Kunreuther 1999).

Flood insurance can also provide incentives for individuals to limit losses by, for example, excluding coverage for damage to carpets or wooden floors, which fosters the use of tiled floors or water-resistant timber floors. Another strategy to reduce flood losses and claims to insurance companies is to inform policyholders about flood-adapted building use and materials, as well as damage-reducing measures that individuals can undertake once a flood occurs (Thieken et al. 2006). With the use of deductibles, policyholders are motivated to be creative in limiting potential losses *ex ante*, *ex post* and during floods. For example, survey analysis indicates that insured individuals spent more time on flood mitigation measures than uninsured individuals at the advent of the 2002 flood in Germany (Thieken et al. 2006).

Another valuable characteristic of insurance is that insurance companies generally monitor the activities of their policyholders. This monitoring is performed to determine whether policyholders operate in a manner consistent with underwriting standards. For example, if insurance arrangements provide premium discounts when loss-reducing measures are undertaken, then insurance companies will monitor policyholders to verify whether the specified standards have been met. Therefore, monitoring insured parties beforehand ensures that loss-reducing measures are actually implemented and adhered to. It should be recognized that, over time, the increased losses that could result from climate change will, at least partly, be shifted to policyholders through increased premiums and altered policy regulations. In this way, increased risks will partly be carried by those who are insured (Tol 1998).

4.4 Problems associated with private flood insurance arrangements and suggestions for possible solutions

Although it can be desirable to extend private insurance coverage to flooding, several problems exist that make it difficult to establish a purely private market. Two conditions must be met before a risk is regarded as insurable (Freeman and Kunreuther 2003; Kunreuther and Michel-Kerjan 2007). First, it must be possible to estimate the probability of occurrence of the event, as well as the extent of losses that the insurance company will incur under different levels of coverage. This may be difficult for flood insurance because floods occur with a very low frequency, and potential losses are difficult to estimate, due to the absence of extensive loss data. Therefore, it might be necessary to share some of the risks faced by insurance companies with individuals and maybe the government (Pearce and Smale 2005). This problem of ambiguity of risk is exacerbated by climate change (Tol 1998). Uncertainty associated with climate change scenarios is significant, and the influence that climate change has on the frequency of flooding is uncertain as well. Second, insurance companies must be able to set premiums for each customer or class of customer. This can be problematic due to ambiguity of risk, adverse selection, moral hazard and correlated risks, as explained below. However, these problems are alleviated by the flexible nature of the insurance sector, since premiums and policies can be adjusted on an annual basis, which limits risks for insurance companies in the face of a changing risk landscape (Bouwer and Vellinga 2005).

4.4.1 Risk and uncertainty

The difficulty in assessing flood probabilities and estimating potential damage hinders the setting of adequate premiums and risk management by insurance companies. Empirical evidence indicates that the larger the uncertainty about the probability of an event and the magnitude of losses, the higher will be the insurance premium charged (Kunreuther et al. 1995). Premiums are higher to account for the risks taken by insurance companies caused by uncertainty about the frequency and magnitude of claims. This implies that insurance premiums for flooding will be higher than, for example, fire insurance premiums, even if expected losses would be the same; this is due to the greater uncertainty associated with the frequency and severity of floods. Uncertainty associated with floods is more significant, since there are fewer historical records and less scientific information available on flood losses compared with fires because floods are low-probability events, while fires occur regularly.

Assessments of natural disaster risks, such as floods, have improved considerably over recent decades due to the advent of catastrophe models. Insurance companies can use catastrophe models to gain insight into loss exposures to catastrophe events, which is useful information for determining premiums and planning the financial risk management of the insurer (Grossi and Kunreuther 2005). Also, in the Netherlands several recent studies have employed catastrophe models to assess flood risk, which can provide a useful basis for determining flood insurance premiums (Aerts and Botzen 2011a; Aerts et al. 2008b; Wouters 2005). The expertise of hydrologists, engineers and risk modellers can be used to obtain an insight into the likelihood of flooding and potential damage, but this is unlikely to completely resolve uncertainty. Uncertainties remain especially large around the extreme risk, i.e. the very low probability of suffering catastrophic damage.

4.4.2 Information asymmetries: Adverse selection and moral hazard

The fear of problems relating to adverse selection was officially one of the main reasons why Dutch insurers resisted government pressure to extend flood insurance coverage in the period shortly before the drafting of the WTS (de Vries 1998). Adverse selection occurs if high-risk individuals are more likely to demand insurance coverage than low-risk individuals. As a consequence, insurance companies will suffer losses if premiums are based on the average probability of a loss. Adverse selection is caused by information asymmetries between insurance companies and policyholders. This problem can arise if individuals are able to determine their individual risk characteristics and insurance companies have difficulty in distinguishing good-risk from bad-risk individuals. If information asymmetries did not exist, insurance companies could simply charge higher premiums to high-risk individuals. Adverse selection can result in very high premiums if insurers are risk-averse, which could explain a missing market.

Adverse selection may arise when flood insurance is available, because individuals living in flood-prone areas are more likely to demand insurance coverage against flooding. Distinguishing low-risk from high-risk individuals and charging the latter higher premiums can overcome this problem. Therefore, permitting price discrimination can mitigate the adverse selection problem. In practice, flood risks can be audited with the use of geographical information systems (GIS) and catastrophe models, allowing insurers to adjust premiums accordingly (Pearce and Smale 2005). The relation between the degree of risk exposure and geographical characteristics is very high. Therefore the adverse selection problem can be overcome for flood insurance, since insurance companies are able to distinguish good-risk from bad-risk individuals. It also seems unlikely that individuals would have better knowledge about flood risks than insurers, given the expertise required in assessing climate-related risks, which would limit the adverse selection problem (Henriet and Michel-Kerjan 2006). Indeed many individuals have difficulty with accurately assessing the flood risk that they face, as will be shown in Chapter 9.

Although premium differentiation can solve problems with adverse selection, it might be politically unfeasible when the difference in premiums for individuals living in low-lying areas compared with high-lying areas becomes too great. A solution to this issue of equity could be to provide support to low-income households who live in areas with a high flood risk. Kunreuther et al. (2011) point out that such support should come from general public funding and not from subsidies of insurance premiums, because incentives to reduce flood damage are distorted if premiums do not reflect risk. Another solution is to bundle flood insurance with other kinds of disasters, such as earthquakes which (partly) occur in different regions in the Netherlands than floods do (Kok 2005). Alternatively, flood insurance could be bundled with property or home contents insurance. This bundling could be made compulsory as is, for example, government policy in France, and this basically solves any problem with adverse selection because individuals both with a low and high risk must purchase the insurance.

Information asymmetries between insurance companies and policyholders can also result in moral hazard. Insured individuals may behave less carefully when they have insurance coverage, although this is not observed by the insurer. This information asymmetry results in an inefficient level of risk prevention (Gollier 2005). For example, where vehicle insurance is not compulsory, people with vehicle insurance are, in general, likely to drive less carefully than people without insurance. The probability of losses increases after the individual has bought insurance coverage due to the behaviour of the policyholder, which is defined as moral hazard (Freeman and Kunreuther 2003). Moral hazard can arise when individuals are covered for flooding. However, problems with moral hazard can be limited in practice by an adequate design of insurance contracts (Thieken et al. 2006). In addition, the moral hazard of insured individuals is argued to be a minor problem for flood insurance because the insured agents do not have control over the catastrophe event (that is, the probability that damage will occur), although policyholders do have control over the extent of damage during and after a flood (Jaffee and Russel 1997).

Monitoring the insured is an option to change asymmetric to symmetric information according to their risk characteristics. However, in practice, monitoring is a costly activity, making perfect monitoring an unattractive option. Incentives to limit losses can be suboptimal if insurance contracts are improperly designed. Other solutions to moral hazard are introducing deductibles, co-insurance, and upper limits on coverage. Sufficiently large deductibles are an incentive for individuals to undertake loss-reducing measures, since individuals will have to pay for part of the damage. This also applies to placing upper limits on insurance coverage. Another advantage of deductibles is that they lower transaction costs, since only insured individuals with a larger amount of damage than the deductible will submit a claim (Gollier 2005). This is particularly relevant for insurance against natural disasters, since transaction costs can be relatively high for these types of insurance. This is the case because catastrophes such as flooding result in many claims at the same time. Under co-insurance, a certain percentage of the losses are borne by the insurers and a certain percentage of the losses are carried by the insured. Therefore, individuals have an incentive to limit damage and the undertaking of precautionary measures is encouraged with this risk-sharing arrangement. Another measure to protect the insurance company from moral hazard is raising insurance premiums in order to reflect the higher risks. A disadvantage of raising premiums is that this decreases the demand for insurance coverage and might be resisted by regulators.

4.4.3 Correlated risks

Insurer insolvency is possible if many losses occur simultaneously due to a single event and the insurer needs to pay out on a very large number of claims (Vellinga et al. 2001). This situation can arise when the risks of individual policyholders are highly correlated. This generally applies to flooding; if a damaging flood occurs, for example due to a breach of a dike, then many individuals will suffer damage at the same time. Moreover, natural disasters such as flooding can have devastating consequences and cause very serious damage. If a major flood damages many homes and companies in a concentrated area at the same time, then insurance companies are suddenly faced with significant losses which may not be matched by premium revenues if the risks are not well diversified geographically. Major flood losses can be beyond the carrying capacity of insurance companies, due to a lack of capital. Spreading of insurance over many different river basins, coastal areas and polders is necessary to achieve a sufficient degree of independence of flooding events (Kok et al. 2002). Insurance companies will charge higher premiums if risks are highly correlated in order to compensate for the possibility of insolvency due to catastrophic losses, and to cover the cost of holding large amounts of capital. A related problem is the high variability in the annual pattern of losses for catastrophe insurance, which is due to the low-probability, high-impact nature of these events. This creates the problems of balancing catastrophe losses over time and having sufficient liquid capital available on a continuous basis.

A solution to excessive concentration of risks for insurance companies is to buy reinsurance and diversify into different types of insurance and different geographical regions. The latter option is limited due to efficiency gains in specialization and also to regulatory as well as practical limits on firm size, explaining the observed demand for reinsurance coverage. In general, two kinds of reinsurance contracts can be identified: proportional reinsurance and excess-of-loss reinsurance. Under a proportional reinsurance contract, the reinsurer assumes a share of the risks in exchange for a share of the premiums. With an excess-of-loss reinsurance contract, the primary insurer retains risks up to a specified amount and the reinsurer above that amount. The problem with reinsuring catastrophe risks is that the reinsurance market may sometimes lack sufficient insurance capacity, especially after the occurrence of a huge disaster that drains the reserves of reinsurers (Verbond van Verzekeraars 2005; Weiss and Chung 2004). Reinsurance is often not available at premiums close to the expected value of the loss because of the high cost associated with attracting large amounts of capital and the need to achieve an adequate return (profit) on capital in the reinsurance business. Reinsurers may, moreover, face capital shortages due to capital market imperfections and may suffer from adverse selection. Moreover, the main reinsurance companies exhibit market power, which partly explains their high prices (Cutler and Zeckhauser 1997; Froot 2001; Jaffee and Russel 1997). In addition, reinsurers (as well as primary insurers) suffer from 'timing risk' if they insure events that have the potential to engender very large claims. Timing risk refers to the possibility that reinsurers that cover catastrophe risks will have to pay large claims before they are able to collect sufficient premiums (Litan 2006). This is especially relevant when new insurance products are introduced. Nevertheless, a promising trend of increased reinsurance industry capital has been observed over time (Kunreuther 2006b), and reinsurance prices have fallen after the destructive hurricane seasons in 2004 and 2005 (Kunreuther et al. 2011).

Insurance companies can also hedge their risks in the capital market directly, instead of relying on reinsurance. Financial instruments that hedge catastrophe risk can provide liquidity to insurance companies to pay for catastrophe claims. These alternative risk transfer (ART) instruments can also be attractive to investors, because their returns are perceived to be almost unrelated to general economic activity. Bond instruments can provide capital to insurance companies in advance of a catastrophe. These (catastrophe) bonds are designed such that the obligation to make interest and principal payments (at least partially) ends once a specific catastrophe event occurs or when specified industry-wide or insurer-specific losses are suffered. In this way, catastrophe bonds can enhance the degree of intertemporal diversification. Other financial instruments available to hedge against catastrophe risks are catastrophe options and futures. These financial instruments are designed such that they pay out a specified sum of money once catastrophe losses have been suffered. However, practical difficulties exist with pricing these instruments and ensuring contract performance (Jaffee and Russel 1997). Pricing catastrophe options is complicated by the low-probability, high-impact nature of catastrophes. Traditional measures to guarantee contract performance, such as marking to market, are inadequate for catastrophe options, since large price changes only take place once a catastrophe occurs, with little warning in advance. Problems with catastrophe bonds are that their trading frequency is low, they are not treated as equivalent to reinsurance by regulators, and correlation of their return with stock markets is higher than initially anticipated (Litan 2006). Overall, catastrophe bond markets have expanded significantly during the last decade and may grow further in the future (Kunreuther 2006b; Kunreuther et al. 2011).

The practical problems associated with reinsurance and financial instruments suggest that there is a role for governments to play in overcoming the liquidity shortages of insurance companies that offer flood insurance, by providing coverage for large catastrophe losses. This is especially relevant for flood risks in the Netherlands, which are highly correlated and potentially catastrophic. In the worst-case scenario, floods can cause damage worth €100 billion. Instead of providing complete public insurance cover, the government could provide public reinsurance to overcome problems with correlated risks and the limited capacity of the insurance sector to cover catastrophic damage. An arrangement where insurance companies are liable up to a certain maximum amount and the government thereafter may be desirable, especially since governments generally have the ability to raise capital by issuing debt. A major strength of the government as a reinsurer is that governments are able to diversify risks over the entire population as well as over future generations (Michel-Kerjan and de Marcellis-Warin 2006). Lewis and Murdock (1996) proposed that governments offer catastrophe reinsurance by auctioning excess-of-loss contracts, which would pay out to their holders when aggregate or industry-wide catastrophe losses from a specified event or over a certain time period exceeded an attachment point. These contracts could complement private market reinsurance for large-magnitude losses. Alternatively, setting up a public or private fund to cover catastrophe losses could ensure the availability of capital. Insurers could pay a fee to the government for the provision of reinsurance capital which reflects risk which can be passed on to policyholders, so that the premiums they pay are risk-based. Furthermore, the government could encourage capital accumulation by insurance companies by providing tax incentives, and stimulate the use of financial instruments to hedge flood risks by making them equivalent to reinsurance from a regulatory perspective.

The above suggestions are similar to a multilayered insurance programme, as has been proposed by Litan (2006), Kunreuther (2006b) and Kunreuther and Pauly (2006) for insuring catastrophe losses in the USA. Such a public–private partnership should be administered by private insurance companies, meaning that they sell insurance, collect premiums, estimate damage and pay claims. It is likely that private insurance companies are more efficient in the day-to-day management of an insurance programme than government agencies are (Michel-Kerjan and de Marcellis-Warin 2006). The first layer of (small) flood losses is borne by the victims themselves, in order to provide loss-reducing incentives and prevent moral hazard. Private insurance companies cover the second layer of flood losses with risk-based premiums and coverage depending on the surplus of the insurer, its current portfolio and its ability to diversify across risks. The third layer of losses is covered by reinsurance and ART instruments, which are to be acquired by the primary insurers. A prefunded government reinsurance programme covers the fourth layer, consisting of extreme losses. Alternatively, this fourth layer could, in due course, be covered by a national fund, which could be publicly or privately financed.

4.5 International experience of flood insurance arrangements

This section compares existing arrangements concerning compensation for flood damage in the USA and three European countries: the UK, Germany and France. This will allow desirable and undesirable characteristics of these arrangements to be identified, which then serves as a basis to suggest possible directions for setting up new flood insurance arrangements, such as in the Netherlands.

4.5.1 United States of America

Since 1968 the federal government has provided flood insurance through the National Flood Insurance Program (NFIP) in the USA. Flood insurance can also be bought in the private market but private insurance is more expensive and its market penetration among households is very low. An exception is commercial flood insurance, which is commonly bought in the private market by large companies. The NFIP was established in response to large payments of federal disaster

Table 4.1 *The stakeholders of the National Flood Insurance Program (NFIP) and their main responsibilities*

Federal Emergency Management Agency (FEMA)
- Set flood insurance premiums
- Identify flood hazard areas
- Make flood risk maps
- Design and review construction criteria and floodplain management by communities
- Fund mitigation projects
- Provide flood insurance

State governments
- Authorize and assist building regulation of local governments

Local governments
- Regulate building in floodplains

Private insurers
- Market insurance policies
- Handle the process of paying out claims

Source: Aerts and Botzen (2011b).

relief in the 1960s and the refusal of the insurance sector to cover flood damage. The NFIP operates like a public–private partnership. Table 4.1 lists the main stakeholders of the NFIP and their responsibilities. The Federal Emergency Management Agency (FEMA) administers the programme and sets flood insurance premiums, identifies flood hazard areas, makes flood risk maps and provides design standards for the construction of buildings in floodplains. The state governments authorize, and assist with, building regulations in flood-prone areas administered by local governments. In principle, communities can decide, on a voluntary basis, whether or not they want to join the NFIP. Insurers are the agents that sell the policies and underwrite the risk, for which they are compensated, but the NFIP has the final financial responsibility and bears the risks. The NFIP operates like a national type of risk pool, and has never been reinsured by private insurance. The government operates like a reinsurer of last resort and Congress can lend money to the NFIP if it experiences deficits and, in principle, this borrowed money should be repaid in the future. Reserves are built up by premiums. Apart from the NFIP, the federal government may provide federal disaster assistance to areas that have experienced a severe natural disaster, such as a hurricane or earthquake, but this ad hoc relief is not directly tied to the flood insurance programme.

The standard flood insurance policy of the NFIP covers direct material damage caused by floods, flood-related erosion as a result of waves or currents of water, and mudslides. The general coverage limits are $250,000 for residential buildings[3] and $500,000 for non-residential buildings such as shops and businesses. Individuals and businesses in the 1/100-year floodplain in NFIP communities are required to have NFIP flood insurance as a prerequisite for receiving any type of federal financial assistance, including federal disaster assistance, and federally backed mortgages. Dixon et al. (2006) have estimated that about 49 per cent of single-family homes in the 1/100-year flood zone carry NFIP flood insurance, while market penetration is very low (about 1 per cent) in flood zones without the mandatory insurance purchase requirement. FEMA determines the insurance premiums based on historical losses averaged for the entire USA. The insurance rates are divided into a few groups and depend on a flood zone classification system. Similar standard premiums are charged per flood zone across the USA, regardless of whether a community has suffered repeatedly from very high flood losses or, at the other extreme, has never experienced major flooding. The existing premium levels are high enough to ensure that the NFIP can pay its claims in an 'average historical loss year' but not a catastrophic year. This implies that, in general, premiums are too low. Existing buildings in floodplains that were erected before the creation of the NFIP flood risk maps pay lower subsidized premiums. This applies to about a quarter of all insured properties (Wetmore et al. 2006).

In principle, communities decide for themselves whether they want to join the NFIP, but the NFIP has established the eligibility requirements related to flood risk reduction that communities have to meet. For example, the ground floor of new constructions in the 1/100-year flood zone needs to be elevated to the estimated 'baseline flood elevation' (BFE) so that damage during flooding is minimized. Policyholders are encouraged to elevate their homes above the minimum requirements; they can be eligible for a premium discount if they

[3] The standard residential policy covers, for example, structural damage to furnaces, water heaters and air conditioners, the clean-up of flood debris, and damage to floor surfaces such as carpeting and tiles.

elevate above the BFE. Policyholders can, moreover, obtain premium discounts through the Community Rating System, which rewards communities who invest in risk reduction with premium discounts of up to 45 per cent of the full rates defined by FEMA. Moreover, FEMA stimulates damage mitigation by communities by providing financial assistance via several grant programmes. It has been estimated that the NFIP standards and mitigation programmes have saved, in total, over $1 billion of flood losses per year (Sarmiento and Miller 2006). Even though the NFIP requirements are quite successful in flood-proofing new buildings, the programme is generally evaluated as being ineffective in limiting new developments in high-risk areas (Burby 2006).

Several studies have evaluated the NFIP and come up with recommendations to improve the programme (Aerts and Botzen 2011b; Burby 2006; Michel-Kerjan 2010; Michel-Kerjan and Kunreuther 2011). Premiums should be appropriately set, according to risk, in order to provide adequate incentives to households to limit damage. Multi-year insurance policies could be introduced in order to prevent policyholders from cancelling their policy, which at present often happens if they have not suffered a loss for a few years, and to provide better incentives for risk reduction (Michel-Kerjan 2010; see also Chapter 7). Also, the elevation requirements of the NFIP could be made stricter so that they do not only protect buildings from current flood risk, but also take into account that future water levels during floods may increase as a result of sea level rise (Aerts and Botzen 2011b). Moreover, an additional layer could be added to the NFIP in which private insurers cover part of the risk, and this layer could be augmented by reinsurance and ART instruments (Michel-Kerjan and Kunreuther 2011). More private sector involvement in covering flood risk could be beneficial because it would reduce the need of the NFIP to borrow money from the US Treasury. During the publication phase of this book, a bill has been passed by the US Senate to reform the NFIP and, in particular, to require FEMA to move to fully risk-based premiums.

4.5.2 United Kingdom

Private insurance companies cover flood risks for households and companies in the UK. The government does not provide compensation if flood damage occurs. Allowing insurers to accumulate tax-exempt reserves promotes capital accumulation by the insurance industry.

Coverage against flood damage is usually included in building or home contents insurance, implying that it is bundled with other risks. Through an agreement between the Association of British Insurers and the UK government, flood cover is included as standard in domestic property insurance policies. Homeowners are obliged to have this insurance, since a mortgage is only provided in the case of full insurance coverage, including flood risks (Hubert 2004). Insurance companies differentiate premiums based on geographical risk characteristics, which may reward settlement in low-risk areas. Companies can also insure themselves against damage that results from business interruption due to floods. In general, potential flood damage in the UK is lower than in the Netherlands (Kok and Barendregt 2004), but several multi-billion-pound flood events have occurred during the last decade.

A drawback of the British system is that public investments in flood protection are considered to be too low (Baan 2004; Crichton 2005). Moral hazard on the part of the government seems to be present, because incentives for the government to invest in protection measures are lower when it is not the government but the private insurance sector that provides flood compensation. Underinvestment in flood defences is especially problematic if flood risks rise because of climate change. Coordination of investments in flood protection between the government and the insurance sector is insufficient (Hubert 2004). The insurance sector threatened to remove flood coverage after major flood events in 1998 and 2000. As a response to these threats, the UK government has increased its spending on flood protection. UK insurers have agreed that they will continue to provide flood insurance cover, but not for homeowners who have a very high risk of flooding. Lately, some homeowners in flood-prone areas have had difficulty in obtaining flood insurance (Crichton 2005; Lamond et al. 2009). Insurers are now less willing to provide flood insurance in the UK because they are dissatisfied with safety standards in high-risk areas and also because of the high cost of recent floods and the projected increase in flood risk as a result of climate change (Lamond et al. 2009). An exception to this is in Scotland, where risks for insurance companies have not increased, due to stricter building codes (Baan 2004). A survey of 400 homeowners in the UK by Lamond et al. (2009) shows that insurers seldom actively engage in encouraging their policyholders to adopt mitigation measures, for example by providing premium discounts. In the face of increasing flood risk due to climate change, persuading policyholders to 'flood-proof' their property can be a useful strategy for insurers to reduce flood claims. Another drawback of the arrangements in the UK is that insurance coverage is not mandatory, which leads to the ethical problem that insurance coverage is low (30 per cent) for poor households, compared with a total market penetration of approximately 95 per cent (Crichton 2005).

4.5.3 France

In France, flood insurance arrangements are based on a public–private partnership. Insurance coverage against flooding and other natural hazards is mandatory and is included in building and home contents insurance (van Schoubroeck 1997). This reduces problems with adverse selection and ensures that insurance penetration is practically 100 per cent. Coverage is also available for the commercial sector. Private insurance companies collect the insurance premiums and handle claims and payouts when damage occurs. The government provides a state guarantee if flood damage exceeds the ability to pay of insurance companies. Low-priced reinsurance for flooding with unlimited coverage is obtainable via a publicly owned reinsurer, the Caisse Centrale de Réassurance (CCR), which again benefits from an unlimited state guarantee (de Marcellis-Warin and Michel-Kerjan 2001). The government sets the 'natural catastrophe' surcharge and premium differentiation based on natural disaster risks is not allowed, which implies that incentives to reduce losses are suboptimal (van den Bergh and Faure 2006). The natural disaster insurance has been combined with what are called 'risk prevention plans' (PPRs). PPRs should limit construction in areas that are vulnerable to natural hazards and are meant to guide and encourage the undertaking of damage mitigation measures by households. Moreover, insurance arrangements include deductibles to stimulate loss-reducing measures. In principle, communities can decide for themselves whether to implement a PPR, but if a community without a PPR suffers from repeated flooding then the deductible of its inhabitants can be increased until the community applies for a PPR. This provides an incentive for the adoption of risk mitigation plans and should stimulate the undertaking of damage reduction measures in disaster-prone areas (Poussin et al. 2011).

In the French system, flood damage is compensated if the government officially recognizes the flood as a disaster and the area where the damage occurred as a disaster area. The French insurance arrangement lacks transparency because the definition of what constitutes a disaster is unclear. It is also possible to obtain private insurance coverage against flooding separately from the public–private partnership. With such an insurance policy, compensation does not depend on governmental recognition of a disaster, as the public reinsurance system does not come into effect. Overall, the French system has proved to be quite efficient and, in general, insurers and insured are quite satisfied with the performance of the system in practice (Kok and Barendregt 2004). Nevertheless, there is scope to further improve efficiency of the natural disaster insurance. Incentives to apply and implement PPRs can be improved, since there are still many communities without such a plan for risk reduction (Poussin et al. 2011). Moreover, risk reduction at the policyholder level can be further encouraged by charging risk-based insurance premiums and rewarding policyholders who invest in damage mitigation measures with premium discounts.

4.5.4 Germany

In Germany, flood damage is privately insurable as a package with other natural disasters as supplements to home contents or building insurance. Damage resulting from storm surges is excluded. The government does not participate in this insurance by providing reinsurance, as in France. Although building insurance has a large market penetration of about 90 per cent, market penetration of flood insurance is only about 10 per cent for home contents and 4 per cent for residential buildings in most parts of Germany (Thieken et al. 2006). Market penetration for industrial risks is also less than 10 per cent. Adverse selection is argued to pose a major problem, despite the bundling of flood insurance with earthquake coverage. A zoning system called ZÜRS has been created to classify flood risk. ZÜRS defines the following four flood zones:

- exposure zone 4, with a flood probability of once in 10 years
- exposure zone 3, with a flood probability of once in 10–50 years
- exposure zone 2, with a flood probability of once in 50–200 years
- exposure zone 1, with a flood probability less than once in 200 years.

Buildings that are situated in hazard-prone areas (zone 4) are often excluded from flood insurance, or they can only be insured for very high premiums. Flood insurance arrangements include deductibles to stimulate loss-reducing measures by individuals. Apart from these deductibles, promotion of damage-reducing measures by insurance companies is minimal (Thieken et al. 2006).

In 2002, a large flood occurred in Germany. The resulting damage was mainly compensated by the government and through donations. Total direct flood damage amounted to €9.1 billion, which was almost completely compensated. Insurance companies compensated approximately €1.8 billion (Baan 2004). Public pressure on the government to provide damage relief was considerable because of the low insurance penetration and the poor economic conditions in the affected areas. However, this public compensation may crowd out the private insurance market and reduce individual incentives to limit damage. Although private insurance coverage is available in Germany, the greater proportion of the flood risks is still carried by the government, as well as by households and businesses if the government decides not to grant (complete) damage relief. Following these floods, insurers and the government have discussed the introduction of a public–private flood insurance system with mandatory coverage, in which private insurers cover the flood risk and the government provides a state guarantee for extreme losses. In the end, this proposal was resisted, mainly by politicians who viewed the ad hoc compensation of flood damage as an effective way of attracting votes and because of conflicts between different levels of government about who should pay the costs of the system (Schwarze and Wagner 2007).

4.5.5 Summary of risk-sharing arrangements

The characteristics of the (public or private) arrangements for flood damage in the USA, the Netherlands, the UK, France and Germany are summarized in Table 4.2. The heterogeneity in risk-sharing arrangements suggests the relevance of identifying desirable and undesirable characteristics which, in turn, aids the design of a more efficient flood insurance arrangement. The involvement of the public sector is substantial in the USA. It has been proposed that a more prominent role of private insurers in covering flood risk in the USA may be desirable in order to reduce the need for federal loans to cover deficits of the

Table 4.2 *Characteristics of financial arrangements for flood damage in different countries*

Kind of arrangement	USA	Netherlands	UK	France	Germany
Private coverage	Yes, but very limited	No	Yes	Yes	Yes
Public insurance	Yes	No	No	No	No
Premium differentiation	Yes	n.a.[b]	Yes	No	Yes
Public reinsurance	No[a]	n.a.[b]	No	Yes	No
Public compensation scheme[c]	Yes	Yes	No	No[d]	Yes

Notes: [a] There is no public reinsurance but the NFIP can borrow money from the US Treasury if it experiences deficits.
[b] Not applicable because private coverage is not generally available.
[c] Does not involve a right to compensation.
[d] The public reinsurance scheme is (partly) financed through taxes.

NFIP. The Netherlands is the only country of these five in which flood insurance is currently not available. The flood insurance market in the UK is closest to a purely private market, because premiums are risk-based, the government is not involved as a reinsurer, and a public compensation scheme is absent. According to the discussion in Section 4.3, this type of arrangement is likely to be the most efficient, due to optimal loss-reducing incentives. However, in practice, insurers in the UK seldom actively engage in encouraging policyholders to invest in mitigation measures. A more coordinated approach may be useful, as has been applied in the USA and France, where the insurance arrangements have been combined with requirements and plans for risk reduction at the community level. Moreover, risk reduction at the policyholder level is encouraged in the USA by providing premium discounts to homeowners who elevate their homes above minimum elevation requirements. Nevertheless, incentives for risk reduction can be improved, in both the USA and France, by better differentiating premiums in accordance with the risk faced by policyholders. The French system is characterized by a high degree of solidarity, since premium differentiation is not allowed and the insurance scheme is kept affordable with public reinsurance. The public reinsurance also solves problems associated with high correlated risks. Although private insurance coverage is available in Germany, insurance penetration remains low. This could be explained by the voluntary nature of the insurance and a belief in public compensation for flood damage. In addition, coverage can be regarded as being too expensive, which is possibly the result of public reinsurance being absent. The negligible insurance penetration in Germany implies that the merits that insurance coverage can bring as an adaptation instrument are not being fully utilized.

4.6 Conclusions

Insurance arrangements can be a useful instrument in adapting to the increase in natural disaster risk that climate change is expected to cause in several regions around the world. The ability of private insurance companies to spread risks, segregate risks, encourage loss-reducing incentives, and monitor and control policyholders can be a rationale for extending insurance coverage to climate change risks that are currently not insured, such as flooding in the Netherlands. Individual uncertainty associated with flood losses can be reduced when private insurance is available, which improves social welfare. Most importantly, private insurance arrangements are able to limit total economic losses if they provide incentives to undertake loss-reducing measures. In addition, premium differentiation can decrease the desirability of settling in risk-prone areas. However, several problems exist with offering private catastrophe insurance, such as ambiguity of risks, adverse selection, moral hazard and correlated risks.

The existence of private insurance schemes in the UK, France and Germany suggests that a partly private market can be viable, despite the aforementioned problems. Adverse selection can be limited by allowing for premium differentiation or by making coverage of flood insurance compulsory on, for example, home contents or building insurance, as is government policy in France. Extensive insurance penetration can also be established by requiring building insurance with flood coverage to be purchased when applying for a mortgage. Moral hazard on behalf of the insured can be limited by monitoring and through adequate design of insurance contracts. International experience further indicates that it may be difficult to ensure adequate insurance penetration in a purely private market. Insurance coverage for poor households is low in the UK and overall penetration is very low in Germany. There might be a role for governments to ensure insurance affordability by, for example, providing public reinsurance. Alternatively, coverage against flood could be made compulsory, as is the case in France.

Problems with correlated risks and with the need for large amounts of capital to pay claims in the event of a catastrophe can be overcome by encouraging reinsurance or capital accumulation by insurance companies, as is government policy in the UK. Alternatively, the state can act as reinsurer and provide a state guarantee, as happens in France. Another option is to set up a private or public fund to cover flood losses and to provide tax exemptions for capital reserves against catastrophes. In addition, it might be desirable to stimulate the use of alternative risk transfer (ART) instruments that hedge flood risks on capital markets, such as catastrophe bonds and options. An advantage of keeping the government partly liable for flood losses is that moral hazard on the part of the government, as experienced in the UK, can be limited. Adequate investments in flood protection measures are especially desirable in a changing climate. A different approach has been taken in the USA, where flood risks are completely covered by an insurance programme that is run by the federal government. However, a greater role of private insurers in covering risks has been proposed in the USA in order to alleviate the pressure on public budgets that arises after major catastrophe losses need to be compensated.

This study indicates that several future directions for risk-sharing mechanisms for flood damage are possible. The multilayered insurance programme seems to be a promising public–private partnership that can provide adequate incentives to limit flood losses and overcome capital shortages in insuring large catastrophe losses. In such a programme, insurers cover a layer of middle-sized losses and charge risk-based premiums that act as a price signal of risk. The government covers a layer of very large losses, for example through a public reinsurance scheme. Increased flood risks and catastrophe losses in the future are likely to heat up the debate about the appropriate compensation arrangements. An economic rationale suggests that extending private coverage against flood risks in the Netherlands could be a fruitful strategy to cope with climate change risks, although an active role of the government in overcoming capital shortages, investing in risk prevention, setting up strict regulations on building codes and land use, and stimulating adequate coverage and market penetration remains desirable.

5 *Dealing with uncertainty in flood risk management*

5.1 Introduction

Climate change and associated sea level rise are projected to increase flood risk in several regions of the world. Flood risk is usually defined as *flood probability*flood damage*, and it is both of these components of flood risk that are expected to increase over time as a result of climate change, as well as socio-economic change which affects the exposure and vulnerability of communities to floods (see Chapters 1, 2 and 3). This requires investment in adaptation measures that limit the expected increase in flood risk and/or ameliorate the societal impacts of flooding. Two main distinctions can be made between measures that can be implemented to limit increased risk: namely, measures that reduce the probability of flooding, such as flood protection infrastructure, or measures that limit the damage that results if a flood occurs, such as damage mitigation measures that reduce the vulnerability of property to floods. Most studies on flood risk management conduct a cost–benefit analysis that evaluates such measures separately on the basis of the (discounted) cost of implementing the measure (including maintenance cost) and its (discounted) benefits in terms of risk reduction (e.g. Levy and Hall 2005; Penning-Rowsell et al. 2005). Such an approach provides useful insights into what measures are likely to be most cost-effective.

A challenge with evaluating flood risk management measures is the uncertainty associated with flood risk and the performance of measures under different flooding conditions. Moreover, an evaluation of specific measures in isolation may hide their combined effects on the complete probability distribution of flood damage. Such combined effects may provide important information for water management. Policy makers may be concerned with the prevention of catastrophic losses and may therefore prefer to have a variety of measures in

Parts of this chapter (in particular Sections 5.3 and 5.4) were published in revised form as Aerts et al. (2008a).

place that are complementary to each other and work as a system to reduce risk under different uncertain flooding conditions. Such a policy maker may be reluctant to rely on only flood protection infrastructure, such as dikes, as a strategy to limit flood risk, since it is uncertain how this flood protection infrastructure will perform under different storms, surge heights and river discharges. Instead, this policy maker may prefer to implement measures that limit flood risk at times when the flood protection fails, such as damage mitigation measures that prevent the occurrence of a catastrophic loss. Such a strategy would be consistent with the concept of adaptive (water-) management that has been introduced in the social sciences, in which dealing with uncertainty has been stressed as a key issue (Downing et al. 2005; Folke et al. 2002; Pahl-Wostl et al. 2005). A diversified flood risk management strategy would be consistent with the creation of a robust flood risk system which can be defined as a system that can 'cope with floods of the full range of plausible magnitudes without the area at stake being affected beyond recovery' (Klijn et al. 2012). Also, in the fields of climate and vulnerability science, reducing current vulnerabilities and dealing with uncertainty has been put forward as an important theme (e.g. Adger et al. 2004; Füssel and Klein 2006; O'Brien et al. 2004).

In the area of financial investments, there is a lot of experience of how to deal with the uncertainty of investment returns and how to follow a strategy that diversifies investments in order to minimize risk. In an approach formulated as the 'modern portfolio theory' (MPT), risk is referred to as the standard deviation of the return on an investment (Markowitz 1952). MPT aims at finding combinations of investments that diversify risks and reduce the overall risk of the total portfolio of investments. There are several studies that use the portfolio concept to guide investment decisions outside the financial domain (e.g. Costanza et al. 2000; Figge 2004; Fraser et al. 2005; O'Brien and Sculpher 2000; Tonhasca and Byrne 1994), but its usefulness for analysing investment decisions in the field of climate change adaptation has hardly been explored.

This chapter applies MPT to water management. The task of a water manager, analogous to a financial manager, would be to construct a portfolio of flood management measures that generates the highest investment return under an acceptable risk. Moreover, a water management portfolio should be developed in such a way that the risk–return ratio will be optimized through the diversification of activities in the portfolio; hence, choosing to implement measures in the portfolio that have returns that are (at least partly) uncorrelated under different flooding scenarios. The terminology of MPT is adopted in this chapter and risk is hereafter referred to as the standard deviation or variance of either the return on a particular investment, or the standard deviation (or variance) of the return on a portfolio of investments. This chapter provides a numerical example that applies MPT to flood risk management in a case study area in the Netherlands. The framework of this analysis is also broadly applicable to other countries and other risks. The objective of this chapter is to examine the effectiveness of a diversity of water management options in limiting flood damage and coping with risk, including damage mitigation measures and investment in dikes.

The remainder of this chapter is structured as follows. Section 5.2 discusses flood risk management policy in the Netherlands. Section 5.3 describes the concept of diversification as formulated in MPT and provides the model framework. Section 5.4 presents the case study and discusses the results. Section 5.5 concludes the chapter.

5.2 Flood risk management in the Netherlands

The Netherlands is one of the most densely populated countries in the world with 395 inhabitants per square kilometre and about 16.5 million inhabitants. The Netherlands consists of a large river delta that is vulnerable to coastal flooding from the North Sea and to floods from major European rivers (the Rhine and the Meuse). About half of the Netherlands (the western part) is below sea level and about 9 million people live in these areas, which are protected by dikes. Also, alongside the Rhine and Meuse rivers there are areas that lie lower than average river water levels. The roots of this situation are historical. Many low-lying parts (which are often called ‘polders’) have been reclaimed from former lakes. Also, subsidence of land induced by agricultural practice is one of the processes that cause increased exposure to floods in already low-lying areas. This situation is further exacerbated by sea level rise.

The low-lying areas in the Netherlands are protected by a system of dikes and embankments along the main rivers and coastal areas. The country is divided into 53 so-called ‘dike-ring’ areas, which are geographical units that have their own closed flood protection system that consists of dikes, dams and sluices. A dike-ring is also a separate administrative unit under the Water Embankment Act that was enforced in 1995. The Water Embankment Act aims to guarantee a certain level of protection against flooding for each dike-ring area. A dike-ring area should be protected against floods by a system of primary embankments, and each dike-ring has been designed such that it meets a safety norm. These safety norms are based on potential high flood levels with a certain probability. For example, a dike-ring with a safety norm of 1/10,000 means that it has been designed such that it can withstand a peak water discharge that occurs, on average, once in 10,000 years (probability 0.0001). These numbers have been derived from extrapolations based on historical data. Some relatively small floodplains in the Netherlands are unprotected and have a higher probability of flooding than dike-ring areas. Figure 5.1 shows the different dike-ring areas in the Netherlands and their safety norms, which are either 1/10,000, 1/4,000, 1/2,000 or 1/1,250. Safety norms have been determined with the use of cost–benefit analysis and vary throughout the country (e.g. van Dantzig 1956). Safety norms near coastal dike-ring areas are higher (western Netherlands) due to higher economic values and population density. Dikes at the coast also have higher safety norms because storm surges are more difficult to predict, which increases the probability of casualties. Moreover, seawater floods are likely to cause more damage than freshwater floods. Coastal dike-ring areas are designed at standards between 1/2,000 and 1/10,000.

Flood risk management in the Netherlands relies strongly on technical engineering capacity. This is a historically created situation in which water managers have developed highly sophisticated flood protection systems with the highest safety standards in the world (Vellinga 2003). The major storm surge of 1953, which flooded large coastal areas in the south-west Netherlands (about 200,000 hectares), initiated a boost in technical innovations in flood protection measures (Aerts and Droogers 2004). The Dutch government stated that a major storm surge flood must never happen again and, therefore the Delta Plan was initiated. This plan led to the construction of the Deltaworks, which consists of a series of dams, sluices, dikes and storm surge barriers that have been constructed between 1958 and 1997. With over 10,250 miles of dikes (1,500 miles designated as primary dikes and 8,750 miles as secondary dikes) and 300 structures such as sluices and bridges, the project is one of the most extensive engineering projects in the world. The Deltaworks reduced the length of the dikes exposed to

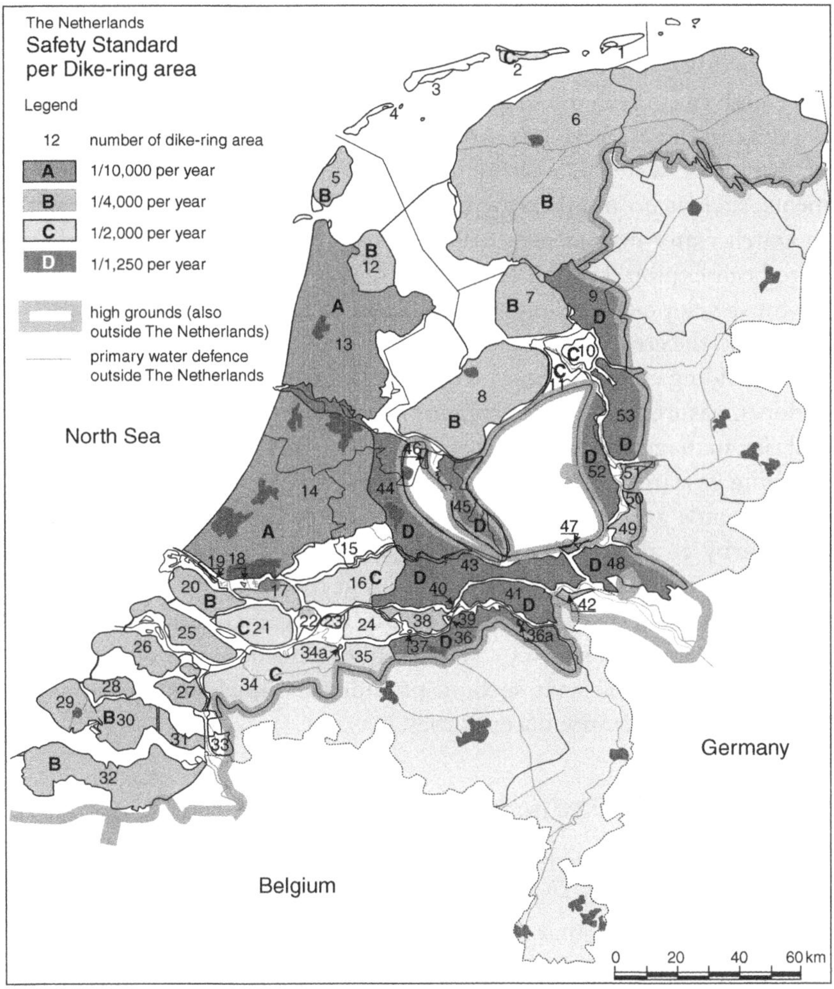

Figure 5.1 Safety standards for dike-ring areas in the Netherlands

storm surge by approximately 400–450 miles, by constructing barriers and dams (Aerts and Botzen 2012b).

The current safety standards in the Netherlands have been under discussion because increased exposure during the past few decades and the projected increase in flood risk as a result of climate change could warrant higher safety standards in some areas or require additional measures that limit potential flood damage (Klijn et al. 2012). Although the prevention of floods will always remain the dominant strategy in Dutch water management policy, other measures than technical investments in dikes and pumping stations have received increasing attention since high river discharges in 1993 and 1995 almost caused large-scale flooding in the Netherlands. Another explanation for the exploration of alternative water management strategies is that long-term developments, such as economic growth and climate change, pose a challenge to Dutch water management (IPCC 2001). Alternative strategies that have been explored in the Netherlands, as well as in other countries, are the development of flood insurance, mapping of current and future flood risk, the creation of evacuation plans and early warning systems, flood building code and flood zoning policies, and other damage mitigation measures (see also Chapters 6 and 7). Moreover, more general risk management approaches which are commonly used in spatial planning research that specifically address the probability of certain future trends are gaining increasing attention in water management (e.g. Burby et al. 1999). A challenge with anticipating the effects of these long-term trends on flood risk is that they are inherently uncertain, which complicates the design of effective risk reduction strategies (DWW 2005a; Gleick 2003). Modern portfolio theory (MPT), which is introduced in the next section, provides a framework for coping with current and future uncertainties.

5.2.1 Climate change and safety standards

For the area of the Netherlands that covers most of the delta of the rivers Rhine and Meuse and is located on the North Sea coast, climate change would mean that the probability and consequences of flood risk may increase (e.g. Aerts et al. 2006; Middelkoop et al. 2001; Ward et al. 2008b). Sea level rise may increase flood risks from storm surges in low-lying delta regions that are already vulnerable to flooding, such as the Netherlands. This could have disastrous socio-economic consequences (Bouwer and Vellinga 2007; Olsthoorn et al. 2008).[1] Moreover, the risk of river flooding may increase if climate change results in an increase in (extreme) precipitation and peak water discharges.

[1] About 60% of the landmass of the Netherlands is located below sea level, while 70% of GDP is generated there (Kabat et al. 2005).

At present, safety standards in the Dutch part of the River Rhine are designed to withstand a peak water discharge that occurs on average once in 1,250 years (1/1,250). The peak discharge, also called 'design discharge', for the Rhine at Lobith, associated with an incidence of 1/1,250, is estimated to be 16,000 m^3/s. As the primary dikes along the Rhine are currently designed for a maximum discharge of 15,000 m^3/s, additional measures in the area are currently needed, as climate change may severely affect the hydrology of the river (IPCC 2001). New scientific results concerning climate change are one of the reasons behind a revision of the flood protection law, and additional measures are required to protect the land against peak discharges of 16,000 m^3/s (Aerts et al. 2006). Middelkoop et al. (2001) estimate that peak flows of the lower Rhine may increase by about 5 per cent to 8 per cent by the year 2050. The maximum peak flow estimated by RIZA (2003) is slightly higher: peak flows are estimated to increase by 5 per cent for 1 °C of warming and 10 per cent for 2 °C of warming by 2050. These estimates consider only increased precipitation. The pattern of precipitation, which encompasses the location of precipitation in the river basin and the timing of precipitation, is assumed to remain constant. Currently, the peak discharge for the Rhine at the gauging station of Lobith, associated with an incidence of once in 1,250 years, is approximately 16,000 m^3/s. In this case study (and in Chapter 8), it is assumed that a flood will occur with certainty once this critical level is reached. Therefore, the probability of observing a peak flow of 16,000 m^3/s and higher can be assumed to be the same as the probability of flooding. According to the RIZA estimates, a discharge with an incidence of once in 1,250 years may increase in 2050 to 16,800 m^3/s under the low climate change scenario (+1 °C) and to 17,600 m^3/s under the higher climate change scenario (+2 °C). An increase in discharge to 18,000 m^3/s by 2100 is possible if mean temperature has increased by 4 °C.[2]

Figure 5.2 shows the distribution of peak water discharges at Lobith and the corresponding incidence in years. The repetition time in years that is shown on the horizontal axis can be interpreted as a probability measure (ten Brinke and Bannink 2004). The probability of

[2] Actually, a discharge of 19,000 m^3/s is possible with a temperature increase of 4 °C. However, maximum discharge at Lobith is set at 18,000 m^3/s to account for decreased water discharge due to floods in Germany that are expected to occur at these high water levels (RIZA 2003).

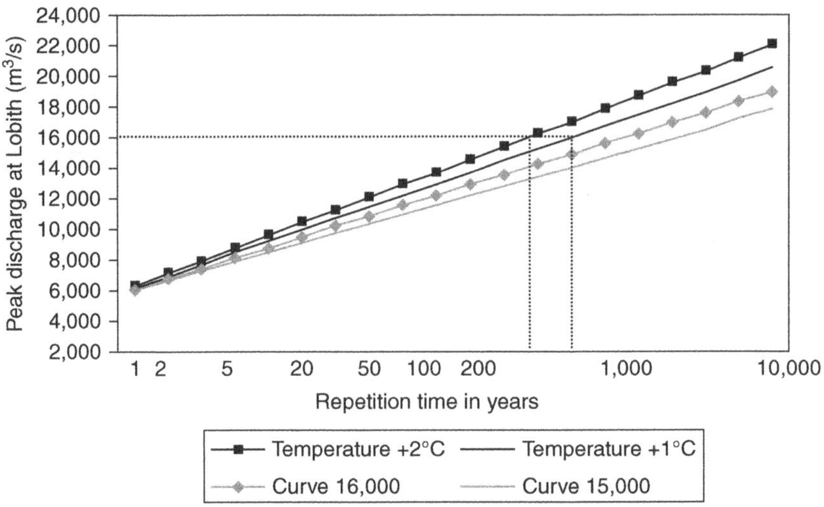

Figure 5.2 Peak discharge at Lobith (m^3/s) and the probability of dike failure

observing a specific water discharge or higher is 1 divided by its repetition time in years. The shift of the 15,000 curve to the 16,000 curve corresponds to an adjustment of the norm which took place in 2001. Climate change may increase peak river discharges, which would shift this curve upward. The upper and second-highest curves represent climate change scenarios. The second-highest curve corresponds to an increase in peak flows caused by 1 °C warming, and the upper curve corresponds to an increase in peak flows as a result of 2 °C warming. The new probability of dike failure can be found by approximating the probability that corresponds to a discharge of 16,000 m^3/s and higher on the new (higher) curve. As is shown by the dotted lines in the figure, the probability of flooding may increase to approximately 1/750 in 2050 if temperature increases by 1 °C, and it may increase to about 1/550 if temperature increases by 2 °C. These probabilities should be interpreted with caution since they are merely indications of how flood risks of the Rhine may change in the future. Flood probabilities will be lower if the government invests in protection measures.

5.3 The framework of modern portfolio theory

The benefits of diversifying investments have been widely recognized by financial economists. Investors rarely hold a single financial asset;

instead they hold portfolios of financial assets. In this way, investors diversify risks and become less sensitive to price changes of individual assets. For example, total returns for an investor will be higher when low returns on an individual stock in a certain period are partly offset by higher returns from other stocks during the same period. Diversification is possible if stock returns are less than perfectly correlated. Modern portfolio theory (MPT) (Markowitz 1952) addressed the question of what potential set of portfolios investors should select. The main criteria in MPT for selecting portfolios are their expected return and risk. The latter can be measured by the variance or standard deviation of the portfolio's return. A portfolio with a relatively high variance or standard deviation is riskier, because the probability of yielding an unfavourable return is higher. According to MPT, investors should first identify the most efficient sets of portfolios from all feasible portfolios. This means finding portfolios that have the highest possible expected return for a given risk or the lowest possible degree of risk for any given mean rate of return. Subsequently, investors can choose a portfolio among the efficient ones according to their individual risk and return preferences (Elton and Gruber 1995).

The expected, or mean, return R_i of an individual asset A_i can be estimated by summing products of the actual return on that asset in a specific state of the economy and the probability that the corresponding state occurs. This can be represented by the following formula:

$$\overline{R_i} = \sum_{s=1}^{n} p_s R_s \tag{5.1}$$

where $\overline{R_i}$ is the expected return on an individual asset, p_s is the probability that state s occurs, and R_s is the actual return in that state s, with a total number of states equal to n.

The variance of the return on an individual asset is the average squared deviation of the actual return on that asset from its expected return. The variance V_i of an asset A_i can be defined as:

$$V_i = \sum_{s=1}^{n} p_s (R_s - \overline{R_i})^2. \tag{5.2}$$

Another measure of dispersion is the standard deviation of an individual asset, SD_i, which can be calculated by taking the square root of the variance; thus $SD_i = \sqrt{V_i}$. The expected return $\overline{R^p}$ of a portfolio can easily be obtained after calculating the expected returns of individual assets. Consider a portfolio consisting of individual assets $A_1, A_2, \ldots, A_n$ with corresponding shares in this portfolio $x_1, x_2, \ldots, x_n$, where $0 < x_i < 1$ and the sum of all x_i equals 1. The expected return on such a portfolio can be estimated by adding the products of the expected return on the individual assets $\overline{R_i}$, and their shares in the portfolio x_i, which can be represented as:

$$\overline{R^p} = \sum_{i=1}^{n} \overline{R_i} x_i \tag{5.3}$$

where the expected returns of individual assets $\overline{R_i}$ are defined by Eqn 5.1.

The covariance between individual assets in a portfolio has to be estimated in order to estimate the variance of a portfolio. The covariance between assets A_i and A_j corresponds to the expected value of the deviation of the actual return R_i of asset A_i from its expected return $\overline{R_i}$, multiplied by the expected value of the deviation of the actual return R_j of asset A_j from its expected return $\overline{R_j}$. This can be represented by:

$$\sigma_{ij} = E\left\{(R_i - \overline{R_i})(R_j - \overline{R_j})\right\} \tag{5.4}$$

or equivalently by:

$$\sigma_{ij} = \sum_{s=1}^{n} p_s (R_{is} - \overline{R_i})(R_{js} - \overline{R_j}) \tag{5.5}$$

where p_s corresponds to the probability that state s occurs, R_{is} to the actual return on asset A_i in that state, and R_{js} to the actual return on asset A_j in that state, with a total number of states equal to n.

The covariance between two assets is positive if returns between assets are positively related, and negative if returns between the assets are negatively related. The interpretation of the actual covariance value is difficult. Therefore, it is suggested that the correlation between two assets should be calculated, which always has a numerical value between −1 and 1. The correlation between two assets A_i and A_j is defined as:

$$\rho_{ij} = \sigma_{ij} / (SD_i * SD_j) \tag{5.6}$$

where SD_i and SD_j are the standard deviations of the individual assets A_i and A_j.

The risk of a portfolio V^p can be represented by the variance of its returns, which can be estimated with the formula:

$$V^p = \sum_{i=1}^{n} x_i^2 V_i + \sum_{i=1}^{n} \sum_{\substack{j=1 \\ j \neq i}}^{n} x_i x_j \sigma_{ij} \tag{5.7}$$

where V_i represents the variance of the individual assets A_i, as can be estimated with Eqn 5.2, and σ_{ij} represents the covariance as estimated with Eqn 5.4. From Eqn 5.7 it follows that the portfolio variance is less than the weighted sum of the variances of the individual assets if the correlation between the assets is less than 1. In other words, diversification is possible as long as there is less than perfect positive correlation between the returns of assets.

Portfolio diversification for a two-asset case is illustrated in Figure 5.3, which shows different sets of portfolios composed of two assets A and B for different correlations ρ between these two assets. The curved lines represent opportunity sets or feasible sets also called 'efficient frontiers'; points on these curves can be obtained by selecting a mix of these two assets. Only one of these curves can exist in the real world, since either $\rho = 1$, or $\rho = 0.5$ etc. The investor can only choose between different points on a curve that have distinct risk and return characteristics for a given correlation between assets. Varying the proportions of assets A and B in the portfolio can develop different portfolios. Points located more to the left represent portfolios with higher proportions of asset A, which has a smaller expected return and risk than asset B. The straight line between the two assets represents possible return and risk characteristics of a portfolio composed of two assets (A and B) with a correlation of unity. The diversification effect applies to the curved lines, where the correlation is smaller than unity. The smaller the correlation between the two assets, the more bent is the curve, indicating that higher returns can be earned for the same *SD* of the portfolio. Alternatively, the lower the correlation the lower the *SD* of a portfolio is for a constant expected return. The point MV, which is actually located on each of these curves, represents the minimum variance portfolio. This backward bending always occurs if $\rho \leq 0$, but may or may not occur if $\rho > 0$. Obviously, no investor wants to hold a portfolio with an expected return below the minimum variance portfolio. Therefore, the efficient set lies between MV and B.

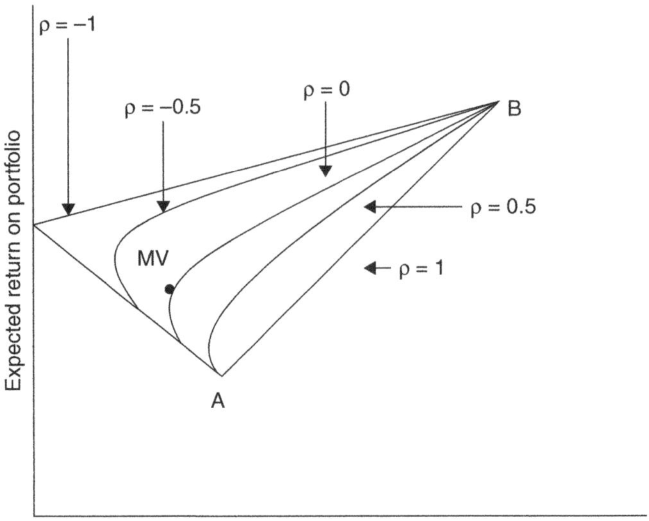

Figure 5.3 The curved lines are 'efficient frontiers' showing the most optimal risk–return values against the variation in return for different portfolios of two assets

Note: Each curve represents a different correlation ρ between the two assets. The minimum variation point (MV) represents an example of a minimum variance portfolio.
Source: Adapted from Ross et al. (2002). With permission from McGraw-Hill.

The above-described concept of MPT can provide additional value to current flood management practices. Most flood management investments have tended to focus on acceptable or tolerable risk, defined as *probability*damage*, and focus on the evaluation of a single flood management measure. An MPT approach in flood management may add new information on the robustness of investments in a variety of flood management measures, not only with respect to their return in terms of net benefits (avoidance of flood risk), but also with respect to the risk (variance of avoided flood risk) in achieving this return.

5.4 Application of modern portfolio theory to flood risk management

The principle of diversification is illustrated in this section with a numerical example that applies modern portfolio theory (MPT) to

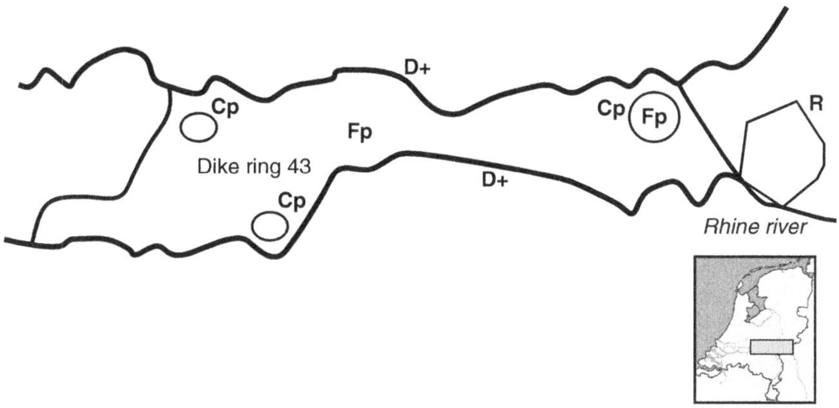

Figure 5.4 Schematic overview of the possible flood protection investments in dike-ring area 43 in the central part of the Netherlands

Note: The four possible investments are: Asset D+, main dike reinforcement; Asset Cp, compartments around urban areas; Asset Fp, flood protection of residences; Asset R, upstream retention area.

four different flood management investments that are proposed for a representative dike-ring area, 'number 43'. The area is located in the central part of the Netherlands (see Figure 5.4). The maximum potential damage in this area is estimated at €17,993 million (DWW 2005b). The hypothesis is that investing in a portfolio of flood protection measures has the potential to increase the expected return on these investments and lower risk compared with investing in individual strategies. Investing only in primary dikes will result in a relatively high expected return (i.e. prevented damage), but also in a large variance (i.e. risk) on this return. Combining dike investments with other investment strategies may lower the variance of returns, because these investments can prevent damage in situations of dike failure, in which the return on the dike is actually zero. In other words, the probability of suffering extreme flood damage may be reduced if water management investments are diversified.

The objective of this application is to provide an illustration of the potential benefits of diversification of water management investments. The analysis is indicative in the sense that the returns on investment strategies are merely approximations. This application focuses only on prevented economic property damage and damage to land use, including crop damage, within the dike-ring area. Macroeconomic costs and casualties are not considered here. In this application, four states of nature are defined, each representing different levels of river discharge of the Rhine at Lobith, which is where the Rhine enters the Netherlands. The probability of observing these discharges is likely to increase as a result of climate change (see Section 5.2). In order to explore the robustness of the effectiveness of water management strategies under climate change, one baseline scenario and two climate change scenarios are defined, which are associated with different probability distributions for each of the four states of nature. Four assets will be evaluated which correspond to water management strategies that can prevent flood damage in the different states of nature. The probabilities of river discharges are listed in Table 5.1. The discussion below defines the states of peak discharges, climate change scenarios and assets that are used in the application.

5.4.1 States of nature

Four states of nature are defined, which correspond to different peak discharges of the River Rhine. One state (state D) is the situation where no damage will be caused (all discharges below 16,000 m^3/s), because the existing dike infrastructure is sufficiently strong to withstand these discharges. The other three states (A, B and C) are other intervals of peak discharges above the current safety level of 16,000 m^3/s, which will all cause damage. Three intervals of discharge are selected for these flood states: namely, C: 16,000–17,000 m^3/s, B: 17,000–18,000 m^3/s and A: >18,000 m^3/s (Table 5.1). The probability of each of these states under each climate scenario can be derived from the curves presented in Figure 5.2. The probability of state A is directly calculated from Figure 5.2 as 1 divided by the repetition time corresponding to 18,000 m^3/s that is shown on the horizontal axes. The probabilities of the intervals C and D are obtained by subtracting the probability of observing the maximum discharge or higher in the interval from the probability of observing the minimum discharge or higher in the interval. For example, the probability of the occurrence of a discharge between 17,000 m^3/s and 18,000 m^3/s (state B) is calculated by subtracting the probability of observing a discharge of 18,000 m^3/s or higher from the probability of observing a discharge of 17,000 m^3/s or higher. The probability of state D (discharges <16,000 m^3/s) is calculated as 1 minus the probabilities of states A+B+C, so

Table 5.1 ***Probabilities of the different states of nature, i.e. possible river discharges, under three different climate scenarios: current, low climate change, and high climate change***

	States of nature			
	A	B	C	D
Discharge [m^3/s]	> 18,000	17,000–18,000	16,000–17,000	< 16,000
Climate scenarios	Probability			
Current	0.000125	0.000250	0.000800	0.998825
Low climate change scenario (CC1)	0.000167	0.000667	0.001333	0.997833
High climate change scenario (CC2)	0.000667	0.001333	0.001818	0.996182
Assets	Actual returns [€million]			
D+: Higher dike-ring	0	0	17,993	0
Cp: Compartments	0	8,000	12,000	0
Fp: Flood-proofing houses	3,000	3,000	3,000	0
R: Retention areas	−4,000	14,000	14,000	0

Note: The returns for each investment are shown in the various states A, B, C and D, expressed as prevented damage (from DWW 2005a, 2005b).

that the probabilities add up to unity. The probability of state D can be interpreted as the probability that no flood damage will occur in a given year.

5.4.2 Climate change scenarios

Table 5.1 shows the probability of observing a particular state under different climate change scenarios. The probability of each state – A, B, C and D – is calculated according to three climate scenarios: current climate, climate change low (CC1) and climate change high (CC2). The probability of observing a particular state under the CC2 scenario can be derived from the upper curve in

Figure 5.2, and the probabilities under the CC1 and current climate scenarios can be derived from, respectively, the second-highest and third-highest curves. The probabilities for each state under a specific climate scenario should add up to unity. The purpose of these climate change scenarios is to examine how portfolio returns and variances are affected by changes in flood probabilities caused by climate change.

5.4.3 The assets: Four flood protection measures

Within and around the case study area, four different flood protection investments to cope with the projected increase in future peak discharges will be considered. These measures are targeted towards lowering water levels under peak discharges, thereby reducing the risk of dike failure, increasing protection by building dikes, or limiting economic damage if a flood occurs. The four investment measures can be characterized by their actual returns in terms of damage saved under the different states of nature (Table 5.1). The protection measures that are being considered are shown in Figure 5.4 and briefly discussed below. Investment costs are not taken into account. The analyses' focus is on returns measured as the 'prevented damage'.

Asset D+: Reinforcement of dikes

This investment concerns the reinforcement of the primary dikes of the dike-ring area. This refers to heightening the river dikes so that they can withstand the design discharge level of 17,000 m^3/s. After having invested in this option, no damage can be expected at discharges from 17,000 m^3/s or lower. At higher discharges, the damage will be equal to the maximum potential damage, which is estimated at €17,993 million. The damage prevented in state C (or expected return) equals the maximum potential damage.

Asset Cp: Dividing a dike-ring into compartments

This strategy leaves the current primary dikes at the safety level below the design discharge of 16,000 m^3/s, but involves developing new 'internal dikes' within the dike-ring to protect the most economically valuable areas, such as urban and horticultural areas. In this example, it is assumed that three additional dikes will be built: two that protect urban areas and one that protects horticultural areas. The two new internal dikes for urban areas can withstand floods in state B: 17,000–18,000 m^3/s. The new internal dike for horticultural areas can withstand floods in state C: 16,000–17,000 m^3/s. The potential damage for urban areas and horticultural areas is estimated at €8,000 million and €4,000 million, respectively (DWW 2001). This therefore results in €12,000 million avoided damage for asset Cp in state C and €8,000 million avoided damage in state B.

Asset Fp: Reducing vulnerability to flooding of individual properties
This investment strategy consists of the flood-proofing of residences. It comprises developing additional flood protection measures for individual houses including, for example, the development of floating houses and elevated houses (see Chapter 6). These measures limit economic damage once a flood occurs. It is estimated that, through additional flood protection measures for all houses in the area, about €3,000 million damage may be prevented in the case of any of the three flood states from 16,000 m^3/s and higher discharges considered here.

Asset R: Creating upstream retention areas
This option consists of the development of upstream areas that are designed to temporarily store water in case of a peak discharge. During a peak discharge, the area will be deliberately flooded to lower the peak and, thereby, lower the water levels downstream. In this example, it is expected that using a 'flood storage area' will prevent damage to dike-ring area 43 at discharge levels between 16,000 and 18,000 m^3/s. Although the retention area only holds a few urban settlements which can be quickly evacuated in case of a flood, flooding of the retention area will cause economic damage to that area (for example, to crops or nature areas) of about €4,000 million.

5.4.4 Calculation of the portfolio return and variance

Using Eqns 5.1–5.7, it is possible to calculate optimal sets of flood protection portfolios. For this purpose, first the expected returns and variances of the returns on individual assets have to be calculated. Subsequently, the expected returns and variances of portfolios can be estimated. The expected return $\overline{R_i}$ on each individual asset (Eqn 5.1) can be calculated using the actual returns and probabilities that are

Table 5.2 *Expected return, variance and standard deviation (SD) per asset under the current climate scenario*

	Expected return	Variance	SD	SD/ Return
	(€million per year)			
Asset D+	14.39	258791.24	508.72	35.34
Asset Cp	11.60	131065.44	362.03	31.21
Asset Fp	3.53	10562.57	102.77	29.16
Asset R	14.20	207598.36	455.63	32.09

provided in Table 5.1. For example, the expected return on asset D+ under the current climate scenario is calculated as follows:

$$\overline{R_{D+}} = 0.00013 * 0 + 0.00025 * 0 + 0.0008 * 17993 + 0.99883 * 0 = 14.3944 \quad (5.8)$$

The variance V_i of each individual asset can be calculated with Eqn 5.2, which yields the values shown in Table 5.2.

The estimated expected returns and variances in Table 5.2 indicate that the expected return on investment strategy D+ is the highest, but the variation in returns is also the highest for this investment. This means that, on average, the prevented damage is high if the government decides to invest in heightening primary dikes. The disadvantage of this strategy is that risk, defined as variance, is very high as well, since the dike only prevents damage up to a certain level of river discharge, and for higher levels of discharge the full potential damage is suffered. The other investment strategies have lower expected returns, but the variance in these returns is also lower. Expected returns are lower because these investment strategies do not prevent all of the maximum potential damage in a state. The advantage of these investment strategies is that risks are smaller, since they prevent damage in more than one state. The individual risk and return characteristics of these assets suggest that a combination of these investments may be desirable in order to reduce overall risk. This will be examined by calculating the expected returns and risks of different portfolios of assets.

The expected portfolio return can be calculated using both the expected returns of the individual assets as displayed in Table 5.2 and the

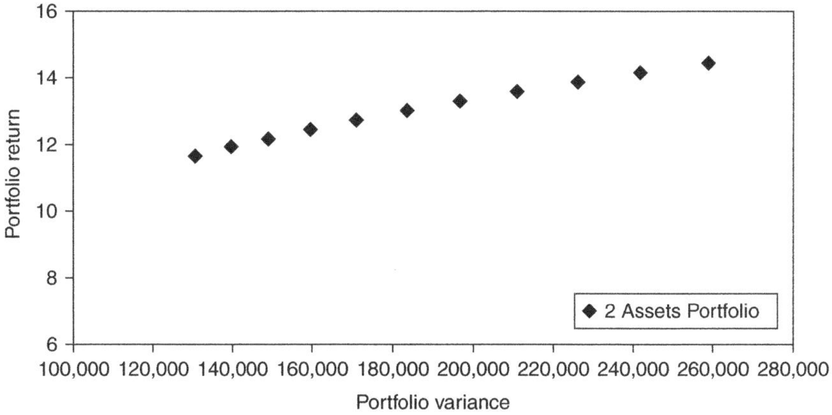

Figure 5.5 Portfolio return and variance for portfolios consisting of two assets (D+ and Cp) under the current climate scenario

shares x_i of each asset A_i in a portfolio. By choosing a variety of shares, different portfolios can be constructed. These shares can be interpreted as the percentage share of total government budget available for water management investments that is spent on a specific investment strategy. Obviously, the expected returns of investments depend on investment shares. For the sake of simplicity, it is here assumed that by lowering the investment share in heightening primary dikes, the resulting primary dike will be lower. As a consequence, the prevented damage of the D+ investment strategy will be lower as well. Lowering the investment share in asset Cp implies that the area protected by compartments will be smaller and the damage prevented by these compartments will be lower. Decreasing the investment share of flood-proofing houses means that fewer houses will be flood-proofed and, correspondingly, damage prevented by this strategy will be lower compared with the 100 per cent flood-proofing scenario. Investing less in retention areas implies that the areas that are deliberately flooded are smaller, which again lowers the damage prevented in dike-ring area 43. Obviously this reasoning is reversed if investment shares of specific investments are increased.

By filling in expected and actual returns per asset using Eqn 5.4 and calculating the portfolio variance using Eqn 5.7, the expected returns and variances of portfolios can be estimated. These results for several portfolios are presented in Figures 5.5 and 5.6. Figure 5.5 shows two-asset portfolios and Figure 5.6 shows two-, three- and four-asset

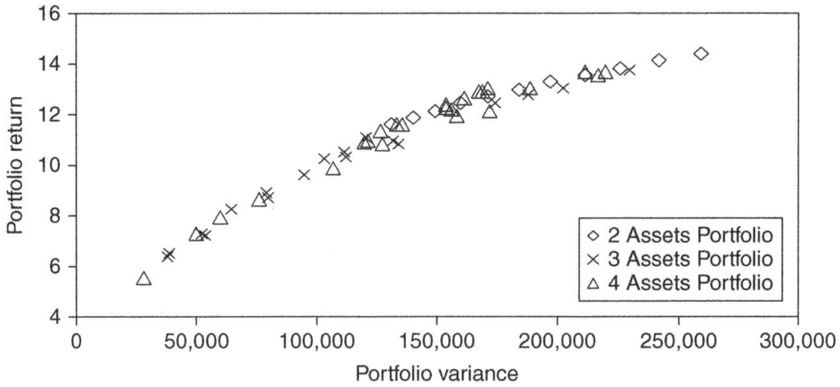

Figure 5.6 Portfolio return and variance for two- (D+ and Cp), three- (D+, Cp and Fp) and all four-asset portfolios under the current climate scenario

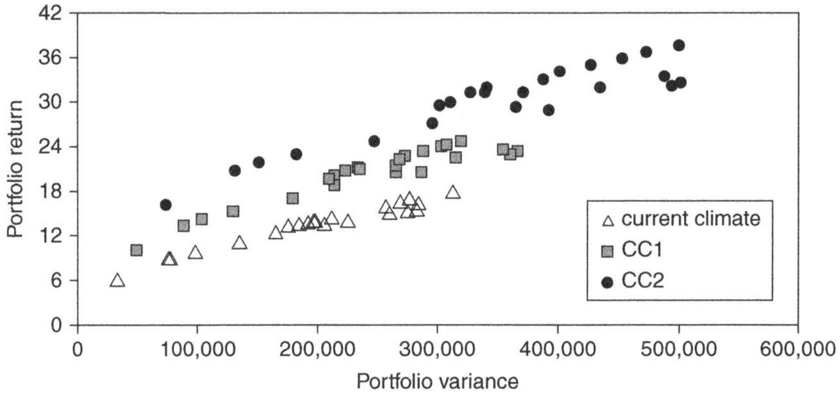

Figure 5.7 Portfolio return and variance for portfolios consisting of four assets under the current, CC1 and CC2 climate scenarios

portfolios under the current climate scenario. Finally, the same calculations are conducted using the probabilities of the climate change scenarios CC1 and CC2 (Table 5.1). These results for portfolios consisting of all four assets are displayed in Figure 5.7.

5.4.5 Discussion and interpretation of results

Figure 5.5 shows the portfolio variance and expected return on several portfolios that consist of the assets D+ (heightening dikes) and Cp (compartments). The upper dot represents the portfolio with a 100 per cent share in D+ and 0 per cent in Cp, which corresponds to a high expected return but also a high risk. Portfolios that are positioned more to the left can be obtained by increasing the investment share of Cp and reducing the share of D+. The middle dot is the portfolio that consists of 50 per cent of D+ and 50 per cent of Cp, and the lower left dot represents a portfolio of 100 per cent Cp. This figure shows that increasing the investment share in Cp reduces overall risk. However, this is only possible at the expense of expected return. This analysis indicates that diversification of investments has the potential to reduce overall risks, compared with the 100 per cent D+ portfolio, or increase the expected return, compared with the 100 per cent Cp portfolio. Thus, water managers can obtain their desired risk and return by choosing specific investment shares.

Benefits of diversification can be larger when more investments are available, as Figure 5.6 shows. This figure adds a three-assets portfolio curve, which consists of the assets D+, Cp and Fp (flood-proofing), and a four-asset portfolio curve consisting of assets D+, Cp, Fp and R (retention areas). From this figure it is apparent that adding assets to the portfolio increases the range of possible return and risk characteristics. The overall portfolio risk can be reduced considerably when the portfolio includes asset Fp, since the three-asset portfolio curve includes variances between 38,000 and 130,000, which cannot be attained by the two-asset portfolio curve. Table 5.3 explains why adding asset Fp to the two-asset portfolio considerably lowers the portfolio variance; it appears that the correlation between the assets Fp and D+ is relatively low, namely 0.82 (Eqn 5.6). The lower the correlations of returns between investments, the larger are the benefits of diversification, as discussed in Section 5.3. This suggests that combining investments in dike reinforcement with the flood-proofing of buildings is especially attractive for diversification opportunities, because the returns on these investments have the lowest correlation.

The portfolio variance can be decreased slightly relative to the three-asset portfolio when the fourth asset, R, is included in the portfolio. The lowest variance portfolio is achieved by having approximately 60–70 per cent of asset Fp, 10–20 per cent of assets D+ and Cp, and a small fraction (5–10 per cent) of asset R in the portfolio. Apparently, asset Fp correlates less strongly with the other assets in terms of their return and hence diversifies the portfolio variance relatively

Table 5.3 *Correlation coefficients between pairs of assets*

Pair of assets	Correlation
Asset D+ and Cp	0.94
Asset D+ and Fp	0.82
Asset D+ and R	0.93
Asset Cp and Fp	0.93
Asset Cp and R	0.99
Asset Fp and R	0.89

well. Furthermore, the four-asset portfolio curve has a slightly higher expected return than the two- and three-asset portfolios for variances between 150,000 and 170,000, whereas expected returns are larger for the three-asset portfolios at variances of 100,000. This suggests that distinct portfolio mixes can be desirable for different risk preferences.

Figure 5.7 shows that climate change may have an important impact on portfolio returns and variances. This figure shows the expected return and variance of several portfolios that consist of all four assets under the two aforementioned climate change scenarios. As the probability of flooding increases in both climate change scenarios CC1 and CC2, the expected portfolio return ('the prevented damage') will rise as well, because for most assets the expected return increases. Obviously, investing in preventing and limiting flood damage has a higher return when the probability of damage increases. Note, however, that the portfolio variance also increases under both climate change scenarios CC1 and CC2. This suggests that the necessity for diversification of water management investments increases due to climate change, in cases where an increase in risk is regarded as undesirable. For example, the individual risk level of the D+ strategy (100 per cent D+) in the current climate scenario can only be obtained through diversification in the CC1 scenario.

This application indicates that diversification of water management strategies has the potential to increase expected returns or reduce risks. The variance of expected returns can be reduced considerably compared with the strategy of investing only in the heightening of primary dikes. Therefore, the probability of suffering extreme flood losses reduces if investments in primary dikes are combined with the other investments discussed in this chapter. The analysis indicates that in most portfolios a trade-off between risk and return exists, which implies that lower risk levels can only be obtained by accepting lower expected returns. Calculations of risks and returns of portfolios under different climate scenarios show that increases in risk (i.e. variance in returns) can be limited through diversification. The costs of these investments should then be subject to an evaluation as well, but this is outside the scope of this chapter. Several issues can be included in future research. Apart from the return and variance of a portfolio, the choice of a particular flood protection portfolio is obviously determined by more factors than those mentioned in the case study. For example, the investment costs of the different assets have been ignored. Another restriction is the budget available for new investments in flood protection. Indeed, due to budget limitations, a less preferred portfolio may be selected. Another topic for further research is to consider multiple goals in water management investments (e.g. Penning-Rowsell et al. 2005). For example, environmental values and social aspects, such as preventing loss of lives, could be optimized in addition to maximizing the return on water management investments and minimizing its variance.

5.5 Conclusions

Climate change is expected to increase flood risk in several regions of the world, which requires the undertaking of adaptation measures aimed at limiting the increase in the probability of, or damage caused by, flooding, or a combination of both. Long-term developments such as climate change are inherently uncertain and therefore it is difficult to predict what implications they have for investments in flood management. Most studies on the effectiveness of flood management measures evaluate the ability of such a measure in isolation from other strategies that could be adopted to reduce risk. This chapter presented modern portfolio theory as a framework for evaluating portfolios of flood management options in terms of their return (avoided damage) and risk (variance of this return). The application of MPT to flood risk management has been illustrated for a case study area in the Netherlands, but obviously this framework has broader relevance and could be applied to other countries and natural disaster risks.

This chapter examined whether damage mitigation measures, such as the flood-proofing of houses, are complementary to a diversity of traditional water management strategies in reducing risks. The application indicates that diversification of different water management investments, such as heightening dikes and mitigating damage, has the potential to increase expected returns or to reduce risks. For example, combining investments in the strengthening of dikes with flood-proofing of houses results in lower variances in return (risk) compared with only investing in dikes, which means that the probability of suffering catastrophic damage decreases. The reason for this result is that damage mitigation measures are able to limit damage in the situation when a dike fails and has a return of zero. The analysis further showed the importance of understanding the correlation of the returns on various flood management investments and that adding partly uncorrelated investments lowers the risk of the total portfolio. It appears that combining risk prevention measures (e.g. dike infrastructure) with damage mitigation measures (e.g. elevating buildings) can be especially attractive because these investments have the lowest correlation of returns. The analysis has, moreover, been conducted under various scenarios of increased flood probabilities caused by climate change. The results indicate that diversification of water management investments is necessary in the face of climate change if an increase in risk is regarded as undesirable. These findings suggest that the implementation of a diversified portfolio of adaptation measures that are able to limit risk in different flooding conditions is likely to be an effective strategy to increase resilience to climate change.

6 Damage mitigation measures at the household level and climate change adaptation

6.1 Introduction

Climate change is expected to increase flood risk in low-lying coastal regions and river deltas due to more frequent extreme river discharges and sea level rise (IPCC 2007). The risk of potential flood damage may be amplified by socio-economic developments; notably, economic growth and urban expansion in vulnerable locations. Adaptation strategies are therefore needed so as to limit the projected increase in flood risk. Such adaptation measures can be beneficial for both insurance companies and policyholders. If effective risk-reducing measures are put in place then insurers may experience to a lesser degree the adverse consequences of climate change on their business (as described in Chapter 3) and experience fewer catastrophic losses, while lower risk, in general, implies lower insurance premiums and/or higher levels of coverage for policyholders.

Chapter 5 shows that a combination of water management strategies, such as reinforcing dikes and flood-proofing structures, is likely to be more effective in reducing the likelihood of suffering extreme damage than investing only in the prevention of floods. Furthermore, research on the effects of promoting flood risk reduction activities through flood insurance, and on building codes and other flood-proofing measures, indicates that preventative action undertaken by households (also referred to as mitigation measures) may be an effective means of limiting residual flood risk.[1] For example, research on the effectiveness of mitigation measures during the 2002 extreme flooding of the River Elbe in Germany shows that damage avoidance can be substantial if households are adequately prepared for floods (Kreibich

Parts of this chapter have previously been published as Botzen et al. (2012).

[1] The residual risk is the flood risk remaining after investing in dike infrastructure.

et al. 2005; Thieken et al. 2006). This chapter provides an overview of several international studies on the effectiveness of flood mitigation measures that are implemented at the household level. This chapter also examines a recent proposal to elevate newly built structures in the Netherlands, and the willingness of Dutch homeowners to invest in elevation. This provides insights into the function that such measures can play in programmes for reducing the impacts of climate change on flood risk.

Adequate policies for adaptation to the impacts of climate change are especially important for a low-lying country such as the Netherlands, since about 53 per cent of the population (total population 16.7 million people) live, and two-thirds of the national income is earned, in areas below sea or river water level. The Netherlands is a delta area and is at risk from storm surges of the North Sea and peak discharges of the rivers Rhine, Meuse and Scheldt (e.g. Aerts et al. 2006; Middelkoop et al. 2001; P.J. Ward et al. 2008). Recent research has found that the consequences of flooding in the Netherlands can be catastrophic both in terms of economic losses and the loss of human lives, despite having the highest flood safety standards in the world (Aerts et al. 2008b). Climate change adaptation strategies, including flood mitigation measures, are being investigated in the Netherlands. For example, the recently executed study 'Attention for Safety' proposes to raise all newly built homes in the Netherlands to several metres above sea level as a viable long-run strategy to accommodate rising flood risk (Aerts et al. 2008b). This strategy is proposed as a complementary strategy to traditional dike reinforcements, and is predominantly geared towards lowering potential flood damage. In contrast to mitigation through insurance, which reduces residual risk, elevating homes completely eliminates the residual flood risk of newly built homes.

Experience in the USA, however, indicates that homeowners rarely undertake damage mitigation voluntarily (Kunreuther et al. 2011). In view of this, a relevant question is whether households would be willing to pay for elevating newly built houses in the Netherlands. This is examined in this chapter by presenting the results of a survey among 473 homeowners in the river delta of the Netherlands. It sought to discover whether homeowners are willing to invest in eliminating flood risk by elevating a new house, and how this decision relates to the perceptions of respondents about the impacts of climate change and risk of flooding, as well as household characteristics. The results of this survey are also compared with another economic valuation study that examined the willingness of homeowners to purchase flood insurance and to invest in more simple risk-reducing measures, such as purchasing sandbags or water barriers, to avoid low inundation floods (see Chapters 7 and 11). This comparison provides insights into the premium which individuals are willing to pay to forego flood risk completely by elevating their home, instead of limiting its damage by purchasing insurance. This provides an indication of the monetary valuation of the welfare loss caused by the emotional stress of floods.

The remainder of this chapter is structured as follows. Section 6.2 provides an overview of international studies of the effectiveness of damage mitigation measures implemented at the household level. Section 6.3 discusses the recently proposed water management strategy to elevate newly built homes above potential water levels in the Netherlands. Section 6.4 presents the survey results dealing with the willingness of homeowners to elevate their home. Section 6.5 concludes the chapter.

6.2 International studies on the effectiveness of flood mitigation measures

6.2.1 United States of America

Kunreuther et al. (2011) estimated the potential effectiveness of damage mitigation measures in reducing hurricane risks in Florida, New York, South Carolina and Texas. They used catastrophe models to assess how much damage could be reduced by mitigation through building codes for hurricanes with different return periods. Their study estimates how much money would be saved if all houses and buildings complied with the most recent building code standards. Obviously this is not a practical or realistic scenario because it would be very costly, maybe even impossible, to subject all older buildings to these standards. Nevertheless, this exercise gives an indication of the potential maximum gain of implementing damage mitigation measures. The analysis of Kunreuther et al. (2011) shows that mitigation can reduce losses caused by a 1/100-year hurricane by 61 per cent in Florida, 44 per cent in South Carolina, 39 per cent in New York and 34 per cent in Texas. In conclusion, the potential gain from hurricane-proofing existing buildings can be substantial.

Table 6.1 *Differences in state requirements for local government building code enforcement and comprehensive plans for urban development that consider natural disaster risks in US states, and mean per capita flood insurance claim payments between 1978 and 2002*

State requirements	States	Mean per capita claim payments
No state local government building code enforcement or comprehensive plan requirements	Alabama, Louisiana, Mississippi, New Hampshire, Pennsylvania, Texas	$299
State local government building code enforcement requirement but no comprehensive plan requirements	Connecticut, New Jersey, New York	$79
State local government comprehensive plan requirements but no building code requirements	Delaware, Georgia, Hawaii, Maine, South Carolina	$137
Both state and local government building code and comprehensive plan requirements	Alaska, California, Florida, Maryland, Massachusetts, North Carolina, Oregon, Rhode Island, Virginia, Washington	$99

Source: Adapted from Burby (2006).

Burby (2006) examined requirements regarding local enforcement of (flood-resistant) building codes and local planning for urban development in the Atlantic, Gulf and Pacific states in the USA, and related this to claims paid by the National Flood Insurance Program (NFIP) in these states between 1978 and 2002. The results of this study are presented in Table 6.1. Six states do not require either building codes or comprehensive plans of urban development that consider natural disaster risks, while three states require their local government to enforce building codes but not comprehensive plans. The majority of the states examined by Burby (2006) require either comprehensive plans but no building codes (five states) or both comprehensive plans and building codes (ten states). The last column in the table shows that the total of NFIP claim payments between 1978 and 2002 averaged per capita were considerably higher ($299) in the states without local government building code enforcement and without comprehensive plan requirements than in states where one or both of such policies were in place. These results suggest that implementing flood-proofing measures at the household level can substantially reduce flood damage in the USA. This is confirmed by Pasterick (1998) who shows that buildings that comply with NFIP flood mitigation requirements, such as elevating new structures, suffer, on average, six times less flood damage than buildings that do not have such measures in place.

An extensive study of the effectiveness of a variety of mitigation measures that have been financed by the Federal Emergency Management Agency (FEMA) has been conducted by the National Institute of Building Sciences (2005). In particular, this study has examined the cost-effectiveness of mitigation activities in several representative communities in the USA. The overall findings are positive, since the benefits of mitigation activities outweigh the costs. Reducing flood risk through mitigation has, on average, a benefit–cost ratio of 4.1. A focal point of the mitigation activities that are required and encouraged by the NFIP is the elevation of buildings. The National Institute of Building Sciences (2005) examined the effectiveness of a project that elevated buildings in the community of Freeport (New York), and showed that the benefits of elevation outweigh the costs by a factor of 5.7 to 1. This is consistent with the findings of Jones et al. (2006), who conducted cost–benefit analyses of elevating single-family homes in the USA up to 4 feet above the expected water level of the 1/100-year flood. The analyses were performed for various discount rates, flood conditions, elevation methods and damage functions. The results show that, most of the time, the benefits of elevation in terms of reduced flood vulnerability exceed its costs, especially in coastal flood zones.

6.2.2 United Kingdom

Defra (2008) estimated how much damage might be reduced by implementing damage mitigation measures in the UK. The results show that temporary resistance measures, such as temporary flood guards and air brick covers, could reduce damage by about 50 per cent. Additional investments in permanent resistance could increase damage savings up

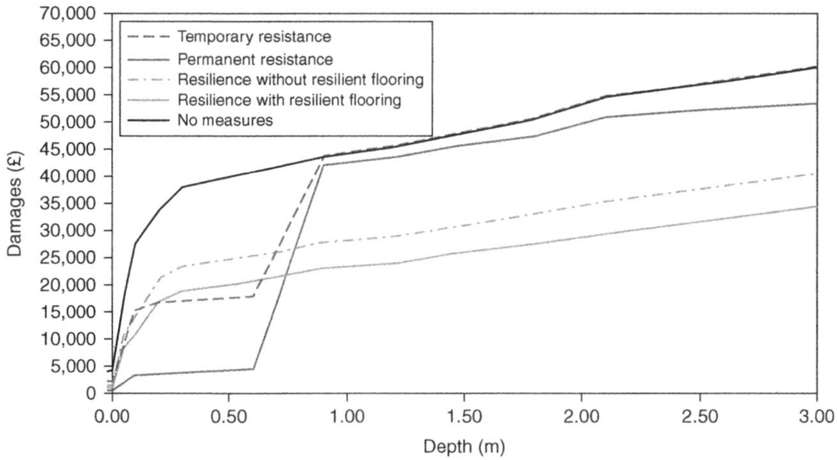

Figure 6.1 Expected flood damage in the UK in relation to flood depth and implemented mitigation measures
Source: Defra (2008).

to 65 per cent or 84 per cent depending on flood conditions, but are more expensive than temporary resistance measures. Resilience measures, which include the use of flood-resilient plaster, resilient kitchens and resilient flooring, are structural flood damage mitigation measures which are especially useful for buildings that are frequently flooded. Figure 6.1 shows the expected flood damage to a building if specific flood mitigation measures have been implemented in relation to the water depth during a flood. This figure shows that resistance measures are more effective than resilience measures at low water levels. However, resilience measures are more effective for floods at higher water levels, which is mainly due to their ability to reduce the costs of restoration, clean-up and alternative accommodation, and because resistance measures are less suitable for blocking high flood waters.

In examining the effectiveness of mitigation measures that will remain functional for a long time in the future, it is appropriate to estimate their ability to reduce risk under scenarios of altered flood risk due to climate change. This was done by Dawson et al. (2011) who examined the effectiveness of implementing non-structural flood mitigation measures in terms of reduction in flood risk in the Thames Estuary under various scenarios of socio-economic and climate change. The benefits of using insurance and planning policies to reduce flood risks are estimated to decrease damage, compared with business-as-usual, between 31 per cent and 50 per cent in 2020, depending on whether or not a large shock to the system (such as a hike in global reinsurance prices) or a major storm surge occurs in the UK. These adverse events would increase insurance prices and thereby provide an incentive to limit new development in flood-prone areas. Moreover, planning policies could be implemented to restrict development in high-risk areas. Of course, these results do not imply that a major disaster is desirable, but it is an interesting observation that large risk reductions may be obtained by steering development towards low-risk locations by means of insurance and planning policies. Furthermore, Dawson et al. (2011) estimated the effectiveness of combining risk reduction through insurance and adaptation policies with the following four types of mitigation measures:

- temporary resistance measures that prevent water from entering a building and then pump it away
- permanent resistance measures that make a building permanently watertight
- resilience without flooring measures, which entails building with waterproof materials and raising electric sockets and appliances
- resilience with flooring, which is the same as above but with concrete/sealed floors.

The results show that implementing these measures could reduce future flood risk by 33 per cent, on average, across climate and socio-economic scenarios of future flood risk, compared with implementing only structural adaptation measures such as flood protection infrastructure.

6.2.3 Germany

Kreibich et al. (2005) interviewed householders from 1,248 households that were affected by the severe River Elbe flood in 2002 in order to assess the level of preparedness of households for flooding, and to estimate the effectiveness of damage mitigation measures that households implemented before and during the flood. Overall, this study shows that the potential gains of implementing mitigation measures at the household level can be substantial. The results show that buildings without a cellar suffer about 24 per cent less building damage and 22 per cent less damage to contents. Water barriers reduced flood damage by about 29 per cent. A stable building foundation or waterproof sealed cellar walls reduced flood damage to buildings by about 24 per cent. The most effective strategies were flood-adapted building use and flood-adapted interior fittings. Flood-adapted building use means that the parts of the building that can be flooded (such as the cellar and ground floor) are not used cost-intensively and do not include expensive constructions such as a sauna. Flood-adapted interior fittings means that only waterproofed building materials and furniture and contents that can be easily moved to higher floors are used in flood-prone parts of the building. Flood-adapted building use reduced damage to buildings and contents by, respectively, 46 per cent and 48 per cent, while flood-adapted interior fittings reduced damage to both buildings and contents by 53 per cent. Placing utility and electrical installations on higher floors reduces flood damage by 36 per cent. These results of the effectiveness of flood mitigation measures in Germany were confirmed by Kreibich and Thieken (2009), who conducted a similar survey after floods in 2005 and 2006 in the city of Dresden. The results of this survey indicate that household preparedness improved before the 2005/2006 floods, compared with the 2002 Elbe flood, and that this improved preparedness resulted in significantly less flood damage. Kreibich et al. (2011) show that the implementation of low-cost mitigation measures, such as the securing of oil tanks and the installation of mobile flood walls, are cost-effective in Germany under a range of flood conditions and discount rates.

6.3 Elevating buildings as a sustainable long-run water management strategy

Until now, investments by the Dutch government in flood risk management have been mainly targeted towards new flood defence infrastructure, such as dikes and storm surge barriers (Kabat et al. 2005). In addition, projects that create more space for rivers are being undertaken, and these also aim at maintaining or lowering the probability of flooding (e.g. Vis et al. 2003). Heightening dikes is effective in terms of its costs and benefits, where the benefits refer to reducing the frequency of floods (Aerts et al. 2008b; Janssen and Jorissen 1997). However, it is increasingly recognized that simply raising dikes is insufficient to achieve a sustainable situation in the long run, because heightening dikes merely reduces the probability that a flood will occur, while the potential damage caused by the flood continues to increase. The latter trend has two reasons. First, most polder areas in the Netherlands are below mean sea level and can in fact be considered as a bathtub. A higher dike lowers the probability of a flood, but also results in higher maximum floodwater levels, and the resulting damage may be potentially higher if it fails. Second, heightening dikes increases the feeling of safety among households and investors, which stimulates further economic and urban development in vulnerable areas (Vis et al. 2003). The latter has been called the 'levee effect'. It is projected that between 500,000 and 1,500,000 new houses will be built in the next 30–40 years in the Netherlands (MNP 2007). These developments, together with economic growth, are projected to increase potential flood damage (at a constant price level) by a factor of 6–7 in the next 80 years (Aerts et al. 2008b). These aspects pose the question whether raising dikes will in itself be sufficient to cope with increasing flood risks, or whether additional adaptation measures should address new spatial planning policies that limit development in floodplains, and new building codes that make houses less vulnerable to flooding (e.g. Aerts et al. 2008b; Olsthoorn et al. 2008).

Aerts et al. (2008b) examined several long-term strategies to cope with the combined effects of climate change and economic trends on flood risk in the Netherlands. In order to reduce flood damage, the study proposed elevating new urban areas with sand from the North Sea to 5 metres above sea level. Given that the Netherlands is a very densely populated country, more efficient use of space through such 'multifunctional land use' relieves pressure on land area. Building new houses on elevated ground does not only eliminate the flood risk for these houses, but potentially confers additional benefits. For example, in elevated areas, infrastructure such as pipelines, powergrids and even railroads can be constructed at lower cost by first developing the infrastructure above ground and then covering the infrastructure with a layer of sand, on top of which new buildings and residential areas can be developed. Another benefit is that new flood protection for existing buildings can be created through elevating new neighbourhoods around low-lying city centres and connecting these higher grounds with dikes. These new dike-rings around old city centres may protect cities once the existing coastal protection has failed during a storm surge. Living in such elevated areas is not very much different from living in low-lying areas, since whole streets or neighbourhoods would be elevated, so that accessibility (for example, for elderly people) is not impaired.

In order to elevate 1,000,000 new houses in the coming decades, about 40 million cubic metres of sand per year needs to be excavated, at a cost of approximately €0.4–1.6 billion per year (Aerts et al. 2008b). This also includes the price of adjusting the infrastructure of these new urban areas, such as roads, pipelines and rail infrastructure. Recent studies show that there is enough sand in the North Sea to meet the demands of this project (see van der Meulen et al. 2007). The cost of elevating an individual new house with a surface area of 150 square metres to a height of 5 metres above sea level with sand has been estimated as between €9,000 and €12,000, depending on the exact amount of sand needed and its price. Although these costs seem quite substantial in an absolute sense, they are only a small percentage of the price of a new house (approximately €220,000 on average in 2008). Future research should focus on assessing precisely how the elevation of structures can be organized as a cost-effective, complementary activity to current water management policies in limiting flood risks induced by climate change. To assess the feasibility of the adaptation strategy to elevate new homes, it is important to know whether households are willing to pay costs of this magnitude. This is examined here with a survey, as explained in the next section.

6.4 Willingness of households to invest in elevating houses

6.4.1 The survey

The willingness of householders to invest in the elevation of new houses was investigated in an extensive online survey that also enquired about the demand for flood insurance by homeowners in the Netherlands (Chapters 10 and 11). The structure of the survey was as follows. The questionnaire starts with questions about the experience of the respondent with flooding and their knowledge about the causes of flooding. In addition, several questions are included about the potential consequences of climate change and the respondent's perceptions of flood risks. These questions are designed to familiarize the respondent with the topic. A section dealing with demand for flood insurance follows, with some questions on the respondent's willingness to pay for the actions involved in the implementation of elevating a house. It is explained that this would completely eliminate flood risks and implies that owners of these homes would not need any flood insurance (see Botzen et al. 2008b for an overview of the questions). The respondents were requested to imagine that they were going to buy a new house in 2015 and asked whether they would be willing to spend €10,000 to elevate their new house to a level that would be safe from flooding.[2] The questionnaire concluded with the usual socio-demographic questions.

Three pre-tests ($n = 88$) were conducted using face-to-face interviews to test the questionnaire, and a final pre-test ($n = 30$) was performed to trial the online survey. The survey was administered over the Internet using Sawtooth CBC software.[3] This computer-based method has the advantage that interviewer effects can be avoided and that a large geographically spread sample can be obtained at relatively low cost. Respondents were selected from the consumer panel of Multiscope.[4] The sample consisted of random draws of panel members living in the river delta in the Netherlands and included only homeowners. The resulting total number of completed questionnaires was 473. Figure 6.2 shows the location of the participants in the survey on a map of the Netherlands. Chapter 7 provides more information about the contents, pre-tests and implementation of the survey. The questions on precautionary measures discussed in Chapter 7 were included in different versions of the questionnaire than the question on elevating homes discussed here.

6.4.2 *Sample characteristics*

The average age of the respondents was 46 years. The proportion of respondents older than 60 years was about 15 per cent, which is less than is the case in the actual Dutch population. Fewer older individuals were represented in the Internet sample, because seniors are generally less active on the Internet than younger people. We do not regard

[2] Elevating houses using sand is not possible in areas with very wet ground, while such areas pose no problem for elevation with poles. Therefore we decided to ask respondents in these areas whether they would be willing to pay this amount for a house on poles instead of a house elevated by sand, so the question was applicable to the living areas of all respondents.

[3] See www.sawtoothsoftware.com.

[4] For more information, see www.multiscope.nl.

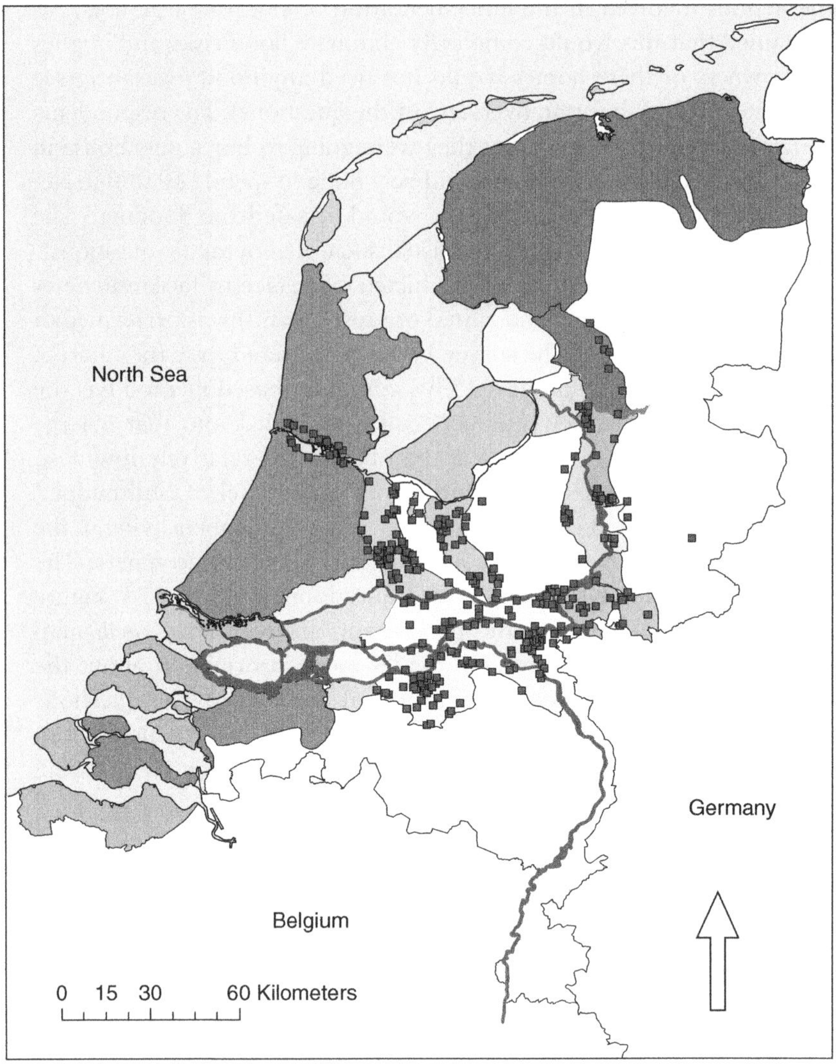

Figure 6.2 Location of the respondents to the survey on a map of the Netherlands

this as troublesome in this application, since it concerns buying a new house in the year 2015. This scenario is likely to be less applicable to older homeowners, who may be less willing to move and are likely either to stay in their current house or move into sheltered accommodation. Furthermore, the increased flood risk posed by climate change is less applicable to older respondents, since it will take several decades before these risks materialize. Our sample had slightly more male (59 per cent) than female respondents, and 45 per cent of the respondents had at least one child. About 39 per cent had a Bachelor's or Master's degree as their highest education level. The median and average after-tax household income fell into the category 'between €2,501 and €3,000 per month'. This is close to the average after-tax income of a household that owns a house in the Netherlands, which is €3,025 per month (Statistics Netherlands 2008).

6.4.3 Willingness to invest in eliminating flood risk through elevation

The answers indicate that slightly more than half of the respondents (52 per cent) would be willing to purchase an elevated house so that flood risks are eliminated, despite the substantial cost of €10,000. These results suggest that this mitigation measure could be very promising. This provides an excellent basis for undertaking future research on the feasibility and return on elevating newly built structures. The results discussed in Chapter 7 indicate that about 68 per cent of homeowners are willing to invest €15 in sandbags that can be used as a water barrier during floods, in exchange for a discount on their flood insurance premium. Hence, the proportion of homeowners who are willing to invest in elevation is not that much smaller than the proportion who are willing to invest in water barriers, despite the much higher cost of elevation (€10,000 compared with €15). This suggests that individuals regard a mitigation measure that completely eliminates flood risk as very attractive. This is consistent with findings in behavioural economics that individuals may place a considerable value on reducing low-probability risk to a probability of zero, which is the so-called certainty effect (Tversky and Kahneman 1986). Moreover, recent theories on individual decision making under risk, such as rank-dependent and prospect theories, indicate that individuals can place a large value on reducing low-probability, high-impact risks, such as flooding in the Netherlands (Chapter 8).

The finding that many homeowners are voluntarily willing to invest in reducing risk is in contrast to practical experience in the USA, where few individuals invest in measures that limit damage, such as the strengthening of roofs with the objective of limiting wind damage

(Kunreuther 2008). The elevation measure examined in this study may be very attractive to individuals because it completely eliminates risk, instead of merely reducing the damage. Individuals may place a premium on the elimination of the total risk compared with simply reducing the potential damage, possibly because of the stress and anxiety associated with flooding. Information on such a premium could be very valuable for cost–benefit analyses of investments in flood risk reduction, but is hardly considered in practice because of a lack of empirical estimates (Kind et al. 2008). As an illustration of the size of this premium, the monthly payments for elevation financed with a mortgage may be compared with the average monthly household willingness-to-pay (WTP) for flood insurance. The WTP for flood insurance indicates the value that individuals place on reducing their financial exposure to flooding, while WTP for elevation can be seen as the value placed on eliminating the emotional stress associated with floods in addition to eliminating potential damage. Suppose that the costs of elevation are financed with an annuity mortgage at the average home mortgage interest rate during the time of the survey of 5 per cent (Statistics Netherlands 2008). In that case, the monthly payment for a mortgage with duration of 20 and 30 years is, respectively, €55 and €67. The average estimated monthly WTP for flood insurance with 100 per cent coverage, as discussed in Chapter 11, is approximately €21 based on estimates of a choice model under current climatic and socio-economic conditions. The comparison between WTP for insurance and WTP for elevation indicates that many homeowners place a considerable premium on eliminating flood risk completely, compared with only financial compensation for damage. This 'safety premium' works out at approximately €35–45 per month.

6.4.4 Factors determining the willingness to invest in elevation

It is useful to examine how the investment decision relates to characteristics of the respondent, expected climate change impacts, and perceptions of flood risks. This information will allow policy makers to target relevant groups of individuals or households, namely those who will potentially be most interested in investing in elevation. In particular, the government could provide information or subsidies to stimulate the elevation of newly built houses or they could require elevation through building codes. The correlation coefficients of the

Table 6.2 ***Correlation coefficients for the decision to eliminate flood risk through elevation***

Variable	Correlation
Negative effects of climate change	0.09*
Perceived flood probability	0.12**
Lower flood risk than average resident	-0.12**
Higher flood risk than average resident	0.11*
Willingness to purchase insurance	0.14**
Close to a main river	0.12**

*Notes:** and ** indicate, respectively, significance at the 5% and 1% levels using a two-tailed test. Estimations were performed with SPSS software.

decision to invest in elevating homes and other variables are shown in Table 6.2 and are discussed in detail below.

6.4.5 Expected impacts of climate change

Perceptions of risk have been found to be an important determinant of individual decision making under risk (Slovic 2000). It has been examined in this chapter whether respondents who expect climate change to have considerable consequences for the Netherlands are more likely to invest in the elevation measure. This was analysed by relating the answers to the question regarding willingness to buy an elevated house with a variable that represents the expectations of the respondent about the negative effects of climate change in the Netherlands. The expected effect of climate change is an ordinal qualitative variable and this was coded according to the method described by Terza (1987). Chapter 7 describes the details of applying this method. The variable has six categories: 'very small', 'small', 'not small/not large', 'large', 'very large' and 'don't know' (coded as missing). Figure 6.3 shows the answers to the question about the size of these climatic effects. The majority of the respondents expected the negative effects to be large or very large (62 per cent), while about 14 per cent expected the negative effects to be small or very small. This suggests that most respondents are worried about climate change. The answers to this question are related positively and significantly (at the 5% level) to the decision to invest in elevation of a new house, as indicated by the correlation

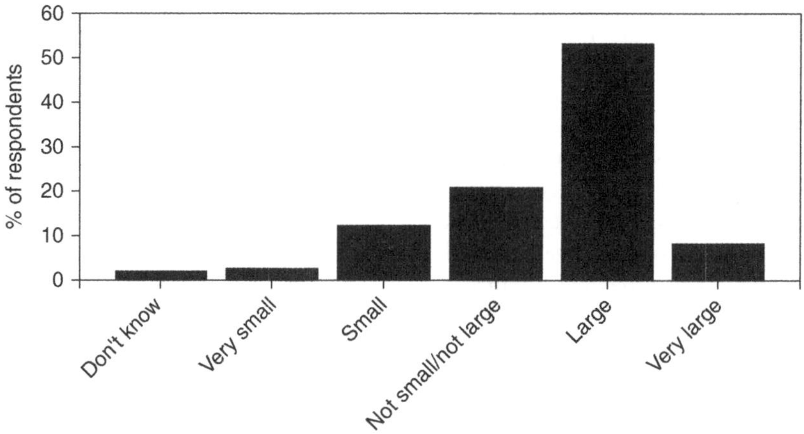

Figure 6.3 Respondents' expected negative effects of climate change for the Netherlands ($p = 0.05$)

statistic shown in Table 6.2. In other words, the larger the perception of homeowners of negative effects of climate change, the greater is the probability they are willing to invest in eliminating flood risk.

6.4.6 Respondents' perceptions of flood risk and geographical characteristics

Individual perceptions of flood risk also relate to the decision to invest in elevated homes. In particular, homeowners are more likely to invest in elevating homes the more they perceive their risk of flooding. Three variables that represent perception of flood risk by individuals are included in the correlation analysis. The first is based on a question that asks respondents to estimate their probability of flooding on a qualitative scale that is coded according to Terza (1987), with the following answer categories: 'I do not have any flood risk', 'very low', 'low', 'not low/not high', 'high', 'very high' and 'don't know' (coded as missing). The correlation coefficient of the relationship of this variable with the decision to invest in elevating houses is positive. The two other variables representing perceptions of flood risk are obtained from a question that asks respondents to rate their flood risk compared with an average resident in the Netherlands, with the answer options 'equal to average', 'higher than average' and 'lower than average'. The correlation coefficients indicate that the probability of investing in elevating a house is higher for homeowners who state that they have a higher flood risk than an average resident, and lower for homeowners who indicate that they have a lower flood risk than an average resident.[5] The three correlations paint a consistent picture.

Next, a variable was constructed that represents respondents who live within 5 kilometres of a major river. This was done with the use of geographical information systems (GIS), and was related to the respondents' zip codes. The correlation coefficient shows that homeowners who live close to a major river are more likely to invest in elevating houses, possibly because they are more aware of the risk of flooding in their residential area. Specifically, 59 per cent of homeowners who live close to a major river were willing to invest in elevating homes, while this was only 46 per cent for homeowners who live further away from a major river.

6.4.7 *Risk attitudes: Willingness to purchase flood insurance*

Risk-averse individuals generally have a positive willingness-to-pay value for insurance against the residual risk of flooding (Chapter 8). It seems likely that individuals who are willing to pay for financial coverage against flood damage by insurance are also more likely to invest in elevating homes to eliminate flood risk than are individuals who are not interested in buying flood insurance. A question was included in the survey that asks whether respondents are willing to pay for flood insurance under the current flood probability. A dummy variable that represents respondents who answered 'yes' to that question correlates positively with the decision to invest in elevating homes. In particular, 63 per cent of homeowners who were willing to pay for flood insurance were also willing to spend €10,000 for elevating their new home. This percentage was only 48 per cent for homeowners who were not willing to purchase flood insurance.

6.4.8 *Socio-economic characteristics*

Finally, we examine whether the decision to invest in elevation is related to socio-economic characteristics of the respondent, such as the value

[5] These two variables were coded as dummy variables.

of their current house, home contents, age, income and education. A positive correlation coefficient can be observed for income,[6] but this is only significant at the 10% level (two-tailed test). The other correlations are not statistically significant and therefore are not reported. In conclusion, perceptions of flood risk and expected effects of climate change, risk attitudes and geographical characteristics are more important determinants in the decision to invest in elevating houses than are socio-economic characteristics of the respondent.

6.5 Conclusions

Climate change may have serious consequences for flood risk in several regions of the world. Socio-economic developments and resulting urban developments in areas susceptible to flooding are an additional factor in increasing potential flood damage. The expected increase in flood risk can be limited by designing adequate adaptation strategies. Effective adaptation measures can be beneficial both for insurance companies, who then experience less adverse impacts from climate change, and also for policyholders because of their ameliorating effects on (risk-based) insurance premiums. Adaptation strategies may involve public investment in flood protection infrastructure and/or measures that reduce the vulnerability of property to flood damage. This chapter has focused on the latter, and has provided an overview of several international studies that provide empirical estimates of the effectiveness of flood damage mitigation measures that can be implemented at the household level. The results of these studies suggest that substantial savings in flood damage can be achieved by mitigation.

In particular, experience during the severe Elbe flood in Germany in 2002 shows that a diversity of measures that were implemented by households significantly reduced flood damage. Studies for the UK arrive at similar results, but also reveal that the effectiveness of flood mitigation measures depends on flooding conditions: measures aimed at preventing water from entering a building are only effective if water

[6] For the question about income, the respondent was asked to mark one of the following categories: <€750; €751–1,000; €1,001–1,250; €1,251–1,500; €1,501–2,000; €2,001–2,500; €2,501–3,000; €3,001–3,500; €3,501–4,000; >€4,000. A continuous income variable was constructed by setting the income of each respondent to the midpoint of the interval (€4,500 was used for the highest category).

depth is less than 1 metre, while other measures can prevent damage at greater water depths. Potential savings due to mitigation could also be substantial in the USA. Building codes can significantly decrease hurricane damage, and elevation of structures above expected floodwater levels is generally a cost-effective strategy in flood-prone areas.

In the Netherlands, most water management investments are currently targeted at controlling flood probabilities through dike reinforcement and heightening. This chapter examined additional adaptation investments that lower potential flood damage. Recent studies have suggested that elevating structures may be a good strategy to limit the impacts of floods, but a remaining question was whether or not homeowners are willing to incur the cost of elevation. This chapter reported the results of a survey that examined the willingness of Dutch homeowners to invest in elevating newly built houses. The results indicate that a considerable proportion of Dutch homeowners (52 per cent) are willing to invest in elevating their houses in order to eliminate flood risks. This suggests that this water management strategy holds promise for limiting potential flood damage.

A comparison between an indicator of the willingness-to-pay (WTP) for elevation and WTP for flood insurance indicates that individuals place a considerable premium on eliminating the risk completely by elevation, instead of only purchasing financial protection against the risk or investing in measures that only partly reduce damage. A reason for this may be the negative welfare effects of the emotional stress and anxiety associated with floods. Moreover, individual perceptions of flood risk and climate change influence the decision to invest in elevation. The results of this chapter show that the willingness of homeowners to invest in eliminating flood risk can be substantial, which provides a basis for further research into the potential role of risk-reducing measures at the household level in climate change adaptation policies.

7 *Insurance incentives for homeowners to invest in adaptation*

7.1 Introduction

The global trend in natural disaster damage and the projected increase in risks of a variety of extreme weather events in several parts of the world, which have been outlined in Chapters 3 and 4, highlight the need to implement measures that ameliorate, and limit, the impacts of natural disasters on human societies. From the point of view of climate change, it is especially important that governments design programmes and implement strategies that prevent natural disasters from occurring and that limit the damage once a disaster happens. In addition, the previous two chapters have shown that damage mitigation measures undertaken by households which reduce the vulnerability of property to natural disasters can perform a useful function in climate change adaptation strategy. This chapter explores how insurance can contribute to achieving risk reduction by encouraging policyholders to implement such damage mitigation measures, with a particular focus on flooding.

The previous chapter showed that international experience suggests that mitigation measures implemented at the household level can be an effective adaptation strategy to limit flood damage. For example, precautionary measures undertaken by households were very effective in limiting flood damage in Germany during the extreme flood event of the River Elbe in 2002 (Kreibich et al. 2005; Thieken et al. 2005).[1] Flood damage could be limited by adapting the use of buildings to flooding, which means that cellars and storeys susceptible to flooding are not used cost-intensively, and by adapting interior fittings to flooding, which includes the use of waterproofed building

Parts of this chapter (in particular Sections 7.5 and 7.6) have previously been published as Botzen et al. (2009a).

[1] The Elbe flood of 2002 is regarded as an extreme flood event. The resulting damage amounted to about €11.6 billion and 21 people were killed (Thieken et al. 2006).

materials and the placing of easily movable furniture on the lower floors of a building. The use of buildings and interior fittings adapted to flooding were especially effective and reduced damage to buildings by 46 per cent and 53 per cent, and damage to contents by 48 per cent and 53 per cent, respectively (Kreibich et al. 2005). Thieken et al. (2006) showed that insured households undertook more mitigation actions during the Elbe flood than uninsured households did. In particular, 28.5 per cent of the insured households undertook at least one of the mitigation measures examined, compared with 20.5 per cent of the uninsured households.[2] Another example is the National Flood Insurance Program (NFIP) in the USA, which by setting (compulsory) mitigation standards contributed to the reduction of flood losses to new structures by about six times (Pasterick 1998). Nevertheless, the NFIP failed to restrain development in floodplains, which may have been because many premiums are not risk-based but are partly subsidized (Burby 2001).[3] For example, during the last few decades, considerable development of new structures has taken place in New Orleans, which increased the damage caused by Hurricane Katrina (Burby 2006).

Even though reducing the vulnerability of structures to flooding can prevent a considerable amount of damage during floods, it has been observed in practice that many households do not invest in mitigation measures, even if they are cost-effective (Kunreuther 2006a). It is therefore of interest to examine how insurance can be applied to provide financial incentives to encourage policyholders to limit risk. Insurance arrangements for flood risk may require households to undertake measures that mitigate damage or stimulate households to voluntarily undertake precautionary measures (e.g. Kleindorfer and

[2] In general, the potential benefits of insurance in mitigating flood damage are not fully used in Germany. Average market penetration is low, namely 10% for household contents and 4% for residences. Moreover, insurance companies put little effort in stimulating mitigation, even though mitigation measures have proven to be effective during the Elbe flood. Only 14% of insurance companies reward voluntarily undertaken flood protection. The higher number of insured households who undertake mitigation, compared with uninsured ones, may be due to the former group being more aware of the flood risk faced (Kreibich et al. 2005; Thieken et al. 2006).

[3] Premiums are risk-based for newly built structures in some areas where flood risk maps are accurate.

Kunreuther 1999; Kunreuther and Pauly 2006; Kunreuther and Rose 2004). For example, households who build their house in such a way that it is more resistant to flooding could be rewarded with premium discounts or higher levels of coverage. These mitigation measures may limit damage during floods and thus be complementary to traditional flood protection. In addition, insurance arrangements serve a useful function in providing financial protection against the residual risk that remains after flood protection infrastructure has reduced risk to optimal levels. Although a number of studies have addressed the issue of how to create incentives for households to control and reduce flood damage through insurance, empirical analyses of the effectiveness of such incentives are rare.

This chapter examines how insurance arrangements can be applied to stimulate investment in risk reduction, and provides a case study of the willingness of Dutch households to undertake voluntary mitigation measures in exchange for insurance benefits. A survey was undertaken among approximately 500 homeowners in the river delta region of the Netherlands. The respondents were asked whether they would be willing to undertake specific mitigation measures in return for benefits on a hypothetical flood insurance policy. The effectiveness of these mitigation measures in preventing damage caused by river flooding and reducing flood risk in the river delta under climate change was estimated. In addition, a statistical analysis examines how perceived risks of flooding, knowledge about flooding, experience with flooding, household and geographical characteristics, and responsibilities for covering flood damage influence the willingness of householders to undertake mitigation.

The remainder of this chapter is structured as follows. Section 7.2 outlines international experiences of the undertaking of voluntary mitigation measures and discusses how insurance arrangements can encourage mitigation. Section 7.3 briefly discusses the role of damage mitigation in Dutch flood risk management. Section 7.4 explains the survey used to elicit the willingness of homeowners to invest in mitigation in exchange for insurance benefits. Section 7.5 provides descriptive statistics derived from the answers to the mitigation questions and estimates the potential of mitigation to limit flood damage. Section 7.6 discusses the estimation results of a statistical analysis of the factors behind the decision of homeowners to invest in water barriers. Section 7.7 concludes the chapter.

7.2 International experience of the undertaking of voluntary mitigation measures and the role of insurance in encouraging mitigation

Several international studies have shown that, in general, few households voluntarily invest in measures that reduce the vulnerability of their property to flooding (Kreibich et al. 2005; Kreibich and Thieken 2009; Kunreuther 1978; Lamond et al. 2009). For example, Kreibich et al. (2005) conducted a survey among 1,248 households to examine, among other issues, the level of preparedness of households before the 2002 Elbe flood in Germany. In general, few people had undertaken damage mitigation measures prior to the flood. About 11 per cent of the respondents had used and fitted their house interior in a way that limits flood damage, 9 per cent had installed their heating and other utilities above the potential water level of a flood, 7 per cent had installed temporary water barriers and 6 per cent had a building structure that reduces vulnerability to floods, such as a building foundation that remains stable during floods or cellar walls that keep flood waters out of a building. Important factors of influence on the decision to take mitigation measures are previous experience of flooding, knowledge about whether the living area of the respondent is prone to flooding, and the perceived effectiveness of the measure. Flood preparedness increased slightly after the flood event. This positive effect of flood experience on mitigation investments has been confirmed by Kreibich and Thieken (2009), who compared how household flood preparedness in Dresden changed after the city suffered floods in 2002 and in 2005/6. Results of a household survey showed a considerable increase in flood preparedness after the first flood event, which implies that households were better prepared for the 2005/6 floods. As an illustration, only 13 per cent of the households had undertaken precautionary measures before the 2002 flood. This increased to 67 per cent before the floods in 2005 and 2006. Because of the severe flood experiences in 2002, many people gathered information about flood mitigation measures and, as a consequence, household preparedness improved.

Similar levels of individual preparedness for flooding have been observed in the UK. Lamond et al. (2009) provide evidence on household flood preparedness, based on a survey of 400 floodplain residents in the UK. Despite the fact that about half of them had registered for flood warnings, fewer than 10 per cent had taken other mitigation measures, including the purchase of temporary or permanent flood barriers, or the installation of resilient fixtures and fittings. Flood experience appears to have a large influence on mitigation behaviour. Of people who had experienced a flood, 73 per cent registered for flood warnings, and these people also removed belongings in advance of the flood that they experienced. Moreover, 20 per cent of the people who had experienced a flood implemented other flood mitigation measures (Lamond et al. 2009). Similar findings were observed by Defra (2008), who conducted a survey of 1,100 floodplain residents in the UK. This research found that 6 per cent of households who have not been flooded previously had undertaken a flood mitigation measure, while this was 27 per cent for previously flooded households.

The observation that few households in flood-prone areas implement flood mitigation measures prior to floods seems to be widespread. In addition to the studies described above about household flood preparedness in the UK and Germany, individuals invest insufficiently in flood mitigation according to experience in other countries, such as in France (Poussin et al. 2011) and the USA (Burby 2001; Kunreuther et al. 2011). The decision of an individual to adopt a flood mitigation measure involves a comparison between the upfront investment costs of installing the measure and the reduction in future flood damage. Homeowners are more likely to undertake mitigation measures than tenants because they receive the full benefit of damage prevented to the home (Kreibich 2011). There are a variety of other factors that affect this decision. An important factor influencing whether people take mitigation measures is their perception of the risk they face. The assessment of a risk by an individual depends on the perceived frequency and the perceived consequences of a disaster, personal experience, anxiety, and media coverage of the hazard or public information campaigns. One reason why individuals do not invest in flood protection measures, even if they are cost-effective, is that their perception of the flood risk is too low. Individuals may adopt an 'it will not happen to me' strategy and ignore hazards with a low probability, and decide not to spend any money on mitigation measures (Kunreuther 1996). Flood risk perception may be especially low in areas behind flood protection infrastructure, where it has been a long time since the last flood happened, such as in the Netherlands.

A review of factors of influence on flood mitigation behaviour by Bubeck et al. (2012) identifies several other factors than risk perception that determine individual flood preparedness. In addition to risk perception, the perceived effectiveness and cost of mitigation measures are important factors of influence on the decision to undertake such a measure (Bubeck et al. 2012; Grothmann and Reusswig 2006). The upfront investment costs of mitigation may loom large in comparison with the future benefits that it might deliver. Individuals may have high discount rates, which has the effect that the future benefits of risk reduction receive little weight when evaluating the mitigation measure (Kunreuther 1996). Such individuals are unlikely to incur high upfront mitigation costs. Moreover, if individuals are not well informed about the effectiveness of mitigation measures, then they may insufficiently appreciate their benefits. Another factor of influence on the observed low adoption rates of mitigation may be that individuals expect that the government will provide compensation for natural disaster damage, which reduces their need to reduce risk themselves (Kreibich et al. 2011).

7.2.1 Encouraging mitigation through insurance

An important principle in actuarial practice and economics is that society's welfare is maximized by undertaking prevention as long as the costs of it are lower than the avoided expected harm of the hazard defined as *probability*damage* over the time frame that the prevention measure is in place (see the so-called 'learned hand' formula in Grossman et al. 2006). However, individuals often lack information about the expected harm before a disaster occurs, making it difficult for them to trade off the expected benefits of investing in mitigation measures and its costs, which may result in a suboptimal level of mitigation investments. Insurance companies with a specialized knowledge of the probability and expected damage of hazards could give a market price signal via premiums about the expected harm that individuals face. This price signal of risk may correct individual biases in judging the probabilities of natural hazards, and overcome the problem that people do not invest in mitigation because they have a too low perception of the risk and benefits of mitigation or ignore the low-probability risk altogether. If premiums are risk-based and insurance markets are competitive, insurance could stimulate the undertaking of damage mitigation measures by offering premium reductions to individuals who invest in mitigation. Individuals could be expected to undertake such investments if the premium reductions outweigh the costs of mitigation over time. Providing premium discounts to individuals who invest in mitigation provides a more tangible reward for mitigation investments than the benefit of, merely, a reduced flood risk in the absence of insurance premium reductions.

Furthermore, if premiums adequately reflect flood risk, then the cost of insurance will be higher in areas with a high risk of flooding than in low-risk areas, which provides an incentive for households to settle in areas with a lower flood risk. Obviously, some individuals may place a high value on living in certain areas near water and still want to locate there, despite a higher flood risk and insurance cost; for instance because of the amenity provided by a beach or riverbed. Nevertheless, overall development in areas with a high flood risk may be restrained if the relative cost of residing in such areas is high due to risk-based insurance premiums. Other ways in which insurance can encourage the adoption of protection measures is to require insurance customers to take specific measures before they become eligible for insurance or to provide higher levels of coverage to policyholders who invest in mitigation.

Few studies have provided a systematic evaluation of the effectiveness of incentives provided by insurance to stimulate damage mitigation in practice. Thieken et al. (2006) have shown that insured individuals in Germany may be more aware of flood risk, because they found that more insured households acquired information about flood damage mitigation measures (48.5 per cent) than uninsured households (33.9 per cent) and that more insured households undertook mitigation measures (28.5 per cent) than uninsured households (20.5 per cent). Nevertheless, Thieken et al. (2006) concluded that insurers could do more to stimulate the adoption of mitigation measures, by better informing policyholders about the benefits of mitigation measures and how they should be implemented. Moreover, German insurers could stimulate risk reduction through financial incentives, such as risk-based deductibles and also by rewarding investments in mitigation with premium discounts.

The National Flood Insurance Program (NFIP) in the USA aims to reduce flood risk, for instance by requiring the elevation of newly built structures in floodplains to the potential water level of the 1/100-year flood. Moreover, policyholders may be eligible for premium discounts if they elevate their building above the minimum elevation requirement. It appears that the NFIP has been quite successful in reducing the vulnerability to flooding of new structures in the 1/100-year flood zone. It has been estimated that the NFIP requirements and mitigation programmes have saved, in total, over $1 billion of flood losses per year (Sarmiento and Miller 2006). Moreover, cost–benefit analyses have shown that the premium discounts provided for elevation are in general cost-effective, but in some high-risk areas the discounts are insufficient (Jones et al. 2006). Even though the NFIP requirements are quite successful in flood-proofing new buildings, the programme is generally evaluated as being ineffective in limiting new developments in high-risk areas and reducing the vulnerability of (existing) older buildings to flood damage (Burby 2006). This may be explained by the fact that NFIP premiums are insufficiently differentiated in accordance with actual flood risk, and that premium levels are too low, at least for policyholders in flood-prone areas. Kunreuther et al. (2011) have argued that premiums reflecting risk is a fundamental principle for natural disaster insurance programmes, in order to provide adequate incentives for mitigation. If premiums do not reflect risk and are, for example, too low, then the price signal to restrain development in high-risk areas is insufficient. Also, if premiums are below actual risk then insurers are unlikely to stimulate mitigation by providing premium discounts, since they are already making a loss on such insurance policies, while the incentive of the policyholder to invest in mitigation is reduced if subsidized insurance is available.

7.2.2 Long-term insurance (LTI) contracts

Recent studies advocate the use of long-term flood insurance contracts on residential properties to provide financial security for homeowners and stimulate investments in risk reduction (Jaffee et al. 2008; Kunreuther 2008; Michel-Kerjan 2010). LTI contracts, for instance with a duration of 5, 10 or 15 years, could be more effective in promoting damage mitigation than annual insurance contracts. Homeowners may be reluctant to undertake mitigation when they have short-term insurance contracts, because of uncertainty both about whether or not they will stay in the same house in the future and whether or not the insurer will charge lower premiums that reflect reduced risk for a long time period. A LTI contract can be tied to the property instead of the individual and provide benefits for the insured, such as premium discounts, for the entire duration of the contract, if individuals invest in risk-reducing measures. With such a contract, investments in mitigation may be financed with mortgage loans, which spread the upfront investment costs for individuals over time. Undertaking a cost-effective mitigation measure would then reduce yearly insurance premiums by more than the yearly increase in mortgage payments used to finance the investments, which benefits consumers.[4] Insurance companies and banks that offer the mortgage also benefit, since the vulnerability of the property to natural hazards has been reduced, so that a win–win situation can be created for all involved parties (Kunreuther et al. 2011). Premiums for annual contracts only reflect current risk and can therefore provide incentives for limiting current risk only. In contrast, LTI premiums should reflect the risk during the entire duration of the contract, which includes any impacts of climate change on this risk. This longer-term price signal may provide better incentives for reducing long-term risks and encourages adaptation to climate change. Moreover, LTI contracts can provide more price stability to policyholders and may reduce search costs for consumers if their annual policies are not renewed and reduce administrative costs for the insurer if consumers commonly cancel their (short-term) policies. This may be an important advantage, since research about flood insurance purchases in the USA has shown that individuals commonly cancel yearly policies; the median tenure of new policies is only 2–4 years (Michel-Kerjan et al. 2012).

Although LTI contracts have important advantages, the pricing of the contract may constitute a problem because socio-economic developments, and possibly climate change, have uncertain effects on future natural disaster risk, which makes it difficult to set premiums that are adequate for the entire duration of the LTI contract. This issue of pricing LTI for flood risk under a range of scenarios of socio-economic and climate change has been examined in a recent study for the Netherlands (Aerts and Botzen 2011a). The premiums for contracts

[4] This is the case because the costs of undertaking mitigation, which are the yearly interest and repayments of the mortgage, are lower than the yearly reduced damage that is reflected in lower yearly risk-based premiums when measures are cost-effective.

with several durations over different time periods up to 2040 under these scenarios are shown in Table 7.1. The premium development for each contract type has been examined over time by estimating the influence on risk-based premiums of the climate change scenarios of 24 cm sea level rise CC24) and 60 cm sea level rise (CC60) and either low or high economic growth (RC or GE).

Overall, climate change has a greater influence on the premium increase than socio-economic changes. This highlights the need for insurance companies to have adequate predictions of climate change in setting long-term flood insurance premiums. In the real world, the actual climate change that will materialize is unknown at the time when premiums are determined, which can result in considerable problems with pricing the long-term flood insurance according to the results in Table 7.1. For example, suppose that in the year 2015 the insurer expects that climate change will develop according to the CC24 scenario until 2040, but that the actual scenario is CC60, while the insurer correctly predicts that the socio-economic scenario is RC. In that case, premiums would be moderately underpriced for the 5-year contract (€16 per year), while premiums would be substantially too low for the 10-year (€30 per year) and 15-year (€36 per year) contracts. Obviously premiums would be even more underpriced if the correct socio-economic scenario is GE instead of RC. The results show that mispricing can be a serious issue even for contracts with a duration of 5 or 10 years. Clearly, in the face of this uncertainty the insurance company is better off keeping contracts as short as possible because premium shortfall, compared with expected payouts on claims, is the lowest with 5-year contracts. For consumers, the opposite is true, and they would be better off with the contract with the longest duration when insurers underestimate the effects of climate change. However, at some point, the insurer will realize that its premiums are too low and will therefore revise its climate change scenario and premiums upwards.

Suppose that the insurer acts prudently and sets the premiums according to the high climate change scenario in 2015, and adjusts its premiums downwards if the low climate change scenario materializes. In that case there would be an advantage for the consumer in choosing the short-term 5-year contract. The longer-term contracts would be less attractive to the consumer because, if climate change turns out to follow the CC24 scenario instead of CC60, then the insurer would

Table 7.1 ***Estimates of flood insurance premiums between 2015 and 2040 in the Netherlands for LTI contracts with durations of 5, 10 and 15 years***

	Scenarios CC24 and RC			Scenarios CC60 and RC		
	5 years	10 years	15 years	5 years	10 years	15 years
Contract duration	[€/ year]	[€/ year]	[€/ year]	[€/ year]	[€/ year]	[€/ year]
Period 2015–20	25	30	33	41	60	79
Period 2020–25	33	30	33	79	60	79
Period 2025–30	39	41	33	118	138	79
Period 2030–35	44	41	50	157	138	196
Period 2035–40	50	53	50	196	215	196
Period 2040–45	56	53	50	235	215	196
	Scenarios CC24 and GE			Scenarios CC60 and GE		
	5 years	10 years	15 years	5 years	10 years	15 years
Contract duration	[€/ year]	[€/ year]	[€/ year]	[€/ year]	[€/ year]	[€/ year]
Period 2015–20	34	42	42	46	69	69
Period 2020–25	50	42	42	91	69	69
Period 2025–30	65	73	42	137	159	69
Period 2030–35	81	73	96	187	159	227
Period 2035–40	96	104	96	227	250	227
Period 2040–45	112	104	96	272	250	227

Note: Premiums are given as fixed yearly amounts and are averages over the dike-ring areas of the Netherlands, estimated under the CC24 or CC60 climate change scenarios and either low or high economic growth (RC and GE).
Source: Aerts and Botzen (2011a).

charge lower premiums for its new contracts, while the consumer would be stuck with a high premium for a long-term contract.

In summary, the estimates of future risk-based premiums, for a variety of contract durations, show that premiums differ significantly depending on the expectations of the insurer about the influence of climate or socio-economic change on flood risk. Given these uncertainties with estimating future risk, insurers are likely to charge a mark-up on LTI premiums. LTI may be promising for providing better incentives for the mitigation of flood damage, but further research is needed to examine how potential problems with the pricing of long-term risk can be addressed before LTI contracts can be implemented on a large scale in practice. Therefore, the remainder of this book focuses on the more usual annual insurance policies.

7.3 Damage mitigation measures and flood risk management in the Netherlands

Most of the investments in flood management in the Netherlands rely strongly on maintaining the safety norms of flood protection infrastructure through dike reinforcements (see Chapter 5). Meanwhile, however, the potential damage has increased sevenfold over the last 50 years due to continuous developments of new urban concentrations in vulnerable areas (Aerts et al. 2008b). Future projections show a gradual upward trend in house construction: by the year 2040 about 500,000–1,500,000 new houses will be constructed, many of them in low-lying areas. This issue has raised the question of whether flood risk management should focus on controlling the likelihood of flooding or also on reducing potential damage. Even if future flood risk, defined as *probability*damage*, is maintained at a constant level by means of higher dikes, the potential damage caused by flooding will increase. Aerts et al. (2008b) show, for example, that additional flood-proofing of new urban areas would lower future flood risk by a factor of two. Elevating a house to eliminate flood risks costs between €10,000 and €20,000, while other measures that limit risk partially can be undertaken at relatively low cost, such as installing central heating and other utility installations on the higher floors of buildings. Chapter 6 examined the willingness of homeowners to invest in the elevation of houses, and the focus of this chapter is on other low-cost measures.

In some areas of the Netherlands, maximum inundation depths of a flood are several metres. At such water levels, simple mitigation measures such as the installation of sandbags are expected to fail (ICPR 2002). However, a recently developed flood risk map (http://www.risicokaart.nl/) shows that despite the low-lying position of dike-ring areas below sea level, inundation depths are not uniformly distributed. Many areas that are located several metres below sea level have an expected maximum inundation depth of only 1 metre. This provides scope for further exploration of the effectiveness of mitigation measures. Moreover, several mitigation measures could be undertaken that limit flood damage even though water levels are high. Examples are moving installations or furniture to higher floors that are safe from flooding.

Until recently, little attention has been paid by policy makers to the role that insurance arrangements can play in damage mitigation in the Netherlands. Private insurance coverage against flooding is not available at this moment. The government may partly compensate damage via the Calamities and Compensation Act (WTS), but it is uncertain whether compensation will be granted, since a right to compensation does not exist (de Vries 1998). The current scheme of government compensation for flood damage, which is not conditional on the risk taken by households who settle in floodplains, can be regarded as undesirable, since incentives to limit damage are minimal (see Priest 1996). Therefore, it is worthwhile to examine how insurance arrangements could encourage mitigation of flood damage.

The advantages and difficulties of introducing flood insurance in the Netherlands are discussed in Chapter 4. The introduction of a multilayered insurance programme in the form of a public–private partnership for insuring flood damage is suggested, as has been proposed for insuring weather risks in the USA (Kunreuther 2006b). In such a programme, a first layer of small losses is paid by households; private insurance companies deal with a second layer of larger losses using risk-based premiums; and the government covers a third layer of very large losses to prevent problems with insurability of highly correlated and potentially catastrophic risks. Availability of insurance is likely to improve the welfare of risk-averse individuals by reducing uncertainty about compensation (see Chapters 8, 10 and 11). Moreover, this insurance scheme may give adequate incentives to limit flood damage.

7.4 Explanation of the survey

The willingness of households to undertake mitigation measures for certain benefits on hypothetical insurance policies was examined using a survey. The four mitigation measures considered were as follows:

1. the purchase of sandbags in order to receive a premium discount
2. the installation of water-resistant flooring if damage to floors vulnerable to flooding is not covered
3. the removal of certain machines (laundry and drying machines) to a higher floor in return for a premium discount
4. the relocation of central heating boiler to a higher floor in return for a premium discount.

These particular measures are examined for the following reasons. The experience during the 2002 Elbe flood and the report of the International Commission for the Protection of the Rhine (ICPR) (2002) indicate that these measures could be very effective in limiting damage from river floods, as explained in more detail in Section 7.5. It is also of interest to examine whether homeowners have different attitudes to measures that prevent damage by keeping water out of the building (with sandbags) or which prevent damage once water has entered the building (e.g. relocating appliances and boilers). Moreover, the incentives provided to encourage homeowners to undertake these measures differ, namely exclusion of coverage in the measure that concerns replacing floors and premium discounts for the other measures. In this way the effectiveness of the 'carrot-and-stick' incentives that can be provided by insurance are examined.

The respondents were selected using area codes that correspond to dike-ring areas near the main river system with safety norms of 1 in 1,250 in the Netherlands (areas indicated with letter D in Figure 5.1). The geographical distribution of the respondents is shown in Figure 7.1, which shows the part of the dike-ring areas that comprise the river delta. Botzen et al. (2008b) give a description of the survey and an overview of the questions.

The structure of the survey is as follows. The questionnaire starts with questions on the experience of the respondent with regard to flooding, flood damage and evacuation due to flood threats. An open-ended question about the causes of flooding was included to test the respondent's knowledge and encourage the respondent to think about the nature of flood risks. In addition, several questions addressed the perception of flood risks and the expected effects of climate change. In particular, respondents were asked to rate their flood risks in comparison with an average Dutch resident (e.g. Viscusi and Zeckhauser 2006). Moreover, respondents were asked to give quantitative estimates of the return

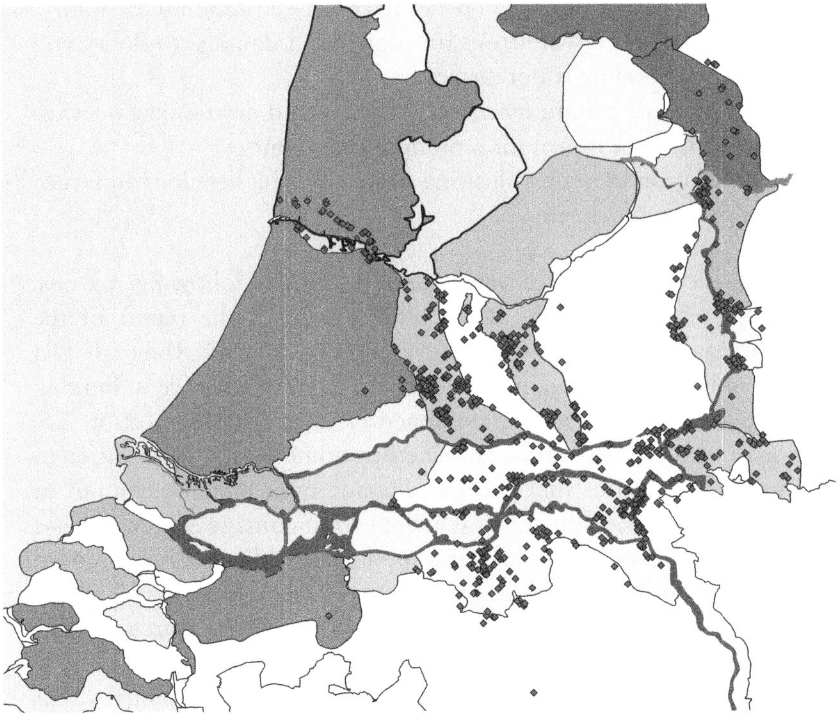

Figure 7.1 Location of the respondents to the survey in the dike-ring areas

period of a flood at their home, using a logarithmic probability scale as a visual aid.[5] These questions were intended to familiarize the respondents with the topic, and the answers may also serve as explanatory variables in modelling the responses to the mitigation questions.

The current regulation of compensation for flood damage was then explained. This explanation differs between two versions of the questionnaire. In one version, it was explained that the government may partly provide compensation for damage suffered by households according to the current legislation and that additional private insurance coverage is available. In contrast, in the other version it was explained that such government compensation would no

[5] Following Schneider and Zweifel (2004) the return period was elicited using a logarithmic scale ranging from 1 to 100,000 years as a visual aid. The legal norm of flooding of 1 in 1,250 years was indicated on top of the scale to help in answering the question.

longer be provided and that only private insurance coverage would be available. This allows for an assessment of the independent effect of the current compensation scheme on the willingness of homeowners to undertake mitigation measures. Moreover, an explanation was included in both versions about the probabilities of flooding and safety standards in the Netherlands to help respondents to understand the risk they face.

The mitigation questions were part of an extensive survey that also included questions on demand for flood insurance (see Chapters 10 and 11). Questions on willingness-to-pay for flood insurance, which are not discussed in this chapter, followed the risk perception questions. Subsequently, the mitigation questions were asked (shown in Appendix 7A), which are the main questions of interest for this study. Each mitigation question and descriptive statistics of the answers are discussed in detail in the next section. The questionnaire concluded with the usual socio-demographic questions.

During the design of the survey, experienced stated choice practitioners, other economists, natural scientists, water management experts and psychologists reviewed versions of the questionnaire. After incorporating their comments, three pre-tests of the questionnaire were conducted between August and October 2007 using face-to-face interviews. Four trained and carefully supervised interviewers (two male and two female) interviewed 88 households. These pre-tests turned out to be useful in checking the understanding of the survey by the respondents, and resulted in several adjustments in the formulation of explanations and questions. A fourth and final pre-test was conducted to test the online questionnaire, which resulted in minor adjustments in layout.

The survey was administered over the Internet using Sawtooth CBC software.[6] This computer-based method has the advantage that follow-up questions can be automated, interviewer effects can be avoided, and a large geographically spread sample can be obtained at relatively low cost. Respondents were selected from the consumer panel of Multiscope and contacted by e-mail.[7] This e-mail did not specify the topic of the survey in order to prevent selection bias. The sample consisted of random draws of panel members who live

[6] See www.sawtoothsoftware.com.

[7] For more information, see www.multiscope.nl.

in dike-ring areas in the Netherlands with a safety standard of once in 1,250 years. The sample was set up to be representative for the Dutch population up to the age of 60 years. Fewer older individuals were represented in the Internet sample because seniors are generally less active on the Internet than younger people. The survey started with a selection question and only respondents who owned a house were allowed to fill out the remainder of the questionnaire. It is more relevant to target mitigation measures to homeowners rather than to tenants, because most flood damage will be caused to buildings, whereas home contents can be moved to higher floors during floods. Therefore, homeowners are more likely to invest in measures that mitigate flood damage, since they fully benefit from the reduced damage following mitigation. Respondents living in apartments higher than the first floor and respondents living outside the sample area were removed from the sample data. The survey was removed from the Internet once the desired number of respondents had been reached.[8] The resulting total number of completed questionnaires was 509.

7.5 Descriptive statistics for willingness to undertake mitigation measures and an analysis of their effectiveness

7.5.1 Mitigation measure 1: Buying sandbags to create a water barrier

Sandbags can act as a water barrier that protects the house from flood damage if they are placed in front of doors or low windows during a flood (ICPR 2002). This measure is only effective if floodwater levels are relatively low, but there is uncertainty about the exact maximum water level at which water barriers are still effective. The maximum height of waterproofing is approximately 1 metre above the ground, according to Kelman and Spence (2003). In contrast, the International Commission for the Protection of the Rhine (ICPR) (2002) states that water barriers are effective during low-depth floods with water levels

[8] The use of the consumer panel of Multiscope does not allow us to calculate the exact response rate to our survey, since the survey was removed from the Internet once a pre-specified quota of completed questionnaires was reached. On average, response rates of the consumer panel are well above 20% (www.multiscope.nl).

of less than 2 metres. A maximum water level of 1 metre may be a safer indicator for determining the effectiveness of sandbags.

According to the ICPR (2002) water barriers can reduce damage by about 60 per cent up to 80 per cent if the floodwater does not overflow the barrier. A good example for assessing the effectiveness of mitigation is the extreme flood of the River Elbe in Germany in 2002. During the Elbe flood, most water barriers were overtopped. Nevertheless, damage was reduced by 29 per cent for buildings with water barriers, which is substantial (Kreibich et al. 2005). This indicates the relevance of examining the willingness to install water barriers in the Netherlands, since the benefits in terms of reduced damage could be large.

The first mitigation question asks whether respondents are willing to purchase 20 sandbags, which has a one-time cost of €20, in return for a €5 discount on their flood insurance premium per year. It was explained to respondents that these sandbags could limit damage during floods, for example if filled sandbags are placed in front of doors. The survey indicates that mean willingness-to-pay (WTP) for flood insurance by respondents who are willing to purchase insurance is about €10 per month or €120 per year based on the contingent valuation method. As an illustration, the discount of €5 yearly is about 4 per cent of WTP for the insurance, which indicates that the discount is relatively small, as are the investment costs. Suppose that respondents view the undertaking of this measure as a standard intertemporal utility maximization problem. This means that the investment is undertaken if the future benefits discounted over time exceed the cost. The ratio of discounted benefits over cost of the investments will be greater than 1 after 6 years using a discount rate of 10 per cent. This indicates that a risk-neutral individual would undertake this investment as long as the individual expects to benefit from the premium reduction for at least 6 years. Individuals with a larger discount rate, i.e. a very short time horizon, as well as very risk-seeking individuals, would be less likely to buy the sandbags.

The results of the survey indicate that about 62 per cent of the respondents are willing to make the investment to buy the sandbags, while the remainder are not willing to buy the sandbags for the discount on the insurance premium. Only 7 per cent of the German households had water barriers available during the Elbe flood, which may have been due to their lack of previous experience with flooding and knowledge of living in a flood-prone area or lack of trust in the effectiveness of private precautionary measures (Kreibich et al. 2005).[9] Our finding of respondents willing to buy the sandbags is large compared with the experience in Germany. The results of our survey suggest that providing benefits on the insurance policy may be useful to encourage respondents to undertake this mitigation measure.

7.5.2 Mitigation measure 2: Installing water-resistant flooring

The second mitigation question examines the willingness of respondents to replace a floor type that is vulnerable to flooding with a tiled floor, which will not be damaged during a flood. Practical experience shows that this flood-adapted interior fitting measure may reduce households' flood losses considerably (Yeo 2002). In particular, the ICPR (2002) states that about 27 per cent of flood damage to buildings consists of damage to floors and floor coverings, which indicates the relevance to examine the willingness of householders to undertake this measure. Damage to parquet, wood, laminate or carpeted floors arises during floods with both very high and low water levels. Therefore, replacing these floor types with tiled floors has the potential to reduce damage during most types of flooding events. A further advantage of this measure is that it can be implemented at relatively low cost, although some people may object to it for aesthetic reasons.

This question consists of two parts. First, it was asked whether respondents have a vulnerable floor type on their ground floor. The vast majority of the respondents have such a floor type (72 per cent) and these respondents are then asked to answer a follow-up question about whether they are willing to buy a tiled floor when their current floor needs replacement in the future if the flood insurance does not cover damage to their current floor type. Approximately 20 per cent of the individuals who have a floor type that is vulnerable to flooding indicated that they will buy a tiled floor if flood insurance

[9] The survey among the residents near the Elbe River indicates that about 25% and 27% of the households had collected information before or during the flood about flood protection. More information campaigns could be run in order to increase knowledge about the effectiveness of precautionary measures, and to encourage households to invest in damage mitigation (Kreibich et al. 2005).

does not cover damage to their current floor. This result suggests that restrictions of coverage in insurance policies can be useful to encourage mitigation.

7.5.3 *Mitigation measure 3: Moving laundry and drying machines to a higher floor*

This measure concerns moving the laundry and drying machines to the first floor or higher if homeowners have installed these machines on the ground floor. In general, moving goods, such as furniture, is a very effective measure to limit flood damage. As an illustration, the flood of the Dutch Meuse in 1995 resulted in 80 per cent less damage to furniture compared with the 1993 flood of the Meuse, due to the appropriate removal of goods, even though flood depth and warning times were comparable for both floods (ICPR 2002). Damage was less during the second Meuse flood because householders were more experienced and better prepared (Wind et al. 1999). A problem is that laundry and drying machines are relatively heavy and are probably only moved to safer floors if flood warnings are issued in plenty of time. Smaller and less heavy furniture is likely to be removed first. For this reason, it is of interest to examine whether households are willing to remove these machines before the flood threat.

This question first asks whether respondents have a washing machine or dryer on a floor that is vulnerable to flooding, which was the case for 43 per cent of the respondents. A follow-up question asks whether respondents are willing to move their laundry and drying machines to the first floor or higher for a yearly discount of €5 on their flood insurance premium. Only 6.8 per cent could not place the washing machine or dryer on a higher floor, because their house consists of only one level. The vast majority (85 per cent) have a first floor or higher, but do not want to move their washing machine and dryer. A small fraction (8 per cent) indicated that they would be willing to move their laundry machine and dryer to a higher floor for the premium discount. This suggests that reasons such as convenience or available space may play a greater role in the decision to move heavy household equipment than monetary incentives given by flood insurance arrangements.

7.5.4 Mitigation measure 4: Relocating the central heating boiler to a higher floor

The final measure concerns moving the boiler of a central heating system to a higher floor, if it is currently installed on a floor level vulnerable to flooding. Of course, this measure can also mitigate flood damage in the case of high water levels (FEMA 1999). Kreibich et al. (2005) have shown that the installation of heating and other (electrical) utilities on higher floors could reduce mean absolute damage by €24,000 per house, based on experience of the 2002 Elbe flood in Germany. Moreover, they estimate that flood damage was about 36 per cent lower for households who had their utility installation placed on a floor safe from flooding. The average damage to the heating installation in particular was €7,836.[10]

This mitigation question again consists of two parts. First, it is asked whether respondents currently have a central heating boiler installed in their cellar or on the ground floor. In total, 19 per cent answered that this was the case. These respondents were then asked to answer the follow-up question: whether they would place their boiler on the first floor or higher when it needs replacement in the future, in exchange for a yearly premium discount of €10 on their flood insurance. A minority of the respondents did not have a higher floor (11 per cent) and were therefore unable to undertake the mitigation measure. Approximately 65 per cent have a higher floor but do not want to move their boiler, and almost a quarter would be willing to place their boiler on a higher floor for the premium discount.

7.5.5 Summary of willingness to undertake mitigation measures

Table 7.2 shows the percentage of respondents for whom the mitigation measures are relevant (column A) and the percentage of those respondents who are actually willing to undertake the measure (column B) and the product of the two, which represents the percentage of total respondents who are willing and able to undertake the measure. We note that the measure to move machines to a higher floor and relocate the central heating boiler to a higher floor is only relevant to

[10] Personal communication with Dr Heidi Kreibich.

Table 7.2 *Responses to the mitigation questions*

Mitigation measure	Percentage for whom the measure is relevant (*A*)	Percentage of (*A*) who are willing to undertake the measure (*B*)	Percentage of total sample willing to undertake the measure (*A***B*)
1) Buy sandbags that can be used as a water barrier	100	68	68
2) Replace current floor with a tiled floor	72	20	14
3) Move laundry and drying machines to a higher floor	43	8	3
4) Install central heating boiler on a higher floor	19	24	5

homeowners who currently have these machines and boiler installed at the ground floor level or below. Also the measure to replace vulnerable floor types with tiled floors is only relevant to respondents who do not already have a tiled floor at the ground floor level.

From the table it is apparent that the sandbag water barrier may be a promising mitigation strategy, since this measure is relevant for all respondents and a large percentage of homeowners are willing to buy the sandbags. The measure that involves replacing a vulnerable floor type with a tiled floor has a large percentage of respondents to whom this measure is applicable, since 72 per cent did not have a tiled floor at the time of the survey. About one-fifth of these respondents indicated that they would replace their floor covering with a tiled floor if damage to their current floor type was not covered by insurance. The measure to move laundry and drying machines to a higher floor is applicable to almost half of the respondents (who currently had placed those machines on the ground floor), but only a very small proportion was willing to undertake it. Therefore, this measure seems the least promising. A smaller proportion had a central heating boiler on a lower floor, so that the mitigation measure was relevant for them, and about a quarter of these homeowners were willing to move it to a floor that is safe from flooding.

7.5.6 Estimating the potential contribution of mitigation in limiting flood damage

It is useful to examine how much damage could be prevented by undertaking the four mitigation measures included in the survey. This in turn allows for the identification of the measure that is most likely to be effective in limiting damage. The last column of Table 7.2 indicates the proportion of homeowners who would undertake a specific mitigation measure. This information is used to estimate damage prevented by mitigation, using two approaches. First, damage prevented by the four mitigation measures is examined in the case where a major river flood occurs in a representative dike-ring area. Second, the reduction in flood risk (defined as *probability*prevented damage*) resulting from the mitigation measures is estimated for all of the dike-ring areas with a 1 in 1,250 norm under different scenarios of climate change. This is of special interest since little is known about the effectiveness of mitigation measures in the Netherlands. We use estimates of damage prevented for each individual mitigation measure based on the 2002 Elbe floods and estimates by the ICPR (2002), because of a lack of empirical evidence about damage prevented by mitigation in the Netherlands.

7.5.7 Case study of dike-ring area 36

Dike-ring area 36 is located in the south of the Netherlands and is part of the 1 in 1,250 safety norm dike-ring areas sampled in our survey. The river Meuse bounds this dike-ring in the north and the east, where dikes protect an area of 740 square kilometres from flooding. Approximately 161,099 houses are located in this dike-ring area (DWW 2005c), of which about 55 per cent, or 88,600, are privately owned (Statistics Netherlands 2008). Wouters (2005) estimates that potential flood damage to houses and their contents due to an extreme flood in this dike-ring amounts to a maximum of €11 billion. Dividing this by the total number of houses and adjusting to end of the year 2007 price levels gives an average expected damage per house of €79,000 caused by an extreme flood. More detailed analyses of potential flood damage in dike-ring area 36 have been made by the Floris study (Rijkswaterstaat 2005a). According to the Floris study, the maximum total flood damage is lower and amounts to €7.5 billion, of which about €4 billion can be attributed to property damage (Rijkswaterstaat 2005b). Using the maximum estimate by Wouters (2005) and the lower estimate by the Floris study results in an upper and lower damage estimate per house of €79,000 and €29,000. These estimates of average flood damage per house will be used in the computation of damage prevented by mitigation.

Table 7.3 shows the number of homeowners who will undertake the mitigation measures as derived from the survey, and an estimate of damage prevented by each measure when a river flood takes place. Multiplying the total number of homeowners in the area (88,600) by the proportion who will undertake the measure (last column of Table 7.2) gives the total number of homeowners who will undertake the measure. The underlying assumption is that the proportions in Table 7.2, which represent a sample of the homeowners in all 1 in 1,250 norm dike-ring areas, apply also to dike-ring area 36 (which is part of the 1 in 1,250 norm areas). Nevertheless, our estimates provide a useful indication of the contribution that mitigation can make in preventing damage and the relative effectiveness of the four measures examined here.

Total flood damage prevented by water barriers is computed by multiplying the number of homeowners who will undertake the measure by estimates of damage prevented per house. It is uncertain how much damage water barriers can prevent, and therefore different scenarios of effectiveness and expected damage will be applied. In particular, an optimistic and a pessimistic scenario of effectiveness of water barriers to mitigate damage will be used in these calculations. The optimistic scenario is based on the ICPR (2002), which states that water barriers can prevent up to 60–80 per cent of total flood damage per house if the flood does not overflow the barrier. We will use the lower estimate (60 per cent) of this range in our optimistic scenario. The pessimistic scenario is based on the effectiveness of water barriers during the 2002 extreme Elbe flood, when many water barriers were overflowed and, on average, 29 per cent of total damage per house was prevented by installing water barriers (Kreibich et al. 2005). We note that above 1.5–2 metres water depth, most mitigation measures are ineffective, as described by Keijzer (2008), and the assumption here is that, on average, the maximum water depth in dike-ring area 36 is below 1.5–2 metres. This is, however, an important variable that should be quantified in more detail in future research, especially because considerable uncertainty exists about the relation between water depth and flood damage (Merz et al. 2004).

The proportions of damage prevented in the optimistic and pessimistic scenarios are multiplied by the potential flood damage per house in dike-ring 36, which is estimated as €79,000 or €29,000 based on Wouters (2005) and the Floris study, respectively. Total damage prevented with this measure is considerable and is between €0.5 and €2.8 billion, as Table 7.3 shows. It should be noted that the damage prevented by water barriers is computed as if it is implemented without undertaking the other mitigation measures. Obviously, damage prevented by water barriers will be lower if floors vulnerable to flooding are also replaced.

The estimate of total damage prevented by changing vulnerable floor types to tiled floors is derived in a similar way using the ICPR (2002) estimate of damage to floors of 27 per cent of total damage per house. It is assumed that this damage is not suffered if tiled floors are used instead of floor types that are vulnerable to flooding. Because of the uncertainty of damage prevented by changing floor types, it is computed using the low (€29,000) and high (€79,000) estimates of average damage per house. Total damage prevented is therefore between €100 million and €270 million, which is substantial but lower than the water barriers measure because fewer homeowners will replace their current floor.

The mitigation measure that involves moving certain machines (laundry and drying machines) to a higher floor is not very effective, because total prevented damage is only about €2 million. Prevented damage is computed by multiplying the number of homeowners who would undertake the measure by the average damage prevented, which gives an estimate of €675.[11] Relocating the central heating boiler to a higher floor is more effective, because more homeowners are expected to undertake the measure and prevented damage per house is larger. Flood damage prevented is based on the experience during the Elbe flood, when average damage to heating installations on lower floors was €7,836,[12] which

[11] The average price of a new washing machine and dryer is about €1,300 in total (www.milieucentraal.nl). It is assumed that the average value of existing machines is about half of this, which results in average damage prevented of €675 per household.

[12] Personal communication with Dr Heidi Kreibich.

Table 7.3 ***Estimates of prevented damage by mitigation in dike-ring area 36***

Mitigation measure	Homeowners who undertake the measure	Total prevented flood damage (million €)
1) Buy sandbags that can be used as a water barrier		
Pessimistic effectiveness and low expected damage	60,248	506.7
Pessimistic effectiveness and high expected damage	60,248	1,427.9
Optimistic effectiveness and low expected damage	60,248	1,048.3
Optimistic effectiveness and high expected damage	60,248	2,855.8
2) Replace current floor with a tiled floor		
Low expected damage	12,758	99.9
High expected damage	12,758	272.1
3) Move laundry and drying machines to a higher floor	3,048	2.1
4) Install central heating boiler on a higher floor	4,040	31.7

is multiplied by 4,040 homeowners who will undertake the measure, resulting in damage prevention of almost €32 million. Table 7.3 summarizes the results, and shows that considerable gains are possible through the first two and the fourth mitigation measures.

7.5.8 Mitigating flood risk in the 1 in 1,250 norm dike-ring areas under climate change

The previous analysis estimated the potential of the mitigation measures to prevent flood damage in a representative dike-ring area. It was then investigated how the mitigation measures could contribute to reducing flood risk in all dike-ring areas with a 1 in 1,250 safety norm, on which our survey sample was based (indicated by 'D' in Figure 5.1), which allows the assessment of the potential benefits of mitigation in a large part of the river delta. It is, nevertheless, very unlikely that all 1 in 1,250 norm dike-ring areas will be flooded at the same time. Therefore, an estimate of the *total value* of damage prevented by mitigation in all these areas is not a very relevant indicator of the benefits of mitigation. The reduction in the *expected value* (*probability*prevented damage*) or in *flood risk* will be estimated instead, which is an important indicator in cost–benefit analysis of water management policy. Moreover, the effects of climate change on the effectiveness of mitigation to limit flood risk are also examined.

The approach to estimating prevented damage per mitigation measure for all 1 in 1,250 norm dike-ring areas, which is an input for the *probability*prevented damage* calculations, is similar to the approach underlying the computations for dike-ring area 36 above. The only difference is that the total number of homeowners and expected damage per house for the 1 in 1,250 norm areas are used as an input for the analysis instead of the values for dike-ring area 36. The total number of houses in these areas is 1,043,758 (DWW 2005c), of which approximately 55 per cent – or 574,000 – are privately owned (Netherlands Statistics 2008).

The average flood damage per house is estimated by adding the total expected damage to houses and home contents caused by a major flood in all of the 1 in 1,250 norm areas, as has been estimated by Wouters (2005), and dividing this by the total number of houses in these areas. Adjusting this for the end of year 2007 price level results in an average flood damage per house of €73,000. Unfortunately, more detailed analyses of expected flood damage, as has been made in the Floris study, are only available for dike-ring area 36 and not for the other 1 in 1,250 norm dike-ring areas. The flood damage per house based on the Floris study was about 37 per cent of the estimate based on Wouters (2005). As an approximation, we assume that the same difference applies to the other 1 in 1,250 norm dike-ring areas and use €27,000 as a scenario of low average flood damage per house. The expected value of prevented damage is then computed by multiplying prevented damage of a mitigation measure by the flood probability.[13]

The expected values of prevented losses are estimated for three scenarios of flood probabilities, namely the current safety norm of 1 in 1,250 and two scenarios where flood risk increases to 1 in 750 and

[13] Computed as *prevented damage*flood probability.*

Table 7.4 *Expected value (EV) of prevented loss (in €1,000) by mitigation under climate change for all 1 in 1,250 norm dike-ring areas*

Mitigation measure	EV prevented loss Current climate P = 1/1,250	EV prevented loss Temperature +1 °C P = 1/750	EV prevented loss Temperature +2 °C P = 1/550
1) Buy sandbags that can be used as a water barrier			
Pessimistic effectiveness and low expected damage	2,445	4,075	5,557
Pessimistic effectiveness and high expected damage	6,610	11,017	15,024
Optimistic effectiveness and low expected damage	5,059	8,431	11,497
Optimistic effectiveness and high expected damage	13,677	22,795	31,084
2) Replace current floor with a tiled floor			
Low expected damage	469	781	1,065
High expected damage	1,267	2,112	2,880
3) Move laundry and drying machines to a higher floor	9	15	21
4) Install central heating boiler on a higher floor	180	300	409

1 in 550 by 2050 due to climate change. These increased flood probabilities are consistent with a 1 °C and 2 °C rise in global temperature, with consequently higher peak discharges on the rivers Meuse and Rhine (see Aerts et al. 2008b) and have been derived in Chapter 5. Of course, flood risk will be lower if the government invests in the heightening of primary river dikes. Note that changes in probability due to climate change differ across each 1/1,250 dike-ring, as shown by Aerts et al. (2008b), depending on varying influences of sea level rise or changes in extreme discharges of rivers. Here we assume, however, that the changes in probability are, on average, the same.

Table 7.4 shows the expected value (EV) of prevented damage for the four mitigation measures under different scenarios of flood probability. The estimates show that, in particular, water barriers and replacing the floor type considerably reduce flood risks. In contrast, the gain from offering premium discounts to households if they move machines to higher floors is small, while stimulating homeowners to relocate their central heating installation is moderately effective. A very relevant insight from the computations in this subsection is that the yearly reduction of flood risk due to mitigation in the entire river delta could be non-negligible even if climate change does not increase flood probability. The (absolute) expected values of prevented damage under the two climate change scenarios indicate that the benefits of mitigation increase considerably if climate change increases flood risk. This suggests that stimulating mitigation with insurance arrangements is an attractive adaptation strategy.

7.6 Statistical model of the decision to mitigate by buying sandbags as a water barrier

The benefits received by the policyholder in exchange for undertaking mitigation measures, for example premium discounts, are likely to be the main determinant in the decision on whether to undertake the measure. However, other factors may also influence decisions to undertake mitigation measures, such as household and geographical characteristics. Insight into these factors is of interest for two main reasons. First, it is relevant from a social welfare perspective to know what types of household are willing to undertake mitigation measures. In particular, whether these households are at low or high risk is important information. Second, it is of interest for insurers to know what type of households they can target or stimulate to undertake such measures. Moreover, the influence of risk perception and availability of compensation for damage by the government in decisions regarding the undertaking of mitigation is also studied. A statistical model is used to examine the factors that influence the decision to undertake mitigation for the first mitigation measure, i.e. the willingness to buy sandbags in exchange for a premium discount. The previous section indicates that this measure may be the most effective because it is applicable to all respondents, a large proportion of respondents (68 per cent) are willing to undertake it, and prevented damage may be large.

7.6.1 Estimation results of the probit model

Appendix 7B explains the probit model used to analyse the mitigation decision and gives an overview of the explanatory variables and their descriptive statistics. The estimation results are shown in Table 7.5 and are discussed in this section. Table 7.5 shows the mean marginal effect (ME) of a change in the explanatory variables, the corresponding standard error and the *t* statistic of the hypothesis that the marginal effect equals 0. This means that the marginal effects of dummy variables give the change in probability of the respondent undertaking the mitigation measure if the variable has a value of 1. The overall fit of the model is reasonable, given that the premium discount is likely to be the main motivation for undertaking the mitigation measure, but cannot be included in the model since it is fixed. The pseudo-R^2 of the model is 11 per cent, which is not the same as the R^2 statistic in an ordinary least squares regression, since a probit model is non-linear. The pseudo-R^2 value of the estimated model corresponds to an R^2 value of about 0.3 (Domencich and McFadden 1975).

Two variables are included to capture actual or perceived responsibility for compensating for flood damage. The first variable represents the version of the questionnaire explaining the current regulation according to which the government may partly compensate for flood damage, compared with the version explaining that such compensation is not available. The variable is significant at the 5% level and the marginal effect indicates that the probability of a homeowner buying sandbags to mitigate flood damage is about 0.09 less if the government provides partial compensation for flood damage. This suggests that the availability of damage relief reduces the incentive to undertake private mitigation efforts, as has been argued in several studies (e.g. Kaplow 1991; Kelly and Kleffner 2003). Furthermore, a variable 'the government is perceived as responsible' has been included to represent respondents who object to buying flood insurance because they feel that the government is responsible for providing damage relief or guaranteeing adequate protection against flooding.[14] It is expected

[14] The questionnaire asked whether respondents are willing to pay for flood insurance. An open-ended question asks for reasons if respondents do not want insurance. This dummy variable takes on the value 1 if respondents

Table 7.5 ***Estimation results of a probit model of the willingness to buy sandbags***

Variable	Marginal effect	Standard error	*t* statistic
Role of government			
Government compensation is available	-0.0899**	[0.0465]	-1.94
Government is perceived as responsible	-0.3094**	[0.1452]	-2.00
Risk perception			
Negative effects of climate change	0.0676**	[0.0314]	2.15
Climate change causes higher flood risk	0.1514***	[0.0619]	2.46
Lower flood risk than an average resident	-0.0938**	[0.0490]	-1.91
Zero expected return period flood	-0.2038**	[0.1499]	-2.25
Expected return period flood	-0.0004*	[0.0002]	-1.74
Experience and knowledge			
Experience of evacuation	-0.1289	[0.0907]	-1.45
Knowledge about floods	-0.1398**	[0.0565]	-2.30
Geographical characteristics			
Elevation house and barrier is lower than water level	-0.1179**	[0.0496]	-2.38
House is close to a main river	0.0857*	[0.0496]	1.71
Rural area	0.3339***	[0.0467]	3.53
Socio-economic characteristics			
Female	-0.0158	[0.0493]	-0.32
Age	-0.0013	[0.0021]	-0.63
Income	0.000004	[0.00002]	0.15
Education	0.0490*	[0.0269]	1.82
Number of observations	494		
Pseudo-R^2	0.11		
Log likelihood	-290		

Notes: *, ** and *** indicate, respectively, significance at the 10%, 5% and 1% levels. Estimations were performed with the statistical software package Stata.

that these respondents are also less inclined to undertake mitigation measures themselves, which is confirmed by the estimation results, since the effect is significant at the 5% level and the size of the marginal effect is large. In particular, the probability of such respondents buying sandbags to mitigate flood damage is about 0.31 lower than for other respondents.[15]

Several variables are included that measure perceptions of the respondent's flood risk and expectations about the effects of climate change. These variables are statistically significant, while the coefficient signs indicate that the larger the respondent expects the effects of climate change to be and the more serious the flood risk is perceived, the greater is the probability of the respondent buying sandbags. In particular, the probability of undertaking mitigation is larger if the negative effects of climate change for the Netherlands are perceived as large (ME = 0.0676), and if it is expected that climate change will increase flood risk (ME = 0.1514). It is less likely that homeowners will undertake mitigation if they perceive their flood risk as being lower than that of an average resident (ME = –0.0938).

The expected return period of a flood is included through two variables: a dummy variable captures the respondents who answered with a zero expected return period and a continuous variable represents the expected return period of respondents with positive answers. The expected flood probability is the inverse of the return period of flooding. The marginal effects are –0.2 and –0.0004 of the zero and positive expected return variables, respectively, which indicates that the greater the respondent perceives the probability of a flood in his or her residential area, the greater is the probability that the respondent will undertake the mitigation measure.

We have included a variable of 'experience of evacuation' in the model.[16] This variable was not statistically significant. Therefore

indicate that the government is responsible for providing damage relief or guaranteeing adequate protection against flooding as reasons not to buy flood insurance, and 0 otherwise.

15 We estimate whether the willingness to undertake mitigation is related to the willingness to purchase flood insurance. The results (not shown in the table) indicate that this relation is not statistically significant, which suggests that the scenario for the mitigation questions proposing respondents to imagine they possess flood insurance coverage is accepted.

16 The questionnaire also asks whether respondents have experienced a river flood in their residential area. About 20% of the respondents answered that

respondents who have experienced a serious threat of flooding, in the sense that they needed to be evacuated, are neither more nor less likely to undertake mitigation. The most recent flood event in 1995 may have been too long ago to affect mitigation decisions. Several studies show that the willingness to undertake precautionary measures first increases after a disaster but then rapidly decreases in subsequent years (e.g. Kunreuther et al. 1985). A variable representing the knowledge of respondents about the causes of flooding was statistically significant. The estimation results indicate that individuals who are able to state at least one cause of flooding have a lower probability of undertaking mitigation, compared with individuals who are not able to name any cause of flooding (ME = –0.1398). This suggests that homeowners who have little knowledge about the flood threat are more likely to buy protection, i.e. sandbags, against the risk.

Three variables in the model reflect objective measures of the flood risk faced by the respondent, based on geographical characteristics. Geographical information systems (GIS), such as a digital elevation map, were linked to the respondents' exact geographical locations using zip codes.[17] A variable indicates whether the elevation of the zip code area of the respondent plus the height of a 1-metre water barrier, for example sandbags, is located below the expected water level of a flood.[18] The other two variables indicate whether the house is located close to a main river and whether the respondent lives in a rural area. These variables are of particular interest for several reasons. The effectiveness of a water barrier depends on the elevation of the area of the house relative to the potential water level of a flood. Therefore, it is of interest to examine whether respondents who live in an area that is too low for additional water barriers, such as

they have experienced a flood. However, only three respondents indicated that they actually suffered flood damage. We regard experience of evacuation as a more relevant measure of experience of flooding than the variable representing experience of flooding in general, since practically all individuals in the latter group did not suffer any flood damage. Therefore, we did not include the experience of flooding in general.

[17] This data is based on four-digit and two-letter zip codes for 494 respondents, which is highly accurate because it gives street location. The data for 15 respondents are based on zip code numbers only because the letters were incomplete.

[18] The elevation of the zip code areas is obtained using the AHN elevation map and potential water levels are based on the '*RWS dijkkruinhoogtesbestand*'.

sandbags, to be effective are more or less likely to buy the sandbags. In this analysis, it is assumed that sandbags are no longer effective for homeowners for whom the elevation of their house plus 1 metre is located below the expected water level of a flood. Furthermore, respondents who live close to a main river may be at higher risk and, therefore, also more aware of the threat posed by flooding, as will be shown in Chapter 10, making them more willing to undertake mitigation. Finally, inhabitants of rural areas with lower population densities and concentrations of economic values may have different attitudes to undertaking flood risk mitigation than inhabitants of cities.

The results in Table 7.5 indicate that homeowners who live in an area that is too low for effective mitigation with water barriers are less likely to undertake the mitigation measure (ME = –0.1179). Even though a large majority of respondents will not know the exact elevation of their house compared with potential water levels, discussions with respondents during the pre-tests of the questionnaire indicated that many households do relate their exposure to flooding with elevation. The results of this analysis indicate that homeowners with very low-lying houses are less likely to undertake mitigation with water barriers.

Respondents with a home close to a main river are more likely to undertake the mitigation measure, but this effect is only marginally significant ($p = 0.09$). Respondents close to a main river are more likely to suffer flood damage; therefore, it is sensible that they are more likely to invest in mitigation. The probability that homeowners in rural areas are willing to undertake mitigation is almost one-third greater than that for respondents in urban areas, and this is significant at the 1% level. This indicates that inhabitants of rural areas have more positive attitudes towards mitigation than inhabitants in cities, which may be because they are more aware of the flood risk (see Chapter 9).

The socio-economic characteristics of sex, age and income have no statistically significant effect on the decision to undertake mitigation. The respondent's education level has a positive and significant effect ($p = 0.07$). In all, the role of the government, risk perceptions and geographical characteristics are more important determinants in the decision to undertake mitigation than the socio-economic characteristics of the respondent. We note that the main incentive for homeowners to choose to invest in water barriers is likely to be the premium discount on the flood insurance policy provided to them in the survey, although the other factors examined in this section also play a role.

7.7 Conclusions

The projected impacts of climate change on flood risks, in various parts of the world, highlight the need to manage the probability of flooding, for example by investing in flood protection infrastructure, and to implement measures that limit flood damage. Chapter 6 showed that experiences in various countries suggest that mitigation measures at the household level can be an effective means of limiting flood damage during floods. Despite the large potential gains through mitigation, international experiences reviewed in this chapter show that many households in flood-prone areas do not undertake damage mitigation measures and are insufficiently prepared for flood events. Several reasons for this have been put forward, such as underestimation of risk by households, a lack of information on mitigation strategies, and high perceptions of the cost of mitigation compared with the potential benefits in terms of risk reduction. It is therefore of interest to explore how insurance can be used, not only as a means of providing protection against residual risks, but also as a tool to encourage policyholders to implement measures that limit the potential damage of natural disasters.

Flood insurance can provide a price signal of risk if premiums are risk-based. This price signal may raise the awareness of individuals about the flood risk they face, and can steer development into areas with a relatively low flood risk because of the lower insurance cost in such areas compared with high-risk areas. Moreover, insurance can achieve risk reduction by requiring individuals to implement damage mitigation measures before they become eligible for insurance, and by rewarding policyholders who invest in such measures with premium discounts or higher levels of coverage. Several leading academics have proposed that long-term insurance contracts, with a duration of several years (e.g. 5, 10 or 15 years), may be especially effective in providing incentives for risk reduction. However, future research is needed to examine how such contracts can be adequately priced in the face of uncertain effects of developments such as climate change on future flood risk.

Few studies have systematically examined the effectiveness of incentives provided by natural disaster insurance in encouraging risk reduction in practice. This chapter presented a study of the potential of flood insurance to achieve risk reduction by homeowners in the Netherlands. Currently, flood insurance is not available in the Netherlands, but increased attention has been paid to the role that insurance arrangements could play in providing financial security against residual flood risks and providing incentives to households to limit potential flood damage. The latter has been examined for four possible mitigation measures: the purchase of sandbags in exchange for a premium discount, the purchase of a water-resistant floor type if damage to other types of flooring is not covered, and the moving of domestic appliances (laundry and drying machines) and central heating boilers to higher floors, in return for a premium discount.

The results of the survey indicate that homeowners can be stimulated to undertake investments to mitigate potential flood damage by offering certain benefits on insurance policies. In particular, offering homeowners a premium discount if they buy water barriers, such as sandbags, before a flood event seems a promising strategy. In addition, the survey reveals that a large majority of homeowners have a floor type that is vulnerable to flooding on the ground floor, indicating that considerable damage could be prevented if flood-resistant floor types replace such floors. About one-fifth of the respondents are actually willing to do this in the face of restrictions on their insurance policy. Moving laundry and drying machines to a higher floor seems to be the least promising measure, while a larger proportion of homeowners are willing to move their central heating boiler to a floor which is safe from flooding. Indicative estimates of the effectiveness of each mitigation measure for a representative dike-ring area confirm that prevented damage is considerable for mitigation with water barriers and replacing vulnerable floors with tiled floors (see Table 7.3). Examination of the benefits of mitigation for all 1 in 1,250 norm dike-ring areas showed that reductions in (yearly) flood risk could be substantial, especially if flood probabilities increase as a result of climate change (see Table 7.4).

A model has been constructed to identify the factors behind the decision to buy sandbags, which can serve as a water barrier. The results provide two main insights. First, the current institutional setting, characterized by the availability of partial compensation for flood damage by the government, reduces the personal incentive to undertake mitigation. Second, perceptions of risk and climate change play an important role in the decision to undertake mitigation. In particular, the higher the risk of flooding is perceived to be, the more likely are homeowners to invest in water barriers. From these results, it follows that homeowners may be stimulated to undertake mitigation investments by abolishing the current scheme of government compensation and by raising awareness of flood risk. The second finding suggests that provision of information about changing climate risks could, through risk perception, influence the mitigation behaviour of homeowners. The analyses presented in this chapter, which show that homeowners can be encouraged to undertake mitigation with the use of insurance policies, which can result in substantial savings in flood damage, provide a basis for designing insurance programmes that aim to limit, or ameliorate, the impacts of climate change on future risks.

Appendix 7A Questions about mitigation and insurance

1. *Suppose that you have insurance coverage against flood damage. Would you spend €15 one time to buy 20 (empty) sandbags if you would then get a discount on your insurance premium of €5 each year?*
   You can prevent damage by placing filled sandbags in front of doors during floods.

□ No
□ Yes

2. *Do you (partly) have parquet, wood, laminate or carpet on the ground floor of your house?*

□ No
□ Yes

Follow-up question 2 (this question is asked if the respondent has answered 'yes' at 2)

Suppose that you have insurance coverage against flood damage. Will you select a tiled floor if you were to buy a new floor and the flood insurance does not cover damage to your current floor?

□ No
□ Yes

3. *Do you have a washing machine or dryer in your cellar, on the ground floor, or in a garage or shed?*

□ No
□ Yes

Follow-up question 3 (this question is asked if the respondent has answered 'yes' at 3)

Suppose that you have insurance coverage against flood damage. Will you move your washing machine and dryer to the first floor or higher, if you were to get a yearly discount of €5 on your insurance premium?

□ This is not possible since my house does not have a first floor or higher
□ I do have a first floor but will not do this
□ Yes, I will do this

4. *Do you have a central heating boiler placed in your cellar or ground floor?*

□ No
□ Yes

Follow-up question 4 (this question is asked if the respondent has answered 'yes' at 4)

Suppose that you have insurance coverage against flood damage. If you need to replace your boiler in the future and you can get a yearly discount on your insurance premium of €10 if you install the new boiler on the first floor or higher, will you do this?

□ This is not possible since my house does not have a first floor or higher
□ I do have a first floor but will not do this
□ Yes, I will do this

Appendix 7B The statistical model and overview of the variables and descriptive statistics

As the dependent variable is binary, a probit model has been constructed to analyse the influence of the explanatory variables on the mitigation decision. This is done by maximum likelihood estimation of the function (e.g. Heij et al. 2004):

$$\log(L(\boldsymbol{\beta})) = \sum_{\{i;y_i=1\}} \log(\boldsymbol{\Phi}(x_i'\boldsymbol{\beta}) + \sum_{\{i;y_i=0\}} \log(1-\boldsymbol{\Phi}(x_i'\boldsymbol{\beta})) \qquad (7.B1)$$

where β is the parameter vector to be estimated, Φ represents the cumulative normal distribution, x_i is a vector of explanatory variables and y_i is the outcome for the ith observation. A detailed description of the variables and their coding as well as their descriptive statistics are given in Tables 7.B1 and 7.B2. Different methods of coding categorical variables have been applied depending on the type of variable. A continuous variable is created from the categorical variable income, which represents monetary classes (e.g. Blumenschein et al. 2008). Ordinal qualitative variables,[19] which are partitioned into J intervals, can be included using $J - 1$ dummies or can be transformed into values on the real axis using an approach proposed by Terza (1987), who advocates the transformation of ordinal qualitative variables as follows:

$$\Phi^j = (n(\theta_{j-1}) - n(\theta_j)) \,/\, (N(\theta_j) - N(\theta_{j-1})) \qquad (7.B2)$$

where n and N are the probability density function (pdf) and cumulative density function (cdf) of the standard normal distribution, respectively, and:

$$\begin{aligned} \theta_1 &= N^{-1}(p_1) \\ \theta_2 &= N^{-1}(p_1 + p_2) \\ &\vdots \\ \theta_{J-1} &= N^{-1}(p_1 + p_2 \ldots + p_{J-1}) \end{aligned}$$

and p_j is the percentage of the sample observed in category J.[20]

[19] These variables are characterized by a continuous unobservable ordinal latent index, while each interval is ranked in increasing order (1 through J) according to its supremum (Terza 1987).

[20] For the lowest and highest categories Eqn 7.B2 reduces to $\Phi^j = -n(\theta_j)/N(\theta_j)$ and $\Phi^j = n(\theta_{j-1})/(1-N(\theta_{j-1}))$.

Table 7.B1 ***Overview of the variables used in the statistical analysis***

Dependent variable	
Sandbags	Binary variable, 1 = respondent is willing to buy sandbags for a premium discount, 0 = otherwise
Explanatory variables	
Government compensation is available	Dummy variable, 1 = it is explained in the survey that it is possible to receive partial compensation for flood damage by the government
Government is perceived as responsible	Dummy variable, 1 = the respondent states that the government is responsible for compensating flood damage or guaranteeing adequate protection
Negative effects of climate change[a]	Categorical variable, respondent expects effects to be 1 = very small, 2 = small, 3 = not small/ not large, 4 = large, 5 = very large
Climate change causes higher flood risk[b]	Dummy variable, 1 = respondent expects that climate change causes higher flood risks
Lower flood risk than an average resident[c]	Dummy variable, 1 = respondent expects that his/her flood risk is lower than that of an average resident in the Netherlands
Zero expected return period	Dummy variable, 1 = respondent expects return period to be zero
Expected return period flood	Continuous variable, return period in thousand years
Experience of evacuation	Dummy variable, 1 = the respondent has been evacuated for threat of flooding
Knowledge about floods	Dummy variable, 1 = respondent is able to state causes of flooding
Elevation house and barrier is lower than water level	Dummy variable, 1 = the area plus the water barrier of 1 metre is below the potential water level of a flood
House is close to a main river	Dummy variable, 1 = the respondent's postcode area is within 4 kilometres of the nearest main river
Rural area	Dummy variable, 1 = the area is a rural area
Female	Dummy variable, 1 = respondent is female
Age	Continuous variable of age

Table 7.B1 (*cont.*)

Income[d]	Continuous variable of the monthly after-tax household income in €
Education	Categorical variable (range 1–7) of the education level, where 1 = elementary education and 7 = university degree

Notes:
[a] Missing values are 'don't know' responses.
[b] The question reads: '*How do you estimate the consequences of climate change for the likelihood of flooding in the Netherlands?*' The answer options are: '*floods will become more frequent, floods will be as frequent as currently, floods will become less frequent, and don't know.*'
[c] The question reads: '*How would you rate your flood risk compares to an average person in the Netherlands?*' The answer options are: '*I have an average flood risk, I have a higher than average flood risk, and I have a lower than average flood risk.*'
[d] For income the respondent could mark one of the following categories: <€750; €751–1,000; €1,001–1,250; €1,251–1,500; €1,501–2,000; €2,001–2,500; €2,501–3,000; €3,001–3,500; €3,501–4,000; >€4,000. A continuous income variable was constructed by setting the income of each respondent to the midpoint of the interval (€4,500 was used for the highest category).

Table 7.B2 ***Descriptive statistics for the variables of the survey***

Variable	Number of observations	Mean	Standard deviation
Dependent variable			
Sandbags	509	0.62	0.486
Explanatory variables			
Government compensation is available	509	0.43	0.495
Government is perceived as responsible	509	0.02	0.152
Negative effects of climate change[a]	494	3.57	0.897
Climate change causes higher flood risk	509	0.65	0.476
Lower flood risk than an average resident	509	0.47	0.499

Table 7.B2 (*cont.*)

Variable	Number of observations	Mean	Standard deviation
Zero expected return period flood	509	0.03	0.164
Expected return period flood	509	19	118
Experience of evacuation	509	0.09	0.287
Knowledge about floods	509	0.82	0.385
Elevation house and barrier is lower than water level	509	0.41	0.492
House is close to a main river	509	0.45	0.498
Rural area	509	0.05	0.216
Female	509	0.43	0.496
Age	509	45	12
Income	509	2,861	1,010
Education[a]	509	5.39	1.404

Notes: [a] The statistics of this variable are in accordance with the coding in Table 7.B1. This original coding has been transformed for the analysis according to Terza (1987), as described above.

An advantage of the dummy approach is that interpretation of the coefficients is straightforward. The transformation of Terza (1987) can result in gains in efficiency and bias, especially if the number of categories is large. For this reason the latter approach has been applied in several studies (e.g. van Praag et al. 2003), even though applications in the environmental valuation literature are scarce. Dummy variable coding has been applied for variables with a small number of categories, such as the rating of the respondent's flood risk compared with that of an average resident and the effect of climate change on flood probability. The transformation is used for variables with a large number of categories, which are the negative effects of climate change and the respondent's education level.

8 Bounded rationality and demand for flood insurance

8.1 Introduction

Scientific uncertainty associated with climate change is considerable, as the recent assessment report of the Intergovernmental Panel on Climate Change indicates (IPCC 2007). Much uncertainty exists about natural climate variability and the consequences of global warming on the natural climate system, which encompasses wind patterns, humidity and rain patterns. Moreover, considerable uncertainty is associated with future impacts and feedbacks of climate change on social and economic structures, which further depend on human adaptation (Tol 2002a, 2002b, 2005).

The uncertainty associated with climate makes it difficult for the public and private sectors to respond optimally to climate change (Dessai and Hulme 2007; Tompkins and Adger 2005). At a governmental level, uncertainty hampers the design of appropriate mitigation and adaptation strategies. At the individual and company level, uncertainty creates difficulties in handling possible future costs arising from climate change. Insurance against natural hazards, such as flooding, has been proposed as a tool for adaptation to climate change (Bouwer and Vellinga 2005; Kabat et al. 2005). Insurance arrangements are a means of reducing uncertainty associated with climate change impacts for individuals by spreading risks geographically as well as temporally. In this way, individual and social welfare is improved if insurance coverage is available for individuals who are risk-averse. Moreover, well-designed insurance arrangements have the potential to provide loss-reducing incentives, which limit climate change impacts, as explored in Chapter 7.

In several countries, including the Netherlands, private insurance against flooding is not readily available, and individuals are dependent on ad hoc compensation by the government in case of flood damage

Parts of this chapter (in particular Section 8.4) have previously been published as Botzen and van den Bergh (2009b).

(see Chapter 2). The risk of flooding is likely to increase in the future, due to increased precipitation caused by climate change. A larger role for private insurance companies may be regarded as desirable by governments that are unwilling to carry these increased risks. However, the collective nature of these risks requires active government involvement, for example in risk prevention, ensuring adequate coverage or perhaps providing reinsurance capital (see Chapter 4).

A purely private market for flood insurance may be difficult to establish due to problems with covariate risks, affordable reinsurance, moral hazard,[1] and ambiguity of risks (Freeman and Kunreuther 2003). These supply-side problems can cause insurance premiums to be higher than under 'normal insurance', such as fire insurance. A role for the government in providing reinsurance capital may overcome problems with covering highly correlated and potentially catastrophic flood losses (see Chapter 4). This is not a sufficient condition, since demand for flood insurance needs to exist as well. Therefore, for policy makers and insurance companies, it is of interest to investigate whether private insurance arrangements are marketable in the sense that individuals are willing to pay for them. Moreover, it is of interest to examine how flood insurance demand may be affected by climate change, given the large projected increase in future flood risk in certain regions as a result of global warming. This is analysed here by estimating risk premiums for flood insurance in the Netherlands using different theories of individual decision making for a number of climate change scenarios. The approach presented in this chapter has, in principle, a wider relevance to decisions and values under extreme and uncertain environmental change, because the analysis of socio-economic impacts of climatic change and climate policy is likely to be very sensitive to the way in which individuals responding to extreme and uncertain climate risks are modelled.

Expected utility theory is commonly used to analyse insurance decisions or measure willingness-to-pay (WTP) to avoid risks and to estimate risk premiums (Brookshire et al. 1985; Heal and Kriström 2002; Kaas et al. 2001).[2] However, the descriptive failure of expected utility

[1] Although individuals have little control over the probability of flooding, individuals are able to influence the extent of flood damage through *ex ante* damage-reducing investments, as well as through protective and ameliorating behaviour during floods.

[2] Applications for insurance against climate change losses are rare.

theory is well documented, making this approach inappropriate for positive analysis such as aimed at here (Camerer 1998). Empirical evidence shows that individual behaviour is better described by a risk-seeking utility function for losses, together with a non-linear transformation of probabilities, which results in risk-averse behaviour for high losses with low probabilities (Kahneman and Tversky 1979; Tversky and Kahneman 1992). This behaviour can be captured by rank-dependent utility theory and prospect theory. For this reason, risk premiums are estimated in this chapter using these alternative theories for individual decision making under risk. Parameter estimates of the value functions used are based on experimental evidence. Rank-dependent utility theory and prospect theory are likely to give a more accurate description of individual behaviour than expected utility theory, especially for high-impact, low-probability events where the expected utility approach is likely to provide considerably biased outcomes (Mason et al. 2005).

Under rank-dependent utility and prospect theories, risk attitudes depend on probability transformations as well as transformations of monetary outcomes (Weber 1994). Traditional measures of risk aversion based on transformations of monetary outcomes only, such as the relative risk aversion index, are not directly applicable to these alternative theories of individual decision making under risk (Cohen 1995). Therefore, an alternative method is used in this chapter, which measures risk aversion indirectly by determining expected utility functions that result in the same risk premiums as rank-dependent utility and prospect theories. This analysis aims to measure the degree of risk aversion, which is implicit in the risk premiums obtained from rank-dependent utility and prospect theories. The application of prospect and rank-dependent utility theories to environmental risks can provide interesting insights, but has hardly received any attention in environmental economics (exceptions are Rosenman et al. 1988 and Quiggin 2005). The interaction between probability weighting and valuing monetary outcomes has non-trivial consequences for insurance demand. Experimental studies indicate that individuals generally overweight low probabilities, while moderate probabilities receive smaller weights. This observation, in combination with a risk-seeking utility function for losses, implies that individuals are either risk-seeking or risk-averse, depending on the characteristics of the environmental risk they face. Indeed, the simulation results of Bargiacchi (2003) in a climate change application indicate that rank-dependent utility theory with an inverted S-shaped probability weighting function does not necessarily result in more risk aversion than given by expected utility theory.

The remainder of this chapter is structured as follows. Section 8.2 outlines experiences with insurance purchases for low-probability risks and relates these to theory. Section 8.3 discusses climate change and the risk of flooding in the Netherlands. Section 8.4 provides estimates of demand for flood insurance in the Netherlands using rank-dependent utility and prospect theories. Section 8.5 concludes the chapter.

8.2 Insurance purchases for low-probability natural hazard risk, and the theoretical implications

Insurance premiums should be high enough to cover the expected value of the loss (*probability*damage*), because otherwise insurers would, on average, lose money from providing insurance coverage. Expected utility theory predicts that individuals who are risk-averse purchase insurance against a risk as long as the insurance premium is not much higher than the expected value of the loss. The explanation for this is that although paying a premium slightly reduces the overall utility level of an individual, the decrease in utility that occurs if a loss is suffered is much more severe in expected value terms, so that purchasing insurance is preferred. However, in practice, insurance premiums are often above the expected value of the loss because of administrative costs associated with providing insurance and because of the profit margin (economic return) required by private insurance companies. Natural disaster insurance premiums, in particular, may be higher than the expected value of the loss, compared with other insurances. This is the case because of the high costs associated with covering correlated risks and holding large amounts of capital or purchasing reinsurance for establishing reserves that are adequate for reimbursing the many claims that can be filed after a disaster. As a consequence, it is not uncommon that many homeowners do not purchase flood insurance if they are not required to do so by regulation. For example, in Germany the market penetration of flood insurance is only about 10 per cent for home contents and 4 per cent for residential buildings (Thieken et al. 2006).

Experiences with flood insurance demand in the USA are especially interesting to examine because, in the USA, flood insurance has been provided by the National Flood Insurance Program (NFIP) at premiums that are close to, or in some cases even below, the expected value of flood damage. In contrast to the prediction of expected utility theory, many individuals do not purchase flood insurance in the USA (Kunreuther 1984). An explanation is that many people have an attitude that 'it will not happen to me' because they perceive that the flood probability is too low to warrant the purchase of flood insurance. This tendency of many people to neglect the low-probability risk cannot be explained by expected utility theory.

An experimental study of McClelland et al. (1993) about insurance purchases against low-probability hazards finds a bimodal distribution of WTP for insurance: many people neglect the low-probability risk and do not purchase insurance, and another large group of people have WTP that is considerably above the expected loss. This latter finding poses another challenge for expected utility theory although it is consistent with the observation, in practice, that many people who purchase insurance against natural disaster risk do so at premiums that have high loading factors, which means that they are significantly above the expected value of the loss. This contradicts expected utility theory, which predicts that people faced with such premium levels will choose not to purchase insurance and will self-insure (Krantz and Kunreuther 2007). An experimental study with real monetary incentives conducted by Laury et al. (2009) finds that individuals are relatively risk-averse to low-probability, high-impact events compared with events that have a higher probability and a smaller loss, and are more likely to purchase insurance for the former. This behaviour can be accommodated by rank-dependent utility theory and prospect theory. In these alternative theories of individual decision making under risk, individuals are, in general, assumed to overweight low-probability risk with severe consequences, such as natural disasters, which can result in more risk aversion and a higher insurance demand than for high-probability risk. Therefore, insurance demand for natural disaster risk may be more accurately modelled with prospect and rank-dependent utility theories, which is why these theories are applied to estimate flood insurance demand in Section 8.4. It should be noted that the parameterizations of both rank-dependent utility theory and prospect theory, as used in this chapter, assume that people overweight low-probability risk, while it has been observed in practice that there are also many people who tend to ignore such risks. This issue of the neglect of low-probability flood risk is examined in detail in Chapter 10.

8.3 Climate change and the risk of flooding

Currently, design standards in the Dutch part of the Rhine River are 1/1,250 per year. This implies that the dikes are designed in such a way that they can withstand a flood event expected to occur once in 1,250 years. These safety norms differ across dike-ring areas in the Netherlands. Design standards per dike-ring in the Netherlands are shown in Figure 5.1. Here we estimate WTP for flood insurance and risk premiums for a representative homeowner living in the river delta in areas with a 1 in 1,250 year safety norm. These dike-ring areas in the river delta with design standards of 1/1,250 are indicated with the letter 'D' in Figure 5.1.

The IPCC (2007) has estimated that the increasing concentration of greenhouse gases in the atmosphere is likely to raise global average surface temperatures between 1.8 °C and 4 °C by the end of the twenty-first century, relative to 1990, according to different scenarios of CO_2 emissions. This may result in a more vigorous hydrological cycle, with changes in precipitation and evaporation rates. Therefore, water availability and runoff are likely to change, which may affect the discharge regime of rivers. For north-western Europe this means that peak flows of the Rhine in the middle and lower sections of the river are expected to increase in winter due to increased rainfall and meltwater runoff from the Alps. Furthermore, the frequency of extreme precipitation events is likely to increase in a warmer climate (Verbeek 2003). In the following section, we estimate demand for flood insurance under current climate conditions and two climate change scenarios, where flood probabilities increase to 1/750 (+1 °C) and 1/550 (+2 °C). These future flood probabilities were derived in Chapter 5.

8.4 Estimating demand for flood insurance

Several empirical studies have estimated the parameters for utility and probability weighting functions. On the basis of these studies, individual WTP and risk premiums for flood insurance can be calculated. Maximum WTP for flood insurance is then derived by computing the maximum (certain) loss that has the same utility value as the utility of a prospect that an individual may suffer flood damage with a corresponding probability. The risk premium is the difference between WTP and the expected value of the loss. It should be recognized that large individual variations in parameter values of utility and weighting function exist. However, in the absence of empirical evidence concerning these parameters for different groups in Dutch society, this chapter estimates risk premiums for an average or representative individual. Indeed, estimates of the insurance demand of a representative agent may provide relevant information for the design of public policy. The parameterization of utility functions for the Netherlands is well researched (Booij and van de Kuilen 2006). While parameter estimates of probability weighting for Dutch individuals faced with extreme (flood) risks are currently not available, experimental results of Donkers et al. (2001), based on a large sample of Dutch respondents, indicate that probability weighting is highly non-linear.[3] This restricts our analysis to a range of parameter values obtained from the current state-of-the-art experimental studies that estimate probability weighting in other countries. A similar approach has been adopted by Bleichrodt et al. (2001); however, these authors restrict their analysis to parameter estimates of the weighting function from Tversky and Kahneman (1992), while our parameter values take a range of experimental studies into account.

Risk premiums are estimated for three scenarios: based on current flood risks of the Rhine River, and flood risks associated with a low (+1 °C) and a high (+2 °C) scenario of climate change. In estimating risk premiums, the average damage caused by flooding per residence is estimated to be €62,600.[4] This damage is estimated by adding total damage to houses that would result from a major flood in 14 dike-ring areas.[5] The flood damage of residences per dike-ring area is obtained from Wouters (2005). Damage per residence is obtained by dividing

[3] In general, the parameterization of utility and weighting functions for these events is currently under-researched.

[4] This estimate represents average flood damage per residence caused by a severe flood. Of course, flood damage may be higher for some residences, for example because they are located closer to the river.

[5] These dike-rings are the areas situated in the Rhine delta with a probability of flooding of 1/1,250 for which these data are currently available. They include dike-ring areas 9, 36, 38, 41, 42, 43, 44, 45, 48, 49, 50, 51, 52 and 53 (Wouters 2005).

this total damage to residences by the total number of residences located in these 14 dike-ring areas.[6]

Certainty equivalents of the prospect (0: (1 – p); –€62,600: p) are calculated for the different probabilities of flooding. A two-outcome prospect is used because it is assumed that a river flood occurs if the central dike-ring of an area fails, which has probability p, while no river flood occurs if the central dike-ring does not fail, which has probability 1 – p. These certainty equivalents are the sure losses (the insurance premium) for which the individual is indifferent between that sure loss and the prospect of suffering flood damage (if no insurance is bought). They correspond to the maximum WTP to avoid the loss of €62,600 with probability p, which can be interpreted as the maximum premium that individuals are willing to pay for flood insurance. Obviously, the WTP estimates reported in this section are absolute values of the certainty equivalents, which are negative. Because the prospect concerns flood damage per residence, WTP estimates and risk premiums should be interpreted as yearly amounts per residence or household. The corresponding risk premiums are estimated by subtracting the expected value of the prospect (*probability of flooding*damage*) from the WTP measure. Risk premiums are an indication of the profit margin for insurance companies when the cost of offering the insurance is close to the expected loss.

Traditionally, risk premiums are estimated with expected utility theory using a concave utility function. However, experimental evidence indicates that the utility function for losses is moderately convex rather than concave. Tversky and Kahneman (1981, 1992) proposed a power utility function that is convex for losses and concave for gains. This utility function is of the form:

$$U(x) = \begin{cases} x^{\alpha} & \text{for } x>0 \quad (8.1a) \\ -\lambda(-x)^{\alpha} & \text{for } x<0 \quad (8.1b) \end{cases}$$

where $0 \leq \alpha \leq 1$ and λ is the loss aversion parameter. Loss aversion means that losses are weighted more heavily than gains in decisions (Tversky and Kahneman 1981). The parameter α can differ between

6 This informtion was obtained from the HIS-SSM information system of Rijkswaterstaat (DWW 2005c).

gains and losses. This function has the property that multiplying the outcome of a prospect by a positive constant multiplies its certainty equivalent by the same constant (Tversky and Fox 1995). This utility function is commonly used in experimental studies (Abdellaoui 2000; Lattimore et al. 1992; Prelec 1998; Tversky 1967; Tversky and Fox 1995; Wu and Gonzalez 1996).

Booij and van de Kuilen (2006) estimated the utility function (Eqn 8.1b) for the Netherlands based on 1,932 respondents who form a representative sample of the population. This study provides a useful basis for this application. Their estimated value of α equals 0.9 when individuals are faced with high loss amounts. This estimate is obtained using the trade-off method proposed by Wakker and Deneffe (1996), which is robust to what is sometimes called subjective probability distortion or probability transformation. Booij and van de Kuilen (2006) concluded from their extensive experimental study that the utility function is concave for gains and convex for losses, as predicted by prospect theory. In addition, their estimates indicate that scaling up monetary outcomes does not significantly alter the degree of utility curvature. The estimated value of α equal to 0.9 is very close to the exponent (α) of 0.88 that was estimated by Tversky and Kahneman (1992). The observed invariance of utility curvature to the scaling of monetary amounts and the convention of applying the Tversky and Kahneman utility representation (Eqn 8.1b) in experimental studies are the rationale for using the functional form shown in Eqn 8.1b with $\alpha = 0.9$ to estimate the risk premiums for rank-dependent utility and prospect theory in the remainder of this chapter. The utility value of the prospect under expected utility theory, and therefore without probability weighting and loss aversion, can be estimated using the formula:

$$EU = \sum_{j=1}^{n} p_j U(x_j) \tag{8.2}$$

where x_j denotes monetary outcomes, p_j the probability of obtaining outcome x_j, and U the utility of receiving outcome x_j obtained from Eqn 8.1b and $\lambda = 1$. The corresponding certainty equivalents for $\alpha = 0.9$ are calculated as $C = -(-EU)^{1/0.9}$. The estimated risk premiums are –€27 under the current climate scenario ($p = 1/1{,}250$) and –€43 and –€57 under the low ($p = 1/750$) and high ($p = 1/550$) climate change scenarios, respectively. These risk premiums are obtained by subtracting the expected value of the prospect from the absolute value of the certainty equivalent. Obviously, these risk premiums are negative, since the utility function is convex for losses, which means that individuals are risk-seeking in the loss domain. This implies that no basis for insurance exists with this convex utility function when the probability transformation is incorrectly assumed to be linear. It is evident that this is not a realistic description of reality, but these estimates serve the purpose of identifying the independent effect of non-linear probability weighting that is introduced below.

8.4.1 Applying rank-dependent utility theory

Rank-dependent utility theory allows for a non-linear processing of probabilities (Quiggin 1982; Schmeidler 1989). In general, the rank-dependent utility value (RDU) can be represented by the formula:

$$RDU = \sum_{j=1}^{n} [w(p_j + \ldots + p_1) - w(p_{j-1} + \ldots + p_1)]U(x_j) \tag{8.3}$$

where p_1 is the probability of outcome x_1 and p_n is the probability of outcome x_n and $x_1 > \ldots > x_n$, and w is the probability weighting function. The decision weight of a negative outcome x_j can be defined as $\pi(x_j) = w^-(p + \ell) - w^-(\ell)$, where ℓ is the loss rank $(p_{j+1} + \ldots + p_n)$ and p_j is the outcome probability. The weighting function used is of the form proposed by Goldstein and Einhorn (1987) and used by Lattimore et al. (1992). Parameter estimates are based on experiments of Tversky and Fox (1995). This weighting function is of the form:

$$w(p) = \delta p^{\gamma} / (\delta p^{\gamma} + (1 - p)^{\gamma}). \tag{8.4}$$

The estimated parameters are $\gamma = 0.69$ and $\delta = 0.77$, using prospects that involve gains. This function reflects the commonly observed inverse S-shaped weighting function (Abdellaoui 2000; Diecidue and Wakker 2001). Probability weighting is assumed to be similar in the loss and gain domain under rank-dependent utility. Under this (restrictive) assumption, empirical estimates obtained from gain prospects should apply also to loss prospects. This assumption will be relaxed in the estimations with prospect theory, where empirical estimates based on loss outcomes will be used. The probability weighting function (Eqn 8.4) gives relatively high decision weights to bad and good ranked outcomes, compared with middle-ranked outcomes.

The rank-dependent utility value (RDU) of the prospect (0: (1 – p); –€62,600: p) is estimated using Eqn 8.3 with the utility function (Eqn 8.1b) and $\lambda = 1$ and the weighting function (Eqn 8.4) using loss ranks. The value of the prospect is easily estimated in this application, which uses the two-outcome prospect (x_1: (1 – p_2); x_2: p_2) and $x_1 = 0 > x_2$. The rank-dependent utility value of x_1 drops out of the equation because its utility value equals zero. The utility value of outcome x_2 is weighted by the decision weight on its own probability p_2, since the probability of receiving a worse outcome equals zero. This results in a utility value of the prospect of $RDU = (\delta p_2^{\gamma} / (\delta p_2^{\gamma} + (1 - p_2)^{\gamma})(-(-x_2)^{\alpha})$. Certainty equivalents are then calculated with the formula $C = -(-RDU)^{1/\alpha}$.

Estimated WTP values and risk premiums are presented in Table 8.1. Note that the flood probability and expected value of the prospect are shown in the upper parts of Table 8.1 and Table 8.2. Risk premiums are positive, implying that individuals are willing to pay more than the expected value of this flood insurance. As an example, the calculation of the risk premium under the current flood probability will be provided. The RDU of the prospect is estimated as $(0.77*0.0008^{0.69})/(0.77*0.0008^{0.69}+(1-0.0008)^{0.69})(-(62600^{0.9})) \approx -115.97$. The corresponding certainty equivalent C is then computed as $-(115.97)^{1/0.9} \approx -197$. The WTP measure shown in the table is the absolute value of the certainty equivalent and the risk premium is obtained by subtracting the expected loss of €62,600*0.0008 = €50 from €197, resulting in a value of €147.

Willingness-to-pay is more than three times the expected value of the insurance, which suggests that a profitable market may exist even if the costs of offering the insurance are considerably above the expected loss, for example due to high transaction costs. As an illustration, individuals who behave according to rank-dependent utility are willing to pay for flood insurance as long as insurance companies do not add more than 294 per cent[7] on top of the expected value of the insurance, under the current probability of flooding. Willingness-to-pay increases more than the expected loss if the probability of flooding increases. As a consequence, risk premiums increase if the probability of flooding rises. This implies that demand for flood insurance is likely to expand in the future, when flooding events become more likely.

[7] Calculated as ((*willingness to pay – expected value*)/*expected value*)*100.

Table 8.1 *Average risk premiums per residence*

	Current climate		Global temperature +1 °C		Global temperature +2 °C	
	p = 1/1,250		p = 1/750		p = 1/550	
	Expected value = 50		Expected value = 83		Expected value = 114	
	WTP	Risk premium	WTP	Risk premium	WTP	Risk premium
Rank-dependent utility	197	147	290	207	368	254
Prospect theory with parameterization of						
Tversky and Kahneman	262	212	385	302	487	373
Abdellaoui	297	247	428	345	534	420
Etchart-Vincent	87	37	141	58	189	75

8.4.2 Applying prospect theory

In prospect theory, the utility and probability weighting functions are different for gains and losses (Kahneman and Tversky 1979; Tversky and Kahneman 1992). This allows for modelling empirically observed differences in probability weighting in the loss and gain domain. Furthermore, loss aversion is explicitly modelled in the utility function. In general, prospects are valued according to the formula:

$$PTU = \sum_{i=1}^{k}(w^{+}(p_i + \ldots + p_1) - w^{+}(p_{i-1} + \ldots + p_1))U(x_i) + \sum_{j=k+1}^{n}(w\ (p_j + \ldots + p_n) - w\ (p_{j+1} + \ldots + p_n))U(x_j) \quad (8.5)$$

with the outcomes $x_1 \geq \ldots \geq x_k \geq 0 \geq x_{k+1} \geq x_n$ and their corresponding probabilities p for $i \geq k$ and $j > k$. Obviously the risk premiums estimated with prospect theory are most relevant for the application at hand, which concerns losses. Relatively few studies have estimated parameters of probability weighting functions for losses (Wakker 2001). In this section, risk premiums are estimated based on experimental results of Tversky and Kahneman (1992), Abdellaoui (2000) and Etchart-Vincent (2004), which are the most well-known studies. The estimated utility function for losses is steeper than the utility function for gains. The loss aversion estimate λ for the Netherlands in the experimental study of Booij and van de Kuilen (2006) is 1.79 for prospect theory estimates with high losses. This estimate is slightly smaller than the parameter of 2.25 estimated by Tversky and Kahneman (1992) but is in line with recent estimates by Johnson et al. (2006). Using the parameter estimates for the Netherlands, the prospect theory utility function is:

$$U(x) = -1.79(-x)^{0.90}. \tag{8.6}$$

Risk premiums for prospect theory of the prospect (0: (1 – p); -€62,600: p) are first estimated using the weighting function proposed by Tversky and Kahneman (1992), who estimated different parameters for gains and losses. Their proposed weighting function is:

$$w(p) = p^c / (p^c + (1-p)^c)^{1/c}. \tag{8.7}$$

Probability weighting is linear for $c = 1$, and the weighting function becomes more curved as c gets smaller. Experiments indicate that $c = 0.69$ is a good representation when individuals are faced with losses. This weighting function results in the commonly observed inverse S-shaped weighting function (Wakker 2001). The value of the prospect (x_1: (1 – p_2); x_2: p_2) is estimated using Eqn 8.5 with the utility function (Eqn 8.6) and weighting function (Eqn 8.7). In this application where $x_1 = 0 > x_2$, this reduces to a utility value of the prospect of $PTU = (p_2^c / (p_2^c + (1-p_2)^c)^{1/c})(-\lambda(-x_2)^\alpha)$. The certainty equivalents are estimated using the function $C = -(-PTU/\lambda)^{1/\alpha}$. As an example, the calculation of the risk premium under the current flood probability will be provided. The PTU of the prospect is estimated as $(0.0008^{0.69}/(0.0008^{0.69} + (1 - 0.0008)^{0.69})^{1/0.69})(-1.79*(62600)^{0.9}) = -268.33$. The corresponding certainty equivalent C is then computed as $-(268.33/1.79)^{1/0.9} \approx -262$. The WTP measure shown in Table 8.1 is the absolute value of C, and the risk premium is obtained by subtracting the expected value of the loss from €262, resulting in a value of €212.

WTP is considerably larger under prospect theory with weighting function (Eqn 8.7) than under rank-dependent utility theory, as is apparent from Table 8.1. Estimates under the current flooding probability suggest that consumers are willing to pay up to 424 per cent on top of the expected value of the insurance. WTP under prospect theory is around 30 per cent above the WTP estimates of rank-dependent utility theory for all three flooding scenarios. This implies that the difference in risk premiums increases if the probability of flooding increases. This difference increases from €65 under the current climate to €119 under the high scenario of climate change. Therefore, future WTP for flood insurance is significantly higher when individual behaviour is consistent with Kahneman and Tversky's prospect theory than with the presented version of rank-dependent utility theory.

Subsequently, risk premiums are estimated using parameter estimates of weighting function (Eqn 8.4) by Abdellaoui (2000). Using medians to estimate the parameters of this experimental study $\delta = 0.84$ and $\gamma = 0.65$ are obtained for losses. The WTP figures are slightly higher according to these parameters than according to those estimated by Tversky and Kahneman. WTP estimates using the Abdellaoui parameterization are approximately 10 per cent higher in all three scenarios. Estimates under the current flooding probability suggest that consumers are willing to pay up to 494 per cent on top of the expected value of the insurance. Finally, risk premiums are estimated using experimental evidence of Etchart-Vincent (2004) on the weighting function of losses. Parameter estimates of this study suggest that probability weighting functions differ slightly for small and large losses, in the sense that overweighting is larger in the latter domain. Nevertheless, the observed curvature of the weighting function and overweighting of probabilities in this study are smaller than the estimations of Tversky and Kahneman (1992) and Abdellaoui (2000). Parameter estimates of the weighting function (Eqn 8.4) are $\delta = 1.179$ and $\gamma = 0.853$ for large losses,[8] which is the relevant domain for the current application. Willingness-to-pay estimates and risk premiums are considerably smaller with this weighting function compared with the other estimates for prospect theory and rank-dependent utility theory. The estimates using the Abdellaoui weighting function are about three times smaller than the estimates obtained with the Kahneman and Tversky function in the current climate scenario. This ratio decreases in the low and high climate change scenarios. Risk premiums are still positive

[8] Estimates for small losses are $\delta = 1.02$ and $\gamma = 0.836$.

with this parameterization; WTP is approximately 70 per cent above the expected value of the insurance in all scenarios.

8.4.3 Implicit risk aversion

Under rank-dependent utility theory and prospect theory, risk aversion results from probability weighting, since the utility function for losses is assumed to be convex. This contrasts with expected utility theory, under which risk aversion can only be modelled with a concave utility function. Frequently used utility functions for expected utility are power functions of the form given in Eqns 8.1a and 8.1b (Wakker 2008). It is interesting to examine which values of α under expected utility theory render the same WTP estimates as rank-dependent utility and prospect theories (shown in Table 8.1). This would give an indication of the degree of risk aversion that is implicitly assumed in the estimates and allows for an assessment of their realism. These results can aid in determining what range of risk premiums is most suitable for policy analysis. The values of α under expected utility theory that result in the same risk premiums as shown in Table 8.1 are presented in Table 8.2. These values differ between the different computed risk premiums, due to distinct parameterizations of the utility and probability weighting functions that result in different degrees of risk aversion. With utility function $U(x) = -p(-x)^{\alpha}$ for $x < 0$ and corresponding (negative) certainty equivalents $C = -(-U)^{1/\alpha}$, the implied coefficients of α can be calculated using the formula (derived in Appendix 8.A1):

$$\alpha = \ln(p) / \ln(C / x). \tag{8.8}$$

These estimated levels of α imply moderate concavity of the utility function; they are all modestly above unity ($\alpha = 1$ is equivalent to risk neutrality). From comparing Tables 8.1 and 8.2 it is apparent that α values are larger for parameterizations that result in higher risk premiums, as is to be expected. Variation in α values over the different climate scenarios is minimal, although there is a tendency towards a slight decrease in risk aversion when probabilities increase. However, overall it can be concluded that risk aversion is rather constant across the three flood probabilities used. This explains the increase in risk premiums when the probability of flooding rises, since overweighting of probabilities is rather constant within this range of probabilities.

Table 8.2 *Values of α for which expected utility risk premiums are equivalent to estimates under rank-dependent utility or prospect theory*

	Current climate	Global temperature +1 °C	Global temperature +2 °C
	p = 1/1,250	p = 1/750	p = 1/550
	Expected value = 50	Expected value = 83	Expected value = 114
	α under *EU*	α under *EU*	α under *EU*
Rank-dependent utility	1.2373	1.2320	1.2282
Prospect theory with parameterization of			
Tversky and Kahneman	1.3018	1.3005	1.2994
Abdellaoui	1.3323	1.3276	1.3242
Etchart-Vincent	1.0841	1.0862	1.0876

Risk aversion is often measured by the index of relative risk aversion, which is $-\ xU''(x)/U'(x)$, and is constant for power functions (Arrow 1971; Pratt 1964). The index is $1 - \alpha$ for power functions of the form given in Eqn 1a, and $\alpha - 1$ for power functions of the form given in Eqn 8.1b (see Appendix 8.A2). The utility function is concave if $\alpha > 1$ for the function in the loss domain (Eqn 8.1b) and if $\alpha < 1$ for the function in the gain domain (Eqn 8.1a). The values of α in Table 8.2 correspond to a risk aversion index of approximately 0.23 for the rank-dependent utility estimates, 0.3 for the Tversky and Kahneman estimates, 0.33 for the Abdellaoui estimates, and 0.09 for the Etchart-Vincent estimates. Apart from the latter risk aversion estimate, these estimates are slightly larger than the values of the risk aversion index that Wakker et al. (2007) estimated for the lower two-thirds income class of the Dutch population. These estimates range between 0.097 and 0.186 and are very close to risk neutrality, but concern significantly smaller losses of up to €91, for which risk aversion is generally assumed to be lower. The risk aversion assumed in the risk premium estimates with the Etchart-Vincent weighting function parameters is even below the estimates of Wakker et al. (2007) and could, therefore, be regarded as the least realistic.

The implied values of α are quite small compared with estimated values in macroeconomics and finance (which are usually greater than 1); however, the latter involve significantly more money than does flood damage per residence (Bliss and Panigirtzoglou 2004). These values of the risk aversion index imply negative power functions for gains of the form $U(x) = -x^{a}$. Commonly used values of the index of relative risk aversion are around 2 in the finance literature (Cubitt et al. 2001; Wakker 2008). It should also be considered that although the monetary loss (€62,600) used in estimating the risk premiums is generally smaller than losses assumed in estimates of risk aversion of investors, the loss is relatively large for households, which usually have a lesser ability to sustain losses. This would be a reason for expecting higher risk aversion in this application. Overall, it can be concluded that the risk aversion implicit in the estimated risk premiums in Table 8.1 is rather low compared with the assumptions commonly made in the economic literature. This is especially the case for the risk aversion assumed in the WTP estimates based on the Etchart-Vincent weighting function.

8.5 Conclusions

An important requirement for the adequate functioning of a natural disaster insurance programme is that demand for the insurance is sufficiently high. A high market penetration of insurance implies that fixed costs can be spread over many policyholders and that risks can be diversified in a large pool of the insured. Moreover, a high market penetration of insurance in combination with adequate insurance coverage reduces the need for ad hoc government compensation of (uninsured) damage after a major disaster. The traditional economic theory of individual decision making under risk – expected utility theory – has been unable to accurately describe individual decision making regarding the purchase of natural disaster insurance. Therefore, this chapter has examined the demand side of flood insurance arrangements in the Netherlands by estimating WTP and risk premiums, using alternative theories of individual decision making under risk. The analysis delivers two main insights.

1. Estimation results suggest that a profitable flood insurance market could be feasible.
2. Climate change has the potential to increase the profitability of offering flood insurance.

With respect to the first finding, the alternative theories of individual decision making under risk indicate that the welfare-improving function of insurance schemes against flood damage can be significant. Although all risk premiums are positive, variation in the estimated risk premiums exists, depending on the parameterization of the weighting function chosen. In order to assess which range of risk premiums is most applicable, the risk aversion is estimated that is implicit in the different parameterizations of the weighting function used. This analysis suggests that the lowest risk premiums (based on the Etchart-Vincent study) may be the least realistic. Under current flood probabilities, individuals are willing to purchase flood coverage as long as premiums charged by insurance companies are not higher than approximately 70 per cent to 500 per cent of the expected value of the insurance. This is in contrast with the intuition of various experts in the field.

With respect to the second insight, peak river discharges in the Netherlands are expected to increase due to climate change. The estimation of risk premiums for three climate scenarios has indicated that WTP increases more than the expected value of the insurance when the probability of flooding rises. According to the prospect theory estimates, risk premiums increase between 40 per cent and 60 per cent in the low climate change scenario, and between 70 per cent and 100 per cent in the high climate change scenario, compared with the current situation.

It should be stressed that the calculations offered in this chapter merely give an indication of the actual WTP for flood insurance. It is possible to raise methodological concerns about the approach taken. Variation in individual probability weighting and risk attitudes exists, and individuals may discount the probability of extreme events altogether, which has not been accounted for in the probability weighting functions applied in this chapter. Nevertheless, the parameter values presented in our study are consistent with recent experimental findings on utility curvature for the Netherlands, as well as with a wide range of parameterizations of the weighting function obtained in a number of studies. Chapters 10 and 11 will examine WTP for flood insurance by Dutch homeowners, using the contingent valuation and choice modelling methods, which provide insights into the individual heterogeneity of WTP for flood insurance, and risk attitudes towards low-probability flood risks in the Netherlands, in particular.

Appendix 8A Mathematical derivations

8.A1 Derivation of Eqn 8.8

Substituting Eqn 8.1b into $C = -(-U)^{1/\alpha}$ for $x < 0$ gives:

$$(p(-x)^{\alpha})^{1/\alpha} = -C$$

Solving this for α results in:

$$\alpha = \ln(p) / \ln(C / x)$$

8.A2 Derivation of the relative risk aversion index

For $U = x^{\alpha}$ and $x > 0$ the index of relative risk aversion is (see Arrow 1971; Pratt 1964):

$$rr(x) = -x(U''(x) / U'(x))$$

$$= -x(\alpha - 1)\alpha x^{\alpha-2} / \alpha x^{\alpha-1}$$

$$= 1 - \alpha$$

For $x > 0$ or $U = -(-x)^{\alpha}$, the index of relative risk aversion is:

$$rr(x) = \alpha - 1$$

9 *Individual perceptions of flood risk*

9.1 Introduction

Individuals often deviate from rational behavioural principles when they make decisions under risk (Kahneman 2003; Tversky and Kahneman 1986). In evaluating hazards, people commonly rely on intuitive risk judgements or risk beliefs, which are often called 'risk perceptions' (Slovic 1987). Such perceptions may differ considerably from expert assessments of risk, as individuals find it difficult to evaluate probabilities of infrequent hazards or may lack adequate information about risk. An understanding of risk perceptions and their determinants and allowing for 'bounded rationality' or limitations in individuals' perceptual and cognitive capabilities are fundamental in correctly anticipating individual responses to natural hazards, as shown in Chapter 8.

Knowledge about risk perceptions of natural hazards may provide important information about individual decisions regarding insurance purchases; decisions to take self-protection measures; and public support for governments' risk reduction or climate change adaptation policies. Household risk judgements can also support the legitimacy of, and compliance with, land-use planning and other risk reduction policies that are undertaken by governments (Peacock et al. 2005). Political support by individuals for risk-reducing investments is stronger if the risk to be reduced is perceived as great by citizens (Viscusi and Hamilton 1999). Detailed information about risk perceptions may further improve communication about risk reduction policies to the public. Individual beliefs about hazards are important factors behind individual decision making under risk with respect to insurance purchases and the undertaking of self-protective measures. As an example, Chapters 10 and 11 empirically analyse demand for flood insurance in the Netherlands along with stated preference

Parts of this chapter have previously been published as Botzen et al. (2009b). Section 4 of the cited article has been shortened here to avoid overlap with other chapters.

methods, and their results show that risk perceptions relate significantly and positively to demand for flood insurance. Often natural hazard risks comprise nature- and government- (protection-) related components, which are exogenous to the individual, as well as individual choice components, such as the decision to live in a particular location and the (lack of) implementation of precautionary measures. Examples of precautionary measures are the anchoring of roofs to limit hurricane damage or elevating houses and installing water barriers to limit flood damage. Individuals' risk beliefs affect the undertaking of damage mitigation measures and the extent to which losses will be reduced by means of precautionary measures (Burn 1999; Flynn et al. 1999). Given the importance of individual risk perceptions in individual protective behaviour towards risk, it is very relevant for the design of climate change adaptation policies to understand how risk perceptions are formed and how they relate to actual levels of risk.

Natural hazards have been an important topic for risk perception research, especially in the psychological literature (Slovic 2000). Nevertheless, there has been little empirical research on the factors shaping individual risk perceptions of specific natural hazards (Peacock et al. 2005). Most risk perception research has focused on explaining risk perception in relation to prior experience, knowledge, and socio-economic as well as demographic characteristics. Especially for flood risk, little research has been conducted on the influence on risk perceptions of geographical characteristics, such as proximity to the hazard. The existing evidence relating risk perception to proximity to the risk is mixed. Palm et al. (1990) and Mileti and Darlington (1997) did not find a relationship between proximity to an earthquake fault line and risk perception. On the other hand, risk assessment of volcanic, toxic gas, or radioactive material releases and proximity to the hazard were found to be positively associated (Lindell 1994). According to Peacock et al. (2005), perceptions of hurricane risks are positively related to living in an area with high potential wind speeds. Brilly and Polic (2005) observed that flood risk awareness is higher in a flood-prone area in Slovenia than in areas where flooding is less common. According to Siegrist and Gutscher (2006), Swiss householders' perceptions of flood hazards are related to the riskiness of a location based on flood risk maps. These authors observed a significant and positive relation between risk perceptions and expert assessment of risk, although in certain regions perceived risks were considerably over- or underestimated by individuals. Some studies have used different types of 'cognitive mapping' methods to elicit relations between perceptions of natural hazard risk and spatial characteristics derived from geographical maps (see, e.g., Gaillard et al. 2001; Kaplan 1973; Ruin et al. 2007). For example, Ruin et al. (2007) related motorists' perceived risk of flash floods on various routes to actual risk derived from geographical maps and found that drivers were more likely to under- than overestimate the risk of flash floods.

The objective of this chapter is to examine individual risk perceptions of river flooding in the Netherlands and assess the relationship between actual or objective risk measures and the shaping of risk perceptions. The research method was a survey of approximately 1,000 homeowners living in the Dutch river delta. The specific nature of flood risk means that it is possible to identify the dependence of risk perceptions on the various geographical characteristics of individuals' residential areas that affect the likelihood and consequence of the hazard. A geographical information system (GIS) was used to obtain information on various indicators of objective exposure to risk derived from the geographical location of households, such as their elevation relative to the potential water level, proximity to a main river, and the degree of protection afforded by dike infrastructure. Particular attention was paid to how these actual measures of flood exposure based on geographical characteristics shape perceived flood risk, in addition to socio-economic characteristics and knowledge about, and experience with, flooding. This approach is supported by dual-process theories in psychology that postulate that risk perceptions are shaped by both experience and analytical reasoning (Slovic et al. 2004). In contrast to many other studies, which often analyse just one indicator of perceived risk, this study has elicited several proxies of risk beliefs in order to arrive at a complete understanding of flood risk perceptions. In particular, the following indicators of risk beliefs were examined:

- the perceived risk relative to an average resident
- the perceived probability of flooding (both in relation to other risks and independently)
- perceptions of whether dike maintenance complies with the safety norms of dike design
- the amount of flood damage individuals expect to suffer.

The determinants of these perceptions were analysed with ordered probit, binary probit and ordinary least squares (OLS) regression models (e.g. Wooldridge 2002).

The remainder of this chapter is structured as follows. Section 9.2 provides information about flood risk and flood risk perceptions in the Netherlands. Section 9.3 discusses the psychology of the formation of risk perceptions. Section 9.4 explains the survey. Section 9.5 provides a detailed analysis of the answers to the questions on individual flood risk perceptions. Section 9.6 presents estimation results for a range of statistical models of the factors that shape flood risk perceptions. Section 9.7 presents conclusions and policy implications.

9.2 Flood risk and perceptions in the Netherlands

9.2.1 Flood risk in the Netherlands

The Netherlands is a relevant case study because the area is very vulnerable to flooding. The country comprises a delta where the rivers Rhine, Meuse and Scheldt flow into the North Sea. About half of the country is situated below sea or river water level, and about two-thirds of the country's GDP is earned in these low-lying regions. Advanced dike infrastructures protect these areas, known as 'polders', from flooding. Currently, Dutch flood safety norms are the highest in the world, ranging from a 1 in 10,000 years flood event near the coast up to a 1 in 1,250 years flood event near rivers. Nevertheless, it is realized that absolute safety cannot be achieved, and residual flood risks remain. The consequences of flooding can be catastrophic, with potential damage costing up to €100 billion (Aerts et al. 2008b). In other words, floods can be characterized as a low-probability, high-impact risk. Moreover, climate change is expected to increase future flood risk as a result of more (extreme) precipitation, higher peak flows of rivers and rise in sea level (IPCC 2007, 2012; Middelkoop et al. 2001).

Several adaptation strategies to limit rising risk have been suggested and are currently being implemented: namely, heightening dikes; creating more space for rivers; and designing flood retention areas. In addition, 'soft' measures are considered, such as insurance and increasing the preparedness of individuals for disasters (e.g. Kabat et al. 2005). At present, private insurance for flood damage is not available in the Netherlands, but the government may grant partial compensation for damage caused by large flood disasters. The introduction of a public–private partnership for insuring flood damage may be useful in order to decrease the uncertainty of post-disaster government compensation and provide households with an incentive to minimize damage (see Chapter 4). Most investments in flood protection are made by the government and are focused on lowering the probability of flooding. Nevertheless, individuals have a certain degree of control over the damage caused by floods. Chapter 7 shows that damage-mitigating measures undertaken by Dutch households can be effective in limiting flood damage. The statistical analysis in that chapter indicates that households are more likely to invest in self-protection measures, the higher they perceive their risk of flooding. This underpins the need to examine individual risk perceptions and their determinants in more detail. Moreover, insights into individual risk perceptions and their determinants are very relevant for the current discussion of opportunities to develop flood insurance in the Netherlands.

9.2.2 Studies on perceived flood risk in the Netherlands

Research on flood risk perceptions in the Netherlands is still in its infancy, and considerable research efforts are needed to provide insights to inform policy. Research on perceptions is unusual, as water management studies in the Netherlands have traditionally focused only on technical analyses of flood risk prevention (Baan 2004). Nevertheless, some studies have been undertaken that provide a relevant starting point and comparison with the present study.

The importance of considering flood risk perceptions in water management in the Netherlands was highlighted by Baan and Klijn (2004). Terpstra et al. (2006) conducted a survey (n = 49) and a focus group (n = 14) to examine flood risk perceptions, and concluded that, overall, individuals found it difficult to imagine a flood really occurring, i.e. the perception of the probability of flooding is low. Terpstra and Gutteling (2008) recently undertook an empirical study on flood risk perceptions in the Netherlands, consisting of a larger sample (n = 658). The authors conducted an Internet survey in the province of Friesland, which is protected by dikes with a safety norm of 1/4,000 years. Respondents were asked to rate:

- their personal risk of flooding
- the likelihood of a flood event in Friesland within the next 10 years
- the severity of personal consequences

- their feelings of fear
- perceived control during a flood
- the frequency of flood risk.

In addition, respondents rated their trust in the expertise and credibility of the authorities responsible for flood risk management in Friesland. Qualitative response scales were used, ranging from four to six items. The results of Terpstra and Gutteling (2008) indicate that, overall, risk perceptions are low, which is reflected in low scores for personal risk, frequency of flooding and feelings of fear. In contrast, the consequences of flood risks were rated considerably higher than the flood probability, indicating an elevated perceived risk of expected damage compared with the flood probability. Individuals showed confidence in their personal ability to survive a flood event. Trust in the expertise of the authorities responsible for preventing flood damage was generally high, but their credibility was rated somewhat lower. Terpstra (2011) showed that Dutch citizens' flood preparedness intentions are related to flood risk perceptions. For example, he found that a higher level of trust reduces individual perceptions of the probability of flooding, which reduces flood preparedness intentions. Moreover, trust reduces the amount of dread evoked by flood risk, which lowers intentions to prepare for flooding; while previous flood experiences increased dread.

9.3 The psychology of the formation of risk perceptions

The thinking, knowing and information processing of individuals is often described according to a dual-process theory (e.g. Kahneman and Frederick 2002; Loewenstein et al. 2001; Sloman 1996).[1] This theory postulates that two systems are at work in the human brain to help people apprehend reality; namely, an experiential system and an analytical system, as depicted in Table 9.1 together with a list of their properties. It should be noted that certain authors have defined these systems somewhat differently. Here, we use the terminology of Slovic et al. (2004), who undertook a study indicating the role of the 'affect heuristic' in shaping risk perceptions. The experiential and analytical systems work in fundamentally different ways. The *experiential system* is

[1] An exhaustive review of the empirical findings on the factors shaping risk perceptions is beyond the scope of this chapter. See Kahneman et al. (1982), Slovic (2000) and Sjöberg (2000).

Table 9.1 *Comparison of the experiential and analytical systems*

Experiential system	Analytical system
1. Holistic	1. Analytical
2. Affective: pleasure–pain-oriented	2. Logical: reason oriented (what is sensible)
3. Associationistic connections	3. Logical connections
4. Behaviour mediated by 'vibes' from past experiences	4. Behaviour mediated by conscious appraisal of events
5. Encodes reality in concrete images, metaphors and narratives	5. Encodes reality in abstract symbols, words and numbers
6. More rapid processing: oriented towards immediate action	6. Slower processing: oriented towards delayed action
7. Self-evidently valid: 'experiencing is believing'	7. Requires justification via logic and evidence

Source: Slovic et al. (2004).

generally labelled as intuitive, automatic, natural, non-verbal, narrative and experiential. Tversky and Kahneman (1983) show the importance of intuitive reasoning related to the experiential system in making probability judgements. In contrast, the *analytical system* is characterized by analytical, deliberative, verbal and rational aspects (Epstein 1994). The two systems operate in a parallel manner and interact so that they simultaneously process information to shape judgements or perceptions.

According to Slovic et al. (2004) and Loewenstein et al. (2001), affective feelings are important in individual risk judgements. Individuals may have a higher risk perception if flood risk is associated with negative feelings, which may have been caused or reinforced by experiences with flooding or evacuation due to a flooding threat (Finucane et al. 2000; Keller et al. 2006). This is related to the relevance of the 'availability heuristic' in risk perception discussed by Tversky and Kahneman (1973, 1974). Heuristics are simple rules that individuals may use in risk judgements. Individuals who use the availability heuristic in forming perceptions judge an event as risky if it is easy to imagine or recall. For example, individuals who have experienced a flood may find it easier to imagine that a flood will happen again in the future and, therefore, indicate a higher perceived risk than individuals without flood experience. From a Bayesian learning standpoint, one would also expect that perceived risk increases after the experience of a disaster (Viscusi and Zeckhauser 2006). In particular, individuals may revise prior risk beliefs upwards after the experience of a disaster if it provides information that risk may be higher than initially anticipated.

Empirical results about the effects of experience on risk perceptions across studies are not entirely consistent. For example, Peacock et al. (2005) found that earlier experience with a disaster can even lower the perceived risk associated with future events. A possible explanation of the latter phenomenon is that some people think that if they have experienced one disaster this reduces the likelihood of experiencing another in the future. The specific nature of the experiences is likely to be important in shaping risk perceptions. For example, all the residents of a region where flooding has occurred may claim that they have experienced a flood, even if not all of them have actually suffered from water near or inside their house. Research has shown that more intense personal experiences, such as suffering damage, results in elevated perceptions of risk (Norris et al. 1999; Perry and Lindell 1990; Raid et al. 1999; Windham et al. 1977). The effects of experiences with flooding on perceptions of flood risk by Dutch households are examined in the present study. The dual-process theory provides a useful basis for this work, since it highlights the relevance of examining the influence of experience with flooding in shaping perceived risk. It also provides a basis for analysing rational decisions in relation to objective information, such as proxies for flood exposure based on the geographical properties of the individual's residential area.

9.4 Explaining the survey and variables used in modelling risk perception

9.4.1 Implementation of the survey

Information on the pre-tests of the survey is provided in Section 7.3. The survey was administered over the Internet using Sawtooth CBC software.[2] This computer-based method has the advantages that interviewer effects can be avoided and a large geographically spread sample can be obtained at relatively low cost. Respondents were selected

[2] See www.sawtoothsoftware.com.

from the consumer panel of Multiscope and contacted by e-mail.[3] This e-mail did not specify the topic of the survey, in order to prevent selection bias. The sample consisted of random draws of panel members who live in 'dike-ring' areas[4] in the Netherlands with a safety standard of once in 1,250 years (see Figures 5.1 and 7.1 for maps of the sample area). Only homeowners could participate in the survey; tenants were excluded since they do not bear the costs of flood damage to their homes. The sample was set up to be representative for the Dutch population up to the age of 60 years. Fewer older individuals are represented in the Internet sample, because seniors are generally less active on the Internet than younger people. The survey was removed from the Internet once the desired number of respondents was reached.[5] A total of 1,140 respondents filled out the questionnaire. Respondents who lived outside the sample area or in apartments higher than the first floor were removed from the data. The resulting total number of completed questionnaires was 982. An extensive description of the pre-tests, structure and implementation of this survey, and an English translation of the complete questionnaire, is given in Botzen et al. (2008b). Several questions were posed to respondents before asking those on risk perceptions. These initial questions addressed their experiences with flooding, evacuation and flood damage, as well as their knowledge about the causes of flooding. Such questions introduce the topic to respondents, and the answers serve as explanatory variables in modelling the responses to the risk perception questions (Section 9.6). In addition, a short introductory text was included before the risk perception questions, explaining that respondents would be asked several questions on how they estimate the probability and resulting damage of river flooding. In Section 9.5 we explain in detail the questions used to elicit risk perceptions. The survey included several questions about willingness-to-pay for flood insurance, which are analysed in detail in Chapters 10 and 11.

[3] For more information, see www.multiscope.nl.
[4] A so-called 'dike-ring' is a geographical unit bounded by a flood protection system of dikes.
[5] The use of the consumer panel of Multiscope does not allow us to calculate the exact response rate to our survey, since the survey was removed from the Internet once a pre-specified quota of completed questionnaires was reached. On average, response rates of the consumer panel are well above 20% (www.multiscope.nl).

9.4.2 Sample characteristics

The sample had slightly more male (58 per cent) than female respondents, which may be because the Internet survey method attracts more male respondents. Moreover, it may be that males are more likely to complete the survey than women because they may predominantly deal with the financial matters of households, such as the purchase of insurance (Chapters 10 and 11). Approximately 50 per cent of the respondents had at least one child still living at home and 39 per cent had a Bachelor's or Master's degree as their highest education level. The average age of respondents was 46 years. The proportion of respondents who were older than 60 years was about 11 per cent, which is smaller than is the case in the actual Dutch population. The median and average after-tax household income was the answer category 'between €2,501 and €3,000 per month', which is close to the average after-tax income of a household that owns a house in the Netherlands: namely, €3,025 per month (Statistics Netherlands 2008).

9.5 An examination of perceptions of flood risk in the Netherlands

Five questions were used to elicit individual perceptions of flood risk. These questions are shown in Appendix 9A. The first three questions asked respondents to rate their flood probability using qualitative answer categories. The last two questions about flood risk perceptions asked respondents to give quantitative estimates of the expected return period of flooding and expected flood damage. These questions and answers are discussed in detail below.

9.5.1 Perceived flood risk relative to other risk

The respondents were asked to rate on a scale from 0 to 10 the probability that their household would suffer financial damage to their property as a result of various risks, including flooding (Appendix 9A). On this scale, 0 represents no probability; 5 is a neutral option; and 10 is an extremely high probability. Similar scales have been successfully used to assess perceived likelihoods of health risks, such as surviving until a specified age (Hurd and McGarry 1995), as well as nuclear risks (Kunreuther et al. 1988). This question allows the assessment of how individuals perceive the flood risk relative to other risks they face, and therefore puts their perceptions of flood risk into perspective. The other risks include a terrorist attack, burglary, a house fire, car theft, a car fire, traffic accident and storm damage. In some countries storm and flood risk can be closely related, such as with hurricanes in the USA. In the Netherlands, river floods are mostly caused by peak water discharges due to rainfall and meltwater contributions in upstream countries, such as France and Germany, and not so much by storms. Windstorms are very common in the Netherlands and often inflict considerable damage to properties without causing any flooding (see Chapter 2). Therefore, it is unlikely that any ambiguity exists between the storm and flood events included in this question.

In practice, the average probability of suffering financial damage caused by flooding is lower than the probability of suffering damage from storms, burglary, a traffic accident, a house fire and car theft, while it is higher than suffering damage from a car fire or a terrorist attack (CBS 2007; Verbond van Verzekeraars 2007). Approximately 82 per cent of the respondents rated flood risk as less likely than storm risk, 79 per cent rated flood risk less likely than burglary, 69 per cent rated flood risk less likely than a traffic accident, 69 per cent rated flood risk less likely than car theft and 67 per cent rated flood risk less likely than a house fire. About 50 per cent thought that the probability of suffering flood damage is smaller than the probability of suffering damage from a car fire, although this is in fact not true for the average citizen (Botzen et al. 2008b). Moreover, 56 per cent of the respondents thought that they are less likely to suffer damage to their property as a result of a flood than from a terrorist attack. In reality, however, the probability of suffering property damage from flooding is likely to be much greater than suffering property damage because of terrorism in the Netherlands. Loewenstein and Mather (1990) found that recent risks often attract considerable concern from individuals (sometimes defined in the literature as 'panic'), which may be due to unfamiliarity with the risk. This may explain why Dutch households are more concerned with risk related to terrorism than with flood risk.[6] Overall, it

[6] Although religion has always been a cause of violent conflict and acts of violence, the 2001 attacks by religious extremists on the World Trade Center that killed almost 3,000 people on a single day raised the ranking of terrorism risk on the political agenda. Therefore, individuals may perceive it as a recent threat.

Table 9.2 *Risk ratings for various hazards (as a percentage of total respondents)*

	Hazard							
Risk rating	Flood	Terrorist attack	Burglary	House fire	Car theft	Car fire	Traffic accident	Storm
0	8.6	14.0	0.1	0.2	6.1	6.4	0.1	0.1
1	20.8	24.8	1.1	5.1	2.3	11.3	1.1	1.5
2	16.1	13.5	4.1	7.8	5.1	12.3	3.2	2.2
3	12.3	9.4	6.1	8.6	8.5	9.3	3.6	4.6
4	10.6	5.5	6.7	9.5	6.2	11.6	4.2	4.6
5	15.2	22.8	23.7	35.3	21.7	28.7	28.7	19.9
6	9.0	5.0	21.9	18.0	20.2	11.2	17.6	21.6
7	4.5	3.3	21.4	10.6	18.2	6.2	22.7	24.7
8	2.3	1.3	11.1	3.3	8.7	2.1	13.4	15.0
9	0.4	0.3	3.1	1.1	2.3	0.6	4.5	4.8
10	0.3	0.1	0.7	0.5	0.7	0.2	0.9	1.0

can be concluded that our respondents judged the level of flood risk to be lower than other types of risks.

Table 9.2 shows the ratings for flood risk and the other hazards. About 9 per cent of individuals rated the probability of suffering flood damage as 0, and most respondents (20 per cent) rated flood risk in the second lowest category. The percentage of respondents who chose a higher level of risk gradually decreased as the risk level rose, apart from the neutral option (category 5), where a small upward jump in the distribution can be observed. Compared with the distributions for the other risks, the risk levels lower than the neutral point were chosen most often for the flood hazard, apart from the categories 0 and 1 for the terrorist risk and category 4 for the risk of a car fire. This confirms that flood risk perceptions are low compared with the other risks.

9.5.2 *Qualitative estimate of the perceived flood probability*

The aforementioned results concerned the rating of the flood probability, together with other risks. These answers are not, however, used in the statistical analysis of the factors underlying the shaping of flood risk perceptions in Section 9.6. This is because the level of the rating of flood probability may have been influenced (so-called 'anchoring') by the rating of the other events in the same table (shown in Appendix 9A). This would imply that the chosen risk level of the perceived flood probability is partly determined by the risk levels assigned by that respondent to the other risks. As a consequence, using the previous question to analyse the factors that determine the rating of the flood risk independently of the other risks would not be very useful. Therefore, we included another question that asked the respondents to rate the flood probability without requiring them to rate other risks, so that the rating of the flood probability itself cannot be biased. The scale differs from the scale of the previous question and has fewer categories so that respondents view the question as being distinct.

This question asked respondents to rate the probability that a flood occurs in their residential area on a qualitative scale with the options: 'I do not have any flood risks', 'very low', 'low', 'not low/not high', 'high', 'very high' and 'don't know'. Table 9.3 provides the rating of the flood probability by the respondents. Overall, individuals perceived their risk of flooding as low. About 11 per cent stated that they do not face any flood risk; 31 per cent of the individuals regarded their flood risk to be very small; and another 31 per cent thought that they faced a small flood risk. Thus, in total, 72 per cent of the respondents rated their flood probability as 'small', 'very small' or 'no flood risk at all'. Few respondents chose the 'high' and 'very high' categories. Only about 0.5 per cent chose the 'don't know' option. The responses to this question are consistent with the results of the previous question, where most individuals rated their probability of suffering damage due to flooding low relative to other risks. A correlation analysis indicates that the perceptions of the probability of suffering flood damage (shown in Table 9.2) are positively and significantly (at the 1% level) related to the perceived flood probability (shown in Table 9.3), as would be expected.

9.5.3 Respondents' perceived flood risk relative to an average citizen

Several studies have elicited individual risk perceptions compared with a national average (e.g. Leiter 2009; Viscusi and Zeckhauser 2006). Our third question follows this tradition in risk perception

Table 9.3 *Respondents' perceived flood probability*

Answer categories	Percentage of responses
No flood risk at all	10.5
Very small	30.7
Small	30.8
Not small/not large	19.3
Large	7.4
Very large	0.7
Don't know	0.6

research and asks respondents to rate their flood risk compared with an average resident, using as answer options 'lower than average', 'equal to average' and 'higher than average'. On the basis of our research, most respondents think that they have a lower than average flood risk (48 per cent); 34 per cent judge that they have an average flood risk; but only 18 per cent think that they have a higher than average flood risk. This finding contradicts the fact that almost all respondents have a higher than average flood risk compared with an average person in the Netherlands because the sample consists of dike-ring areas with the lowest protection norm of the country. Viscusi and Zeckhauser (2006), who used a similar question to estimate perceptions of several risks based on a representative sample of the USA, obtained a similar result. In their study, most people rated the risk of dying because of a car accident, terrorist attack or natural disaster as below average or average. An explanation for this finding may be that people realize that the true distributions of these risks are skewed, in the sense that a small fraction of the population is at much higher risk (e.g. because they live in flood-prone areas), meaning that most people have, in fact, lower than average risk. This is, however, not a satisfactory explanation for the findings in the present survey, since the sample is drawn from individuals who live in high-risk areas, and therefore it is not a representative sample for the whole country. This implies that individuals who indicate that they have a lower flood risk than the national average are in fact underestimating their own risk or overestimating the national average risk. Many individuals seem to be unaware that they live in an area vulnerable to flooding.

9.5.4 The perceived return period of flooding and compliance with the flood safety norm

We further elicited a quantitative estimate of the return period of flooding expected by individuals. This is an indicator of the perceived flood probability, because the inverse of the return period equals the probability of flooding. Asking for a return period rather than a probability may be easier for respondents, as judgements under uncertainty are generally improved when risks are expressed in frequencies rather than probabilities (Gigerenzer and Hoffrage 1995). A logarithmic scale of return periods was used as a visual aid, as in Schneider and Zweifel (2004). It was explicitly stated that respondents could answer any return period, even if it was not on the scale. Kunreuther et al. (2001) found that providing an anchor helps respondents to assess low probabilities. A text in the questionnaire explains that the government is responsible for dike maintenance, and that a dike should be high enough to ensure that, on average, it does not flood more than once in 1,250 years, which is the 'return period' that represents the legal safety norm for flooding. This is indicated on top of the scale presented to the respondents as an anchor to help respondents give an expected return period for floods (see Appendix 9A). This makes it possible to assess whether individuals expect a flood return period higher or lower than the legal safety norm. The answers to this question indicate whether individuals expect that dike maintenance is adequate, i.e. it complies with the legal norm. This is of interest since, in the present situation (year 2008), it appears that approximately 20 per cent of the total 3,500 kilometre length of the main dike-ring system does not in fact comply with the safety standards required by the Dutch law. In areas with insufficient dike maintenance, the flood probabilities are higher, as indicated by a recent study (DWW 2005d). This study shows that for some dike-rings, flood probabilities may be as high as 1/200 due to the early failure of weirs and sluices in the case of extreme events. Investments in the Dutch dike-ring system are ongoing, and it is expected that the system will be upgraded by the year 2015 (Aerts et al. 2008b).

The perceived flood frequencies of respondents are summarized in Figure 9.1. A large majority of the respondents (70 per cent) expect that the return period of flooding is greater than the legal norm of 1/1,250 years. Almost 40 per cent expect that the return period is

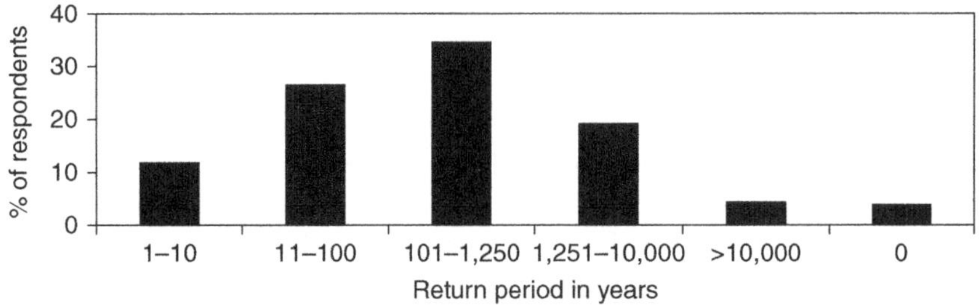

Figure 9.1 Respondents' perceived return period of flooding (in years)
Note: The return period 1,250 has been used in defining the categories 101–1,250 and 1,251–10,000 rather than 1,000, as would be usual, because the legal norm of flooding is 1 in 1,250 and this has been provided as an anchor.

1/100 years or less, which indicates that many individuals perceive the flood probability to be considerably higher than the legal norm. These high perceived frequencies are not consistent with actual risk, even when taking into account that the current dike maintenance in some areas is inadequate. A small fraction of respondents (4 per cent) gave a zero return period of flooding, meaning that they expect that a flood will never occur. A similar fraction gave a recurrence interval larger than 10,000 years, which represents a very small perceived flood probability. Overall, pessimistic beliefs about the flood probability outweigh optimistic beliefs, which is in accordance with the findings of other studies (Camerer and Kunreuther 1989; Schneider and Zweifel 2004). These results are also consistent with findings in the behavioural economics and psychological literature that individuals overestimate low frequency risks when these are expressed as a probabilistic metric (Hurly and Shogren 2005).

9.5.5 Expected flood damage

Flood risk is often defined as *flood probability*flood damage*. An assessment of perceived risk should examine both probability and damage components. While the first four perception questions address the probability of flooding, the fifth and last question asked the respondents to estimate the amount of flood damage individuals expect to suffer once a flood occurs. This provides insight into the perceived consequences of a flooding event. We did not provide a scale as a visual aid for the question on expected flood damage as we

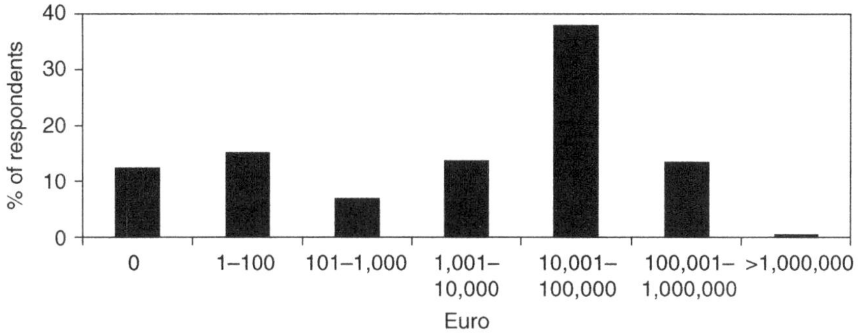

Figure 9.2 Respondents' perceived flood damage (€ per flood event)

did for the question on the expected return period of flooding. The reason is that the pre-tests indicated that respondents found it less difficult to assess the expected damage from a flood (on a monetary scale) than the expected return period of a flood, which is an indicator of the flood probability. This is not surprising because, in general, individuals have difficulty with assessing probabilities (Viscusi 1998). Nevertheless, we note that the open-ended format of this question may result in a large variation in answers to the expected flood damage question.

Figure 9.2 shows the flood damage expected by individuals. It is remarkable that many individuals expected to suffer very little damage; in particular, 12 per cent did not expect to suffer any flood damage and 15 per cent expected to suffer damage valued between €1 and €100. A large proportion (38 per cent) expected to suffer flood damage between €10,000 and €100,000, which is indeed a realistic range of potential flood damage for most people. The mean expected damage is about €70,000 and the median estimated damage is €15,000, indicating that the distribution is positively skewed. A few large damage estimates of between €1 million and €1.5 million caused this skewness. This skewness is not unrealistic as, in practice, a small proportion of homeowners whose properties have a high value will suffer a very large amount of flood damage, causing a right tail in the loss distribution. It is surprising that the expected mean flood damage is equal to the expected mean damage that has been estimated by flood models for the sample area of this survey (see Botzen et al. 2008b). This average estimated flood damage to homes and their contents would cost approximately €70,000 at 2008 price levels.

9.6 Results of statistical analyses of factors determining risk perception

9.6.1 Statistical methods

The role of geographical and socio-economic characteristics in shaping risk perceptions are examined next, using statistical models that estimate the independent effects of several explanatory variables on perceived risk. Different regression methods are required, depending on the type of dependent variable. Table 9.4 shows a description of the dependent variables and their descriptive statistics. The respondents' perceived flood probability and rating of flood risk compared with those of an average resident are ordered categorical variables, and therefore the preferred mode of analysis is an ordered probit model. The variable representing perceptions that the flood return period exceeds the safety norm is a binary variable, and hence a binary probit model is employed. An OLS regression estimates the model for expected flood damage. Standard errors are corrected for potential heteroskedasticity using the Huber–White estimator (Huber 1967; White 1982).

The OLS and binary probit regression methods are commonly applied, and no further explanation is needed here. The ordered probit model may be less familiar and is, therefore, briefly explained. An ordered probit model of the dependent variable y that takes on the ordered values $\{0, 1, \ldots, J\}$ for some known integer J can be derived from a latent variable y^* that is determined by:

$$y^* = x\beta + \varepsilon \text{ and } \varepsilon \mid x \sim N(0, 1) \qquad (9.1)$$

where β is a parameter vector of k number of explanatory variables, x, excluding a constant; and N represents the normal distribution. Observation of a specific value of y is dependent on whether y^* falls between certain intervals of unknown threshold parameters $\alpha_1 < \alpha_2 < \ldots < \alpha_j$ according to:

$$\begin{aligned} y &= 0 \quad \text{if } y^* \le \alpha_1 \\ y &= 1 \quad \text{if } \alpha_1 < y^* \le \alpha_2 \\ &\vdots \\ y &= J \quad \text{if } y^* > \alpha_J. \end{aligned}$$

Table 9.4 *Mean, standard deviation and description of the dependent variables*

Dependent variables	Mean	Standard deviation	Description of the variable
Probability of flooding	2.85	1.13	Categorical, 1 = no risk, 2 = very low, 3 = low, 4 = neutral, 5 = high, 6 = very high
Flood risk compared with that of an average resident	1.71	0.76	Categorical, 1 = below average, 2 = equal to average, 3 = above average
Return period exceeds safety norm	0.34	0.47	Binary, 1 = return period > norm, 0 = return period ≤ norm
Expected flood damage	70,382	185,400	Continuous, damage in €

The maximum likelihood estimation of the log likelihood function for individual *i* gives the parameters of interest:

$$\begin{aligned}\ell_i(\alpha, \beta) = {} & 1[y_i = 0]\log[\Phi(\alpha_1 - x_i\beta)] \\ & + 1[y_i = 1]\log[\Phi(\alpha_2 - x_i\beta) - \Phi(\alpha_1 - x_i\beta)] + \ldots \\ & + 1[y_i = J]\log[1 - \Phi(\alpha_J - x_i\beta)].\end{aligned} \tag{9.2}$$

9.6.2 *Explanatory variables used in the statistical analysis*

The explanatory variables used in the analysis include: respondents' geographical characteristics; their experience with, as well as knowledge about, flooding; and their socio-economic characteristics. Each variable is now explained in more detail.

Three variables reflect objective indicators of the flood exposure faced by the respondent, based on geographical characteristics. Geographical information system (GIS) maps, such as a digital elevation map, have been linked to the respondents' zip codes to obtain these data. These geographical variables indicate: the difference between the elevation of the zip code area of the respondent and the expected water level of a flood; the distance of the house to a main river; and whether the respondent lives in a flood-prone area that is not protected by dikes. The potential flood damage within an area protected by dikes ('dike-ring') is largely dependent on the maximum flood depth and the duration of a flood. The relation between flood depth and flood damage is usually described with what are called 'stage damage functions' (e.g. Aerts et al. 2008b). These functions are based on historical loss data and show the maximum damage for a particular flood depth. The explanatory variable of the relative elevation of a location compared with the potential water level in a dike-ring can be used as a proxy for the potential water level during a flood near an individual's house. This variable was constructed in three steps.

1. The mean elevation of the ground level of a zip code area was computed.[7] The zip codes consist of letters and numbers for most respondents and indicate, with high precision, the location of streets or parts of streets. Therefore they can be used as a proxy for the location of individual houses.
2. The elevation of the flood protection infrastructure, such as a dike, nearest to the respondent's zip code area was computed.[8]
3. The elevation of the flood protection infrastructure was subtracted from the elevation of the respondent's zip code area.

Note that, in this context, we assume that the dike system is overtopped with water in the case of a 1/1,250 flood (or lower probability floods), and that water levels in the flooded area are as high as the maximum dike height.

Flood velocity also plays a role in the damage that occurs because of flooding, but only within the vicinity of a dike breach, where flow velocities can be high. This has been derived from historical loss data from a large flood that occurred in the south-west Netherlands in 1953 (e.g. Keijzer 2008). The variable giving the distance of the individual's house from a main river is another indicator of potential flood damage in addition to the elevation of the house. This has been created by computing the exact distance of the precise coordinates of the middle of the respondent's zip code area to the closest main river in metres using a Euclidian distance measure.[9] This variable may also

[7] The elevations of the zip code areas were obtained using the AHN elevation map.

[8] The elevations of the flood protection infrastructures were obtained using the 'RWS dijkkruinhoogtesbestand'.

[9] The locations of the main rivers were obtained using the GIS map of main surface waters provided by the Dutch government department Rijkswaterstaat.

reflect a higher probability of flooding. Houses near a river are more likely to suffer flood damage than houses far away from a river once a dike is breached or is overtopped by high water levels. This is the case because water may not reach areas located far away from dike breaches because of the presence of natural or man-made obstacles, such as elevated railroads or small hills created in historical ice ages that prevent the complete flooding of a dike-ring area.

Some houses in floodplains are located outside the protection of the Dutch dike-ring system. These houses are very vulnerable to flooding and may already flood with peak river discharges with a recurrence interval of once in 100 or 200 years or even more frequently. The frequency of flooding in unprotected areas is considerably higher than the frequency in protected areas, where dikes are generally designed to prevent overtopping of peak discharges with a recurrence interval of once in 1,250 years. Therefore, the explanatory variable that represents individuals living in unprotected flood-prone areas identifies respondents with a higher flood frequency compared with the remainder of our sample. Finally, a geographical variable was constructed that represents respondents in rural areas. Inhabitants of rural areas with lower population densities and less concentration of economic activity may be expected to have different attitudes towards flood risk.

The questionnaire included three questions on flood experience. We asked whether individuals had ever experienced a flood in their residential area. If the individual had experienced a flood, then we enquired whether any damage was suffered. Finally, we asked whether the individual had ever been evacuated because of flood threat. The most recent flood events in the Dutch floodplain occurred in the years 1993 and 1995. In 1995, nearly 250,000 inhabitants were evacuated. Most unprotected areas were flooded, and considerable damage occurred, including in the upstream countries of the rivers Meuse and Rhine. However, almost none of the respondents (0.4 per cent) had actually suffered damage because of a flood, so this variable cannot be used in the statistical analysis. Therefore, we decided to include a variable representing respondents who indicated that they had experienced a flood in their residential area and were evacuated because of a flood threat. Actual experience with flooding is likely to affect risk perceptions, as a number of studies indicate (discussed in Section 9.3).

A variable was created to represent individuals who could not state the causes of flooding events. It is relevant to examine how this lack of knowledge about floods determines risk perceptions. For example, if less knowledgeable individuals have lower perceptions of flood risk, then a campaign providing information about floods might be useful to increase awareness. Individual risk perceptions may be related to the degree of financial risk aversion of the individual. Here, a risk-seeking index is derived from a question that asks respondents to indicate how similar they are to a hypothetical risk-averse person, described as one who avoids taking financial risk in general and prefers to be well insured. The scale ranges from 1 ('extremely similar to me') to 6 ('not similar to me at all'), where a higher category indicates a higher degree of risk-seeking attitudes.[10] It may be expected that individuals who are generally risk-seeking have a lower perception of risk (Sitkin and Pablo 1992).

Several studies indicate that perceptions of risk vary between different groups of people (Hakes and Viscusi 1997; Leiter 2009). The socio-economic variables included in this study are: the age of the respondent; gender; education level; total value of property; and income.[11] A positive relation between age and risk perception has been found in some studies, while a negative relation has been observed in other research (Peacock et al. 2005). Gender is often found to be an important determinant of risk perceptions (Gustafson 1998). Women perceive risk differently than men and are often more likely to view disaster events and natural hazards as risky (Fothergill 1996; Turner et al. 1986). In addition, individuals with a high income and educational level generally have a lower perception of risk (Slovic 1997, 2000). The education level in this study is a categorical variable which indicates the highest level of education obtained by the respondent. The total value

[10] The question was: 'Some people avoid financial risks as much as possible. They are also well insured. How similar are these people to you?' The answer options were: '1. Extremely similar to me'; '2. Very similar to me'; '3. Similar to me'; '4. A little bit like me'; '5. 'Not similar to me'; '6. Not similar to me at all'.

[11] The effect of the presence of children in the household on risk perceptions in other studies has been mixed, with some studies finding a positive significant effect (Perry and Lindell 1990; Turner et al. 1986), and others reporting no effect (Baker 1991; Lindell and Prater 2000). We have experimented with models that include a variable representing the number of children in the household. Given the unclear theoretical relation between the presence of children and perceived flood risk, the mixed evidence concerning the relation between the presence of children and risk perception in other studies, and the observed non-significance of this relation in the models we estimated, we therefore decided to exclude this variable in our final models.

of property is the sum of the current value of the house and the home contents, which we asked respondents in the survey. The expected flood damage is likely to be related to the total value of property because it is an indicator of the financial value exposed to flooding.

The descriptive statistics of the explanatory variables are shown in Table 9.5. Different methods of coding categorical variables have been applied, depending on the type of variable. Categorical variables with two outcomes (binary) are coded as dummy variables and can be included as such in the regressions without any further transformation. Continuous variables are created from the categorical variables income and value of property, which represent monetary classes (e.g. Blumenschein et al. 2008). Ordinal qualitative variables,[12] which are partitioned into J intervals, can be transformed into values on the real axis using an approach proposed by Terza (1987), as explained in Chapter 7. The variables representing the risk-seeking index and the education level are transformed according to this method. We prefer this method rather than constructing $J - 1$ dummy variables, since the transformation of Terza (1987) can result in gains in efficiency and bias.

9.6.3 *Estimation results*

Table 9.6 reports the results of the ordered probit models of the perceived flood probability and perceived flood risk compared with an average resident; the probit model of whether the expected flood return period exceeds the legal safety norm; and the OLS regression of the expected flood damage. Coefficient values are shown for all models, and heteroskedasticity-robust standard errors are provided in brackets below the coefficients. It should be noted that coefficients do not equal marginal effects (of x on y) in ordered and binary probit models. The estimates indicate whether the explanatory variables relate positively or negatively to the perceived risk, while the size of the coefficient is not directly informative. Asterisks indicate the significance level, and one (*), two (**) and three (***) asterisks represent, respectively, significance at the 10%, 5% and 1% levels. The auxiliary parameters (α_J) of the ordered probit models are statistically significant and of

[12] These variables are characterized by a continuous unobservable ordinal latent index, and each interval is ranked in increasing order (1 through J) according to its supremum (Terza 1987).

Table 9.5 *Mean, standard deviation and description of the explanatory variables*

Explanatory variables	Mean	Standard deviation	Description of the variable
Distance to main river	9.40	11.77	Continuous, distance to main river in kilometres
Elevation relative to water level	-1.48	6.15	Continuous, elevation of area relative to water level in metres
Rural area	0.05	0.22	Dummy, 1 = the area is a rural area
No dike protection	0.11	0.32	Dummy, 1 = the area is not protected by dikes
Experience of flood and evacuation	0.03	0.16	Dummy, 1 = respondent has experienced a flood and evacuation
No knowledge of the causes of flooding	0.19	0.39	Dummy, 1 = respondent cannot state the causes of flooding
Risk-seeking index	3.24	1.16	Categorical (1 to 6), 1 = very risk-averse, 6 = very risk-seeking
Age	45.57	11.98	Continuous, age of respondent in years
Female	0.42	0.49	Dummy, 1 = respondent is female
Education	5.40	1.38	Categorical (1 to 7), 1 = elementary, 7 = university
Value of property[a]	411.70	169.78	Continuous, total value of house and its contents in €1,000
Income[b]	2.86	1.02	Continuous, total after-tax household income in €1,000

Notes: [a] For housing value the respondent could mark one of the following categories: <€100,000; €100,000–150,000; €150,000–200,000; €200,000–250,000; €250,000–300,000; €300,000–350,000; €350,000–400,000; €400,000–500,000; €500,000–600,000; >€600,000. For home contents values the respondent could mark one of the following categories: <€25,000; €25,000–50,000; €50,000–75,000; €75,000–100,000; €100,000–125,000; €125,000–150,000; €150,000–175,000; €175,000–200,000; €200,000–300,000; >€300,000. Continuous values of housing and home contents variables were constructed by setting the housing and home contents value of each respondent to the midpoint of the interval (€650,000 and €350,000 were used, respectively, for the highest housing and home contents value categories).

[b] For income the respondent could mark one of the following categories: <€750; €751–1,000; €1,001–1,250; €1,251–1,500; €1,501–2,000; €2,001–2,500; €2,501–3,000; €3,001–3,500; €3,501–4,000; >€4,000. A continuous income variable was constructed by setting the income of each respondent to the midpoint of the interval (€4,500 was used for the highest category).

Table 9.6 ***Parameters of factors shaping risk perception estimated by four statistical models with varying indicators of flood risk beliefs***

Dependent variable	Model 1 Probability of flooding	Model 2 Flood risk compared with average	Model 3 Return period exceeds safety norm	Model 4 Expected flood damage
Type of model	Ordered probit	Ordered probit	Binary probit	OLS regression
Geographical characteristics				
Distance to main river	-0.0203*** [0.0031]	-0.0226*** [0.0042]	0.0141*** [0.0037]	-1,117.95*** [382.67]
Elevation relative to water level	-0.0407*** [0.0064]	-0.0434*** [0.0070]	0.0305*** [0.0072]	-1,584.10** [653.75]
Rural area	-0.589 [0.1835]	0.4008** [0.1890]	-0.3378* [0.1993]	60,055.67 [42,092.12]
No dike protection	-0.3705*** [0.1187]	-0.3836*** [0.1317]	0.1710 [0.1353]	-16,445.44 [12,094.05]
Experience and knowledge				
Experience of flood and evacuation	0.9604*** [0.2262]	1.7704*** [0.2816]	-0.3976 [0.2796]	-41,718.79** [19,545.66]
No knowledge of the causes of flooding	-0.1720** [0.0876]	-0.2384*** [0.0974]	0.2136** [0.1112]	-24,056.72* [14,132.16]
Socio-economic characteristics				
Risk-seeking index	-0.0623* [0.0365]	-0.0152 [0.0397]	-0.0198 [0.0441]	3,166.30 [8,239.41]
Age	-0.0057* [0.0032]	-0.0019 [0.0036]	-0.0059 [0.0039]	-1,055.27* [585.63]
Female	0.0233 [0.0731]	-0.1098 [0.0807]	-0.1298 [0.0901]	-23,194.99** [11,317.38]

Table 9.6 (*cont.*)

Dependent variable	Model 1 Probability of flooding	Model 2 Flood risk compared with average	Model 3 Return period exceeds safety norm	Model 4 Expected flood damage
Type of model	Ordered probit	Ordered probit	Binary probit	OLS regression
Education	-0.0675*	-0.0451	0.1121**	-21,604.95*
	[0.0399]	[0.0458]	[0.0494]	[12,298.50]
Value of property	0.0004	0.0004	-0.0005*	255.33***
	[0.0002]	[0.0003]	[0.0003]	[87.51]
Income	0.01935	0.0315	-0.0306	-4,278.70
	[0.0412]	[0.0469]	[0.0494]	[12,303.03]
Constant	n.a.	n.a.	0.0799	47,758.16
			[0.2325]	[36,031.59]
Log likelihood	-1408	-937	-603	n.a.
Likelihood χ^2	131.47***	137.52***	55.72***	n.a.
F statistic	n.a.	n.a.	n.a.	2.75***
Number of individuals	975	981	981	980

Notes: *, ** and *** indicate, respectively, significance at the 10%, 5% and 1% levels; n.a. stands for not applicable. Huber–White heteroskedasticity-robust standard errors are given in square brackets below coefficient values. Estimations were performed with the statistical software package Stata.

reasonable magnitude and sign. The likelihood χ^2 statistic of the probit models and the *F* statistic of the OLS regression indicate overall model significance at the 1% level.[13] The results of the four models are discussed in detail below.

[13] Commonly used goodness of fit measures of OLS regressions, such as the R^2 statistic, are not defined for non-linear (probit) models. Sometimes pseudo-R^2 statistics are reported that are derived from the improvement in the log likelihood by the variables added to the model. However, the interpretation of such statistics is not immediately clear (e.g. Train 2003), which makes some authors wary of reporting them (see Hoetker 2007). The reason is that

Model 1 of the perceived flood probability

The results of the ordered probit model of the perceived flood probability indicate the importance of the proxies of actual flood exposure in shaping perceptions. The distance that the individual lives from a main river is negatively related to the expected flood probability, i.e. the further the individual is situated from a river, the lower is the perceived flood probability. This indicates that perceived risk is correlated to the flood exposure of the respondent, which is consistent with the findings of other studies that individuals have a greater perception of the risk of a hazard the closer they live to that hazard. In addition, the perceived risk relates to the elevation of the area relative to the potential water level of a flood in the expected manner. Specifically, the higher the elevation of the area relative to the potential water level, the lower are individual risk perceptions. There is no statistically significant difference in the perceived risk of individuals who live in rural areas compared with individuals who live in urban areas. Individuals living in floodplains that are not protected by dikes have a lower perceived probability of flooding than individuals who live in protected areas, which is contrary to actual flood exposure. This finding suggests that inhabitants of unprotected floodplains are insufficiently informed about the risk they face. Baan and Klijn (2004) argue that people living in unprotected floodplains in the Netherlands may be less frightened of flooding than people in areas protected by dikes, since people in floodplains are more used to the dynamic of the river. Previous experience with flooding and evacuation is related to a higher perceived flood probability. This is consistent with studies which find that personal experience with a hazard increases its perceived probability. The significance of the experience variable supports views prevalent in the psychological literature that the affect heuristic plays a role in individual risk assessment in addition to analytical reasoning. A lack of knowledge about the causes of flooding results in a lower perceived flood probability.

the underlying latent variable y^* is not observed, so that it is impossible to calculate the percentage of its variance that a model explains. The R^2 of model 4 is 0.07 and the pseudo-R^2 statistics of our models are between 0.05 and 0.1, which may be approximated by R^2 statistics between 0.1 and 0.2, according to Hensher et al. (2005). Often R^2 statistics of models of risk perceptions are low (e.g. Peacock et al. 2005), which indicates that it is difficult to model differences in perceived risk across individuals.

The risk-seeking index relates negatively to the perceived flood probability, meaning that risk-seeking individuals expect a lower probability of flooding. Older individuals and individuals with a higher education level have a lower perceived flood probability, which confirms the general stance in risk perception research. Gender, the value of property and household income are not statistically significant.

Model 2 of the respondents' perceived flood risk compared with an average resident

The geographical characteristics are the main determinant in shaping the respondents' perceived flood risk compared with an average resident. All four geographical indicators are statistically significant, while the socio-economic characteristics are non-significant. Individuals living far away from a main river are less likely to rate their flood risk as higher than average. Furthermore, the higher the elevation of the respondent's dwelling relative to potential water level, the lower they perceive their flood risk compared with the national average. Contrary to actual flood exposure, individuals who live in floodplains that are not protected by dike infrastructure are more likely to underestimate their risk, and hence indicate that their flood risk is lower than average. The relations between these geographical variables and perceived flood risk compared with an average resident are similar to their relations with the perceived flood probability (Model 1). A difference from the previous model is that the variable representing the inhabitants of rural areas is statistically significant. Inhabitants of rural areas are more likely to indicate that their flood risk is higher than an average resident, perhaps because they think that rural areas with fewer inhabitants and a lower concentration of economic activities are less well protected than urban areas. Experience with flooding and evacuation increases the likelihood that an individual expects to be at a higher than average risk, while a lack of knowledge about the causes of floods decreases the perceived riskiness. The effects of these latter two variables in Model 2 are consistent with their effects in Model 1.

Model 3 of respondents' perceptions that the flood return period exceeds the legal safety norm of dike design

Estimates show that individuals who live far away from a main river are more likely to expect a flood return period larger than the legal norm. Such individuals have a lower probability of suffering flood damage and are, therefore, more likely to state a flood return period that is higher than the legal norm of dike designs. The positive and significant coefficient of the elevation variable indicates that individuals who live in a high area relative to the potential water level are more likely to expect that the maintenance of dikes is adequate so that dike designs comply with the safety norm. These results are consistent with those of the first two models, which show that individuals living far from a main river and in relatively high areas generally have a lower perception of flood risk. It makes sense that these individuals who, in practice, will have a lower exposure to flooding are more likely to predict a return period of flooding higher than the legal safety norm. On the other hand, inhabitants of rural areas are less likely to expect that the flood return period exceeds the norm, which is consistent with the findings of Model 2 that such individuals are more likely to view their risk as higher than average. The variable representing individuals in floodplains not protected by dikes is not statistically significant. In practice, the safety norm of a 1/1,250 years flood event is not met in those areas, but the results indicate that inhabitants do not realize this.

Experience with flooding and evacuation does not affect expectations of whether the return period exceeds the legal norm. An explanation of this may be that the anticipated flood events in 1993 and 1995 did not actually result in breaches or overtopping of the primary dikes. The evacuations in anticipation of the flood threat were unwarranted from an *ex post* perspective, since major flooding did not take place and almost none of the respondents suffered flood damage. Therefore, these flood experiences may have little effect on expectations of whether dike maintenance is sufficient to meet the dike safety norm, even though they generally increased risk perceptions, as Models 1 and 2 indicate.

Individuals who do not know the causes of flooding are more likely to expect that the authorities are complying with the safety norm. In general, less knowledgeable individuals have a lower perception of flood risk, as the results of Models 1 and 2 show. Respondents with a high education level are more likely to expect that the flood return period exceeds the safety norm. The estimate of the coefficient of total value of property is significant and negative in this model, which indicates that individuals with a higher value of their house and contents are less likely to indicate a return period larger than the legal norm. Perhaps this is the case because such individuals are more financially vulnerable in terms of potential damage due to flooding.

Model 4 of the amount of flood damage that is expected by individuals

In accordance with the previous three models, the variables representing distance from a main river and elevation of the area are statistically significant. Individuals who live far from a main river or in areas that are relatively high expect to suffer less flood damage than individuals who live close to a river or in relatively low-lying areas. These judgements are likely to hold in practice, since flood damage is generally greater for houses close to rivers because of high flow velocities of water. Also, low-lying areas suffer more damage during floods than high areas because of elevated water levels and a higher probability of a building collapse. The expected flood damage is unrelated to living in a rural area. In practice, it is also unlikely that flood damage per house is significantly higher in rural areas than urban areas. The damage per square kilometre in urban areas is obviously higher than in rural areas because of a higher density of houses, but that does not imply that the damage per house differs. Neither is expected flood damage significantly different between inhabitants of areas that are protected and those who live in floodplains that are not protected by dikes. This is consistent with reality, since it is unlikely that a flood event would inflict more damage in unprotected areas than in areas protected by dikes. It may be the case that damage to houses in an unprotected area is greater if the area lies closer to the river. However, the effect of proximity to the river on expected damage is included as a separate variable in Model 4. Its coefficient already captures the higher risk perception of individuals who live close to the river. Obviously, the frequency of flooding will be higher in unprotected floodplains, but this does not necessarily imply that the damage per house for a specific flood event will be higher as well.

Individuals who have experienced a flood and evacuation expect to suffer less flood damage than individuals who did not have such an experience. The results of the first two models indicate that individuals who have experienced a flood and evacuation have a larger perceived flood probability. These findings appear contradictory, but may be explained by the observation that practically none of the respondents who have experienced a flood or have been evacuated have actually suffered any damage. Although these individuals have a larger perception of the probability of flooding, their experience of not having suffered damage during the 1993 and 1995 flood threats seems to have lowered their anticipated amount of flood damage. Expected flood damage is lower for individuals who do not know the causes of flooding. Older individuals and individuals with a higher education level also expect to suffer less flood damage. Surprisingly, females expect to suffer about €23,000 less damage than males. Households with higher property values expect to suffer more flood damage: expected flood damage is about €250 higher per €1,000 value of a home and its contents.

9.7 Conclusions and policy implications

Individual risk perceptions may considerably deviate from expert judgements of risk. Risk perceptions are an important factor which causes individuals to take precautionary measures and give support to policies that aim to limit flood risk. Insights into individual risk perceptions are therefore very important for the design of climate change adaptation policies. For example, low perceptions of natural disaster or climate change risk may result in a lack of support for public investments in climate change adaptation, while excessively low natural disaster risk perceptions are commonly put forward as an explanation for insufficient disaster preparedness of households. The following chapters will show that individual beliefs concerning flood risk are important drivers of household demand for flood insurance. Therefore, knowledge about the factors of influence on risk perceptions provides useful information on the determinants of the demand side for flood insurance.

This study examined individual perceptions of flood risks in floodplains of the Netherlands, where flooding can have catastrophic consequences. A survey was conducted among approximately 1,000 homeowners to elicit their expectations regarding different aspects of the risk posed by flooding. A range of statistical models estimated the factors shaping flood risk perceptions, such as: relations with indicators of actual risk exposure using geographical characteristics; experience with, and knowledge about, the risk; and socio-economic characteristics. The study provides insights into individual perceptions of flood risk that are potentially useful to policy makers and water managers.

An examination of the various indicators of perceived risk shows that individuals generally expect that the probability of flooding is low. The flood probability is underestimated relative to other risks, and most people rate the flood probability as 'very low' or 'low' when it is elicited independently of other hazards. Moreover, many individuals underestimate their flood risk compared with the national average. Most individuals expect a lower flood return period than the safety norm, which suggests that they expect that dike maintenance is inadequate to comply with the safety norm of dike design. Nevertheless, on average, individuals do provide realistic estimates of the damage they expect to suffer because of a flood, since the average damage expected by respondents is similar to estimates derived from expert models.

These findings have important implications for policy makers. Currently, considerable investments have been made and are planned in the future to limit the projected rise in flood risk due to climate change. The low perception of flood risk may undermine the public's perceived legitimacy of these flood risk reduction policies. It has been argued that, as part of an adaptation policy to limit the potential impacts of climate change, households should be encouraged to invest in measures that limit flood damage, such as the 'flood-proofing' of buildings. An enhanced awareness of flood risk may be needed to promote the implementation of such damage mitigation investments. In light of the evidence presented here, a communication campaign could be undertaken to increase awareness of flood risks.

Even though perceptions of flood risk are low in general, it is important to note that variation exists between individuals' risk beliefs. Insights into what type of individuals have high or low risk perceptions is very relevant for the adequate targeting of information campaigns and insurance products that provide coverage against flood damage. The statistical models concerning the shaping of risk perceptions provide five main insights.

1. Differences in expected risk are consistently related to actual flood risk exposure, since individuals in the vicinity of a main river in low-lying areas generally have higher risk perceptions. There is

some evidence that the inhabitants of rural areas are more aware of their risk of flooding than inhabitants of cities.
2. Individuals in flood-prone areas that are unprotected by dikes tend to underestimate their risk of flooding.
3. We found evidence in support of psychological studies that indicate the importance of the 'affect heuristic', which depends on previous experiences with hazards in the formation of perceived risk. It should be noted, however, that flood experience has different effects on distinct aspects of risk beliefs. Individuals who have experienced flooding and evacuation, but suffered hardly any flood damage, have an elevated perception of the flood probability but anticipate lower flood damage.
4. Individuals with little knowledge of the causes of flooding have lower perceptions of flood risk. This suggests that providing information to households about flooding may increase flood risk perceptions.
5. Some support was found for the role of socio-economic variables in shaping perceived risk, as has also been observed in other studies. Education level relates negatively to perceived risk in three out of our four models, and age relates negatively to perceived risk in two of these models, which are the expected effects according to other studies. The risk-seeking index is only significant in the model of flood probability, and more risk-seeking individuals have a lower risk perception. Other studies usually find that women have a greater perceived probability of hazards, which was not observed in our results. The results of Model 4 indicate that women even expect to suffer less flood damage than men do. The value of property relates negatively to perceptions about the return period exceeding the legal norm, while it affects the perceived flood damage positively, which is intuitively clear.

For policy makers, these results imply that it is especially important to improve knowledge of the flood risk of individuals who live in flood-prone areas unprotected by dike infrastructure since these individuals tend to be unaware of or ignore the high risk they face.

Although the focus of this chapter is on the Netherlands, the survey method used and some of the results obtained could also be applicable to other countries. Moreover, the range of statistical models employed can provide examples of how different types of risk perception variables should be treated if their determinants are to be analysed. The statistical methods presented provide an alternative to the simple correlation analyses often used in risk perception research and result in more accurate estimates of the independent effects of explanatory variables on perception. The survey elicited various indicators of perceived risk, which could also be useful to examine vulnerability to flooding in other regions. In particular, our study shows that perceptions and determinants of the two distinct components of flood risk – damage and probability – differ, while this is ignored by several studies that only examine perceptions of probabilities. Our results show the importance of indicators of objective risk exposure based on geographical characteristics in determining risk perceptions. Knowledge about levels of risk perceptions and relations with objective risk may also provide practical information to improve flood risk communication policies in other regions. The results of this survey may be applicable to risk perceptions of households living near the Meuse and Rhine rivers in upstream countries, such as Germany and France.

The effectiveness of information campaigns for increasing the awareness of flood risks is an important topic for future research (Terpstra et al. 2009). Moreover, knowledge on how this translates into flood insurance demand and the willingness of households to invest in damage mitigation measures is important if insurance and strategies aimed at reducing vulnerability to natural disasters are to be applied to ameliorate, or limit, the impacts of climate change. The next chapter examines the influence of risk communication and individual flood risk perceptions on the demand for flood insurance.

Appendix 9A Questions used to elicit risk perception

1. *How high do you estimate the probability that your household will suffer financial damage to property due to the events mentioned below? Rate each possible event on a scale from 0 to 10, where 0 = no probability and 10 = extremely high probability.*

	No probability					Neutral				Very high probability	
Event	0	1	2	3	4	5	6	7	8	9	10
a. Terrorist attack	O	O	O	O	O	O	O	O	O	O	O
b. Burglary	O	O	O	O	O	O	O	O	O	O	O
c. House fire	O	O	O	O	O	O	O	O	O	O	O
d. Car theft	O	O	O	O	O	O	O	O	O	O	O
e. Car fire	O	O	O	O	O	O	O	O	O	O	O
f. Flood/water inside house	O	O	O	O	O	O	O	O	O	O	O
g. Traffic accident	O	O	O	O	O	O	O	O	O	O	O
h. Storm	O	O	O	O	O	O	O	O	O	O	O

2. *How high or low do you estimate the probability that you will experience a flood in your home?*

□ I do not have any flood risks
□ Very low
□ Low
□ Not low/not high
□ High
□ Very high
□ Don't know

3. *How would you rate your flood risk compared with an average person in the Netherlands?*

□ I have an average flood risk
□ I have a higher than average flood risk
□ I have a lower than average flood risk

The government is responsible for the maintenance of dikes. A dike in your region should be high enough so that a flood does not occur on average more often than once every 1,250 years. The scale below depicts different flood probabilities decreasing from once per year to once in 100,000 years. The legal norm is shown at the top of the scale.

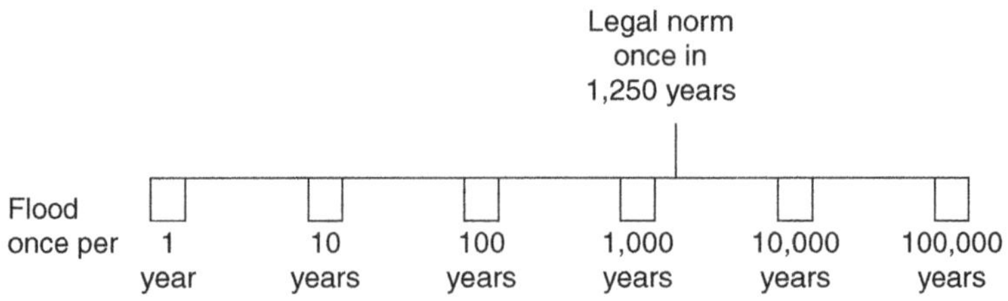

4. *Can you indicate how often you would expect a flood to occur at your home? You can fill in one of the probabilities shown on the scale above or fill in another probability below.*
   I expect that a flood would occur once in … years
5. *How much damage do you expect that a flood would cause to your house and home contents?*
   … euro

10 *Willingness-to-pay (WTP) for insurance against low-probability flood risks*

10.1 Introduction

Several studies have examined the demand for insurance against low-probability, high-impact events, such as natural catastrophes (e.g. Kriesel and Landry 2004; McClelland et al. 1993). A common finding about decision making under risk is that many individuals either neglect or overweight low-probability risks. This behaviour is consistent with prospect theory (Kahneman and Tversky 1979; Tversky and Kahneman 1992). Individuals have a zero willingness-to-pay (WTP) value for insurance if they neglect the low-probability risk. Revealed preference data on insurance demand in the USA indicates that many homeowners neglect flood risk and fail to purchase partly subsidized flood insurance policies (Kunreuther 1978; Kunreuther et al. 2011). In contrast, overweighting of low-probability risk would generally imply that individuals have a positive WTP value for insurance against natural catastrophes, and that WTP may change non-proportionally if the likelihood of the occurrence of a disaster rises. Therefore, these risk attitudes have very different implications for demand for catastrophe insurances. Understanding what kinds of individual risk attitudes are present in the population is relevant for assessing WTP for insurance coverage against low-probability natural catastrophe risk.

Only a few empirical studies have analysed the possible effects of climate change on demand for insurance against natural disasters (Chapter 8). Climate change is expected to increase the frequency and severity of certain natural disasters (Chapters 2 and 3), which is likely to increase demand for financial compensation or insurance against the impacts of such events. Two main issues are of interest in

Parts of this chapter have previously been published as Botzen and van den Bergh (2012b).

examining the effects of climate change on insurance demand. First, it is useful to investigate whether individuals neglect the low-probability risk, and whether this behaviour changes if risk increases due to climate change. Second, an increase in the likelihood of disasters is likely to influence demand for insurance by individuals who do not neglect the risk under current climate conditions and therefore already have a positive WTP value. An important issue is whether WTP rises close to proportionally with the probability, as is roughly the case in the expected utility model,[1] or disproportionally, as predicted by several popular non-expected utility models with probability weighting.[2]

Experience in several countries has shown that a lack of demand for catastrophe insurance may impede an optimal functioning of insurance markets for natural disaster risks (see Chapter 4). The present chapter presents a contingent valuation study that provides a detailed examination of household WTP for low-probability, high-impact insurance against river flooding under current flood probability and after expected increases in the likelihood of flooding due to climate change. The particular focus of this study is on the Netherlands, but the results presented in this chapter are of broader relevance. The overview of international studies of flood insurance demand provided in this chapter shows that few studies have examined the demand for flood insurance using survey data, even though such data may be particularly well suited for unravelling the individual decision-making process regarding insurance purchases.[3] A focal point of this chapter is whether individuals neglect the low-probability flood risk and how WTP relates to the increases in flood probabilities presented to respondents in the questionnaire. Individual risk attitudes towards the low-probability flood risk will be examined. Furthermore, it is investigated how individual perceptions of the flood risk influence insurance demand, while controlling for the relevant flood probabilities that we provided to respondents. Several studies suggest that individuals follow an updating process based on prior perceptions of risk when they are provided with new information about the risk (e.g. Cameron 2005; Delavande 2008; Viscusi 1989; Viscusi and O'Connor 1984). As a consequence,

[1] See McClelland et al. (1993) for a formal derivation.

[2] Starmer (2000) provides an overview of non-expected utility theories.

[3] There are only a few contingent valuation studies which value insurance or mitigation measures against flood risk or changes in flood risk (e.g. Akter et al. 2009; Brouwer and Akter 2010; Brouwer et al. 2008; Daun 2000).

decision making under risk, such as determining whether to purchase flood insurance, may be driven by a combination of prior risk perceptions and new information provided to individuals.

In general, individuals find it difficult to understand differences in probabilities, which complicates the use of surveys to assess WTP for risk changes (Hammitt and Graham 1999). Several studies include visual aids, such as risk ladders, to communicate probabilities in questionnaires aimed at facilitating comprehension of the risk by respondents (e.g. Corso et al. 2001). Communicating probabilities and changes in probabilities with risk ladders can have two effects on stated WTP for flood insurance, and it is these effects which are examined in this study. First, it may influence the level of WTP. Second, it may affect the dependency of WTP on probability changes, which is an indicator of sensitivity to scope in this application. Our study estimates the independent effect of communicating the baseline (current) probability and probability changes using risk ladders on the level of WTP for flood insurance, irrespective of probability changes. In addition, the influence of showing the risk ladders on the relation between probability changes and WTP is examined. This allows us to assess both of the aforementioned impacts of using visual aids to communicate risk in surveys.

This study is of practical interest for insurers and the government. At present, flood insurance is not available in the Netherlands, but under certain conditions the government can, according to the Calamities and Compensation Act (Wet Tegemoetkoming Schade bij Rampen en Zware Ongevallen – WTS), partly compensate for damage caused by large-scale floods. However, a public–private partnership for insuring flood damage could be a more efficient way to spread climate change induced flood risk (see Chapter 4). The current study could help the government and insurers to assess whether there is sufficient demand for flood insurance to warrant its introduction. In addition, this chapter examines the effect on WTP for flood insurance of the current institutional setting, characterized by the availability of partial compensation for damage by the government. This provides a test of the charity hazard, which refers to the crowding out of demand for disaster insurance by individuals who anticipate that the government will compensate them for part of the damage that they may suffer because of disasters (Raschky and Weck-Hannemann 2007).

The remainder of this chapter is organized as follows. Section 10.2 provides a review of several representative international studies on the demand for flood insurance. Section 10.3 gives an overview of the current state-of-the-art of communicating low probabilities in surveys, which forms the basis for the risk-related questions in the Dutch questionnaire. In addition, the set-up of the survey and its implementation are explained. Section 10.4 presents the statistical model and the results about demand for flood insurance in the Netherlands. Section 10.5 concludes the chapter.

10.2 International studies on household demand for flood insurance

This chapter provides a review of the results of several representative studies of flood insurance demand in the USA, Germany, Austria and Bangladesh. These case studies provide interesting comparisons with the survey about demand for flood insurance in the Netherlands and, in general, provide useful insights into the factors influencing demand for flood insurance. Several studies that use extensive data on flood insurance demand have been conducted in the USA. Germany and Austria are, from a geographical and cultural perspective, broadly similar to the Netherlands; the study conducted on flood insurance demand in Germany and Austria focuses on the charity hazard, which may be a problem in many countries where government relief of disaster damage is provided, including the Netherlands. The Bangladesh case studies provide an interesting example of the demand of poor households for micro-insurance, which can be a useful financial arrangement for spreading climate change risks in the developing world (see Chapter 2).

10.2.1 Household demand for flood insurance in the USA

The USA is an interesting case for a study on demand for flood insurance. Flood insurance has been offered at prices that are close to the expected value of flood damage since 1968 by the federal National Flood Insurance Program (NFIP) (see Chapter 4). The NFIP has struggled with a low uptake of insurance despite the low levels of premiums compared with the private flood insurance market. Table 10.1 summarizes the results of several major studies about the factors of influence on the demand for NFIP policies, which are discussed in more detail below.

Kunreuther (1978) interviewed more than 2,000 homeowners in 43 areas throughout the USA about, among other issues, their demand

Table 10.1 *Main studies about flood insurance demand in the USA ('+' indicates a positive relation with demand and '–' a negative relation)*

Study	Data collection	Sample size	Main relations with flood insurance demand	Interesting non-significant relations with demand
Kunreuther (1978)	Household survey	2,000 households	Past flood experience (+) Friend, neighbour or relative has purchased flood insurance (+) Awareness of the availability of insurance and its cost (+) Perceived flood risk (+)	Expectations of federal disaster assistance
Browne and Hoyt (2000)	State-level insurance purchases	50 states between 1983 and 1993	Income (+); Price (–) Number of federally backed mortgages (–) Past flood experience of flood damage (+)	Federal disaster relief
Kriesel and Landry (2004)	Household survey and property data	2,486 properties	Income (+); Price (–); Mortgage (+) Distance of the property from the coast (–) Hurricane frequency (+) Presence of artificial flood protection (+)	

Table 10.1 (*cont.*)

Study	Data collection	Sample size	Main relations with flood insurance demand	Interesting non-significant relations with demand
Dixon et al. (2006)	Property data	75,000 properties in 100 communities	Price (–); Value of property (+); Mortgage (+) Number of single-family homes in a community Number of single-family homes in the 1/100-year floodplain Located near the coast (+); South of the USA (+) Past flood experience (+)	Participation in the Community Rating System (CRS)
Zahran et al. (2009)	County-level insurance purchases	67 counties in Florida	Points earned in the Community Rating System (+) Home values (+); Education level (+) Land area in the floodplain (+) Located near the coast (+) Past flood experience (+)	Density of streams in a county
Michel-Kerjan and Kousky (2010)	Individual NFIP policies in force	7.5 million policies in Florida	Located near the coast (+); Single-family homes (+) Past flood experience (+); 1/100-year flood zone (+) Single-family homes (+) Participation in the Community Rating System (CRS) (+) Tenure of flood policies (–)	

for flood insurance. Their study concluded that most people rely on past flood experience in their decision to purchase flood insurance. Another important factor appeared to be whether the person knew someone who had purchased flood insurance or had discussed insurance with a friend, neighbour or relative. In addition, many uninsured people were unaware of the availability of insurance coverage and its cost. Many people appeared to have difficulty with assessing flood risk and either significantly overestimated or underestimated the flood probability they were facing. Underestimation of flood risk is regarded as an important reason why people did not purchase flood insurance. In contrast with theoretical predictions about the charity hazard, many uninsured people did not expect compensation for flood damage by the federal government.

Browne and Hoyt (2000) estimated how state-level flood insurance purchases between 1983 and 1993 in 50 states were dependent on expenditures on mitigation, disaster relief, the price of the insurance, the income of the state, flood experience, disposable personal income and the number of federally backed mortgages for which flood insurance is compulsory. This data cannot be used for estimating the market penetration of flood insurance in different flood zones because the number of properties in those zones is unknown. Browne and Hoyt (2000) found that flood insurance purchases are positively correlated with income and negatively with price. The number of federally backed mortgages has a negative influence on demand, which is contrary to expectations since homeowners with such mortgages are obliged to carry flood insurance. An explanation for this that the authors put forward is that a high number of such mortgages may be an indicator of low wealth. Moreover, the purchase of flood insurance in a particular state is positively related to the amount of flood damage suffered in that state during the previous year. This study finds no evidence of the charity hazard.

Kriesel and Landry (2004) examined the determinants of flood insurance demand by means of a mail survey that was conducted in nine coastal counties in the USA. A total of 2,486 individual properties were included in their analysis, of which 49.8 per cent had flood insurance which was almost exclusively underwritten by the NFIP. Their regression analysis of the factors of influence on whether people have purchased flood insurance shows that demand is negatively related to price and positively related to income. In contrast to Browne and

Hoyt (2000), Kriesel and Landry found a positive relationship between insurance demand and whether the property was subject to the mandatory insurance purchase requirement by the NFIP. An explanation that Kriesel and Landry (2004) provide for this difference is that the purchase requirements were more strictly enforced during the time that they conducted their study. The authors also estimated the dependence of demand on indicators of actual flood risk, and found that properties that are located further away from the coast are less likely to carry flood insurance and that flood insurance demand is higher in areas where hurricanes have occurred more frequently in the past. The presence of artificial flood protection, such as a seawall, groyne or nourished beach protection, has a positive effect on flood insurance demand. This may be surprising, because the presence of such protection is not reflected in lower insurance prices. Kriesel and Landry (2004) argue that individual risk awareness may be higher if shoreline protection is present, which may explain the positive effect of protection on insurance demand.

Dixon et al. (2006) researched the market penetration of flood insurance, based on information of about 75,000 properties in 100 communities. About 49 per cent of these properties, located in the 1/100-year floodplain, have NFIP policies, while this is only 1 per cent outside the 1/100-year floodplain. Insurance demand was found to be negatively related to the price and positively related to the value of the property and whether or not it has a mortgage. Market penetration was higher in:

- communities with more single-family homes
- communities with more single-family homes in the 1/100-year floodplain
- communities that are subject to coastal flooding
- communities that are located in the South of the USA
- communities that experienced a recent flood.

Zahran et al. (2009) analysed household flood insurance purchases in Florida from 1999 to 2005, at a county level, and estimated, in particular, to what extent these are related to flood mitigation activities by local governments who are involved in the federal Community Rating System (CRS). The CRS encourages local governments to invest in damage mitigation and to improve their flood preparedness by rewarding the policyholders in a community that participates in the

CRS with premium discounts (up to 45 per cent) on flood insurance policies provided by the NFIP. The CRS translates flood risk reduction activities of a community into 'CRS points', on the basis of which the level of the premium discount is determined. Communities that have engaged more in flood risk reduction receive more CRS points and higher reductions in flood insurance premiums. Zahran et al. (2009) found that insurance demand is positively related to the number of CRS points. Moreover, flood insurance purchases in a community are observed to be positively related to house value, respondents' education level, the local land area in the floodplain, whether the community is a coastal community, and the frequency and intensity of prior flood experience (Zahran et al. 2009).

Michel-Kerjan and Kousky (2010) provide a detailed analysis of 7.5 million policies in force of the NFIP in Florida in the period 2000–5. About 40 per cent of the total policies of the NFIP have been sold in Florida. The take-up rates of flood insurance are highest in counties that are located near the coast. Most policies are for single-family homes that are located in the 1/100-year flood zone. Most policies (about 75 per cent) received a premium discount because their community participates in the CRS. The number of policies in force increased after Florida experienced severe hurricanes in 2004. Moreover, past storm experience, generally, resulted in a preference for lower deductibles and higher coverage limits. Such an increase in insurance take-up after a flood may be short-lived. Michel-Kerjan and Kousky (2010) have shown that people may drop their coverage quickly; about half of the policyholders do not renew their flood insurance coverage after 3 years.

10.2.2 Flood insurance demand in Germany and Austria

Raschky et al. (2012) estimated WTP for flood insurance by means of a contingent valuation survey in Austria ($N = 218$) and Germany ($N = 305$). In both of these countries, private insurance and government relief for natural disaster damage co-exist. Raschky et al. (2012) examined what factors influence WTP for flood insurance in both countries. In particular, they showed how demand is affected by the charity hazard. The government in Austria commonly compensates part of the flood damage suffered by households through a national disaster relief fund, and politicians in Germany have, in some cases, provided almost complete relief for flood damage suffered by uninsured households. The main hypothesis of the authors is that an assured partial relief scheme, as exists in Austria, has a stronger effect in crowding out demand for private insurance than an uncertain scheme that provides full relief, as exists in Germany. WTP for flood insurance is found to be positively correlated with income, individual flood risk perceptions and previous experience of flood damage. WTP for flood insurance is significantly lower in Austria compared with Germany. This is even lower for individuals in Austria who consider the provision of compensation for flood damage a governmental duty. The main conclusion drawn from this analysis is that a public compensation scheme that offers assured partial relief for disaster damage (as in Austria) has a stronger effect on the crowding out of insurance than an uncertain scheme of full relief (as in Germany).

10.2.3 Demand for (micro) crop insurance in Bangladesh

Micro-insurance schemes that provide insurance at low cost to poor households in developing countries may be particularly suited to spreading the cost of natural disasters and climate change impacts in the developing world (see Chapter 2). The commercial viability of a crop insurance scheme that covers flood damage in Bangladesh has been assessed by Akter et al. (2009), who estimated WTP for crop insurance by agricultural farm households using the contingent valuation method, and compared this with the expected insurance cost. The data for this analysis were collected by conducting surveys among 3,600 households. This study is of particular importance for Bangladesh, where agriculture is the largest source of employment for its rural population, while farmers' income levels are very volatile because crops are often damaged by severe flooding. Floods occur frequently in Bangladesh due to its very low land elevation. About half of the interviewed respondents indicated that they were, in principle, interested in participating in the proposed micro-insurance scheme. As can be expected in a developing country, income is a major determinant of whether or not households are willing to participate in the insurance scheme. Moreover, unfamiliarity with insurance was observed to be an obstacle to the willingness of households to participate in the crop insurance scheme. WTP for crop insurance was higher for households located in areas that are protected by flood embankments, which makes sense in this context, since households in these areas tend to suffer more damage if a flood occurs. Moreover, household income is an important factor of influence on WTP for crop insurance.

Average WTP was too low to cover the expected claims on insurance in areas that are exposed to coastal flooding and flash floods. An exception is riverine floodplain areas, where average WTP exceeded the expected costs of claims. From these findings, the authors concluded that crop insurance is marginally commercially viable in riverine floodplain areas. However, the administrative costs of the insurance are not accounted for in this study. If such costs are included in premiums, then the observed WTP for crop insurance is unlikely to be sufficient, even in the riverine floodplain. A similar study has been conducted by Brouwer and Akter (2010), who estimated WTP for micro-insurance against flood damage using a choice experiment involving 1,200 households in Bangladesh, and found results similar to those of Akter et al. (2009). Interestingly, Brouwer and Akter (2010) observe that whether or not households had previously received relief for flood damage in the past, and how effective they considered this relief, has no significant influence on WTP for crop insurance against flood damage. Brouwer and Akter (2010) confirm that the affordability of insurance can be a major impediment for flood insurance in Bangladesh. Given the severe income constraints faced by farmers in Bangladesh, some form of government subsidy or international support is likely to be needed to facilitate the development of a micro-insurance scheme for flood damage.

10.2.4 Summary of the international studies on flood insurance demand

It is apparent from the studies of flood insurance demand in the USA that the decisions of many people regarding the purchase of flood insurance are not consistent with the trade-offs between the expected costs and benefits of the insurance in a way that is in agreement with expected utility theory. Even though NFIP policies are offered at prices that are below, or close to, expected flood losses, many people do not purchase flood insurance, although expected utility theory predicts that risk-averse people would prefer to insure. All the studies reviewed concerning flood insurance purchases in the USA find that past flood experience has a large positive effect on insurance demand. However, the increase in flood insurance demand after a flood event may not last very long, as many people appear to cancel their insurance policies if they do not suffer any flood damage in the next few consecutive years (Michel-Kerjan et al. 2012). Prices have, as expected, a negative effect on insurance demand, but most studies reported that price elasticities are below unity, which indicates that demand is relatively price inelastic. Moreover, it appears that flood insurance demand is influenced by several geographical characteristics that are related to actual flood risk, such as whether or not property is located in the floodplain or near the coast. These geographical variables may be proxies for individual flood risk perceptions, which are only explicitly accounted for by Kunreuther (1978) who argues that low perceptions are a major determinant of low demand for NFIP policies.

Contrary to theoretical predictions, the studies of flood insurance demand in the USA find no evidence of the charity hazard. WTP for flood insurance in Germany and Austria is also found to be positively related to past flood experience and flood risk perceptions. In contrast to the US studies, flood insurance demand in Austria is significantly lower due to the presence of assured government relief for flood damage, which points to the existence of the charity hazard. However, the availability of government relief does not appear to influence demand for flood micro-insurance in Bangladesh, which may be due to the uncertainty of obtaining adequate compensation by the government after a disaster. As expected, income is a major constraint for the purchase of flood insurance in a developing country such as Bangladesh. The positive relation between income and demand for flood insurance has also been observed in the studies conducted in the USA, Germany and Austria.

10.3 Household survey on the demand for flood insurance in the Netherlands

10.3.1 Communicating low probabilities in surveys

Research on flood insurance demand in the Netherlands examines, among other issues, the influence on demand of increases in flood probability due to climate change, which has not been addressed by the other studies of flood insurance demand that were reviewed in the previous section. The survey conducted in the Netherlands elicits WTP for flood insurance under the current safety norm of 1 in 1,250 and climate change scenarios where flood probabilities increase up to 1 in 400. In general, individuals find it difficult to understand probability concepts and especially to assess the likelihood of low-probability events (Viscusi 1998). After having low probabilities of disastrous events, such as flooding, carefully explained to them, individuals can choose the appropriate level of protection or insurance coverage against such risks (Kunreuther et al. 2001). There is a substantial literature on the valuation of low-probability risks and the reduction in risks using surveys. Common applications assess the value of a statistical life based on WTP for changes (usually reductions) in environmental, health or accident-related mortality risks (Alberini et al. 2004, 2006, 2007; Bhattacharya et al. 2007; Van Houtven et al. 2008; Vassanadumrongdee and Matsuoka 2005). These studies use several risk communication methods, which provides an excellent basis for the design of the present survey.[4] The main concepts used to communicate low-probability risks in questionnaires and their applicability to this study are examined next.

Some studies that estimate the value of a statistical life using changes in extremely low-probability risks, such as 1 in 10,000 annually or lower,[5] transform yearly probabilities into probabilities that an event happens in a longer time span. This increases the presented probabilities, which may facilitate their interpretation (e.g. Alberini et al. 2004). This approach involving transforming probabilities is not suitable for the current application. This is the case because insurance policies are usually bought for a period of a year. Transforming probabilities to another time scale is thus likely to be confusing and may even be perceived as misleading by respondents. Moreover, the probabilities used in this survey are considerably higher than mortality risks. In particular, they are often similar or smaller in size than the transformed probabilities used in health applications.

It has been shown that visual aids can improve the comprehension of probability concepts by respondents in surveys (e.g. Connelly and

[4] Some surveys fail to explain risk changes that are to be valued by respondents (e.g. Johannesson et al. 1996, 1997). In such studies, WTP for risk reductions should be interpreted with caution, since respondents may not comprehend the change that they are valuing.

[5] Some studies value health risks that are expressed as low as x times in 50 million (Chilton et al. 2006).

Knuth 1998; Corso et al. 2001). Commonly applied visual aids are risk circles, risk ladders and square grids. Risk circles (pie charts) are sometimes used to communicate the current or baseline probability and changes in probability (e.g. Smith and Desvouges 1987). However, this is only suitable for relatively high probabilities. Differences in low flood probabilities are not clearly visible on pie charts. Moreover, individuals often find it difficult to understand pie charts and therefore this method to communicate probabilities has been criticized (e.g. Wilkinson 2005).

Several studies use risk ladders to communicate baseline probabilities, or changes in probabilities (e.g. Hammitt 1990; Vassanadumrongdee and Matsuoka 2005). A risk ladder shows the baseline risk on a probability scale, together with other risks that the respondent commonly faces. The purpose is to show how large the probability to be valued is relative to probabilities of other risks with which respondents may be more familiar. Depicting both the baseline and the new probability on the risk ladder can show changes in risk. An arrow can be used to indicate the change in the probability. Such a risk ladder illustrates the size of the change, and also how the new probability compares with the other risks. In general, Dutch households are accustomed to purchasing insurance for more common risks such as those of fire and theft. A comparison of flood probability with probabilities of events already covered by insurance (and with which individuals are more familiar) on a risk ladder is likely to enhance understanding of the relative size of the probability of flooding.

Square grids, often in the order of 10,000 or 100,000 squares, on which the probability is represented using randomly coloured squares, have also frequently been applied to communicate low probabilities in surveys (e.g. Alberini et al. 2004; Bhattacharya et al. 2007; Jones-Lee et al. 1985; Krupnick et al. 2002). Changes in probabilities can be presented on such a grid by increasing or decreasing the number of coloured squares. An advantage of the square grid compared with the risk ladder is that it reflects a certain degree of randomness in the occurrence of the event. A disadvantage is that the grid does not provide a comparison with other risks, which may be a shortcoming for the insurance context of this survey in which a comparison of the flood risk with other commonly insured risks could be a suitable communication method. In conclusion, risk ladders and square grids are likely to be suitable risk communication devices for this study. Therefore, during the pre-tests of the survey we assessed which device respondents perceived as the clearest and most useful.

10.3.2 The commodity valued in the survey

The insurance product valued in the survey completely covers damage to both buildings and home contents caused by river flooding. Demand for flood insurance is elicited by means of three contingent valuation questions. The first one elicits WTP under the current flood probability, while the other two elicit WTP for scenarios in which the flood probability rises. The effect of risk communication on WTP for insurance is also estimated. In addition, the influence of the availability of government compensation on WTP for flood insurance is examined. The contingent valuation method has been commonly applied to assess household WTP for environmental goods or new products (Mitchell and Carson 1989). This method is appropriate for the focus of this chapter, which assesses whether individuals neglect the low-probability flood risk and have a zero WTP value for flood insurance, or whether individuals have a positive WTP, and how this behaviour changes if the flood probability changes as a result of climate change.

10.3.3 Pre-tests of the questionnaires

During the design of the survey, experienced stated choice practitioners, economists, natural scientists, water management experts and psychologists reviewed various versions of the questionnaire. After incorporating their comments, three pre-tests of the questionnaire were conducted between August and October 2007, using face-to-face interviews. Four trained and carefully instructed and supervised interviewers (two male and two female) interviewed 88 households.

The interviewers were instructed to pay particular attention to the comprehension of the flood probabilities stated in the survey. Different risk communication devices were tested: risk ladders; 10,000 square grids on which baseline risk and changes in risk were represented using coloured squares; and a variety of textual explanations. The results indicated that the square grids were generally regarded as difficult to understand and too abstract, so they were excluded from the final survey. Instead, the risk ladders were perceived as providing very clear and useful information. Therefore they were included in the final survey to communicate the flood probabilities presented with the contingent valuation questions.

Textual comparisons of flood risks for different kinds of households with other risks, such as fire risk, have been tested, since they may increase comprehension, as shown by Kunreuther et al. (2001). However, as these comparisons were not considered to add much additional information to the risk ladders, they were excluded. Nevertheless, a text explaining the current flood probability and causes of differences in flood probabilities across areas was generally perceived as useful by respondents. We further experimented with questions that checked the understanding of probability concepts (e.g. Vassanadumrongdee and Matsuoka 2005), but also decided to exclude these from the survey because they annoyed several respondents, who noticed that their knowledge was being checked. The pre-tests further indicated that it is easier for respondents to express insurance premiums in monthly than in yearly payments. Overall, the interviews showed that respondents were generally comfortable with the valuation tasks. A fourth and final pre-test was conducted to trial the online implementation of the questionnaire, which resulted in minor adjustments in layout.

10.3.4 Structure of the questionnaires

Botzen et al. (2008b) give a full description of the survey and an overview of the questions. The questionnaire opens with questions on the experience of the respondent of flooding, flood damage and evacuation because of flood threats. An open-ended question about the causes of flooding is included to test the respondent's knowledge and stimulate the respondent to think about the nature of flood risks. In addition, several questions address the perception of flood risks. These questions and their answers are discussed in detail in Chapter 9. The assessed perceptions may be important in decision making under risk, as several studies suggest (Viscusi 1989), and may serve as explanatory variables in the statistical model of WTP for flood insurance.

Whereas the previous questions were identical in all versions of the questionnaire, some subsequent explanations and questions differ, giving rise to four versions. In versions 1 and 2, it is explained that households may receive partial compensation for flood damage by the government via the current regulation (WTS), although they have the opportunity to purchase additional private flood insurance. In other words, this states that the government may partially compensate for damage caused by major floods, but this compensation is not granted for smaller flooding events. The uncertainty about receiving relief is mentioned, and several examples of recent floods after which damage was partly compensated are given. This is followed by a question asking whether respondents ever received relief via the Compensation Act, stimulating respondents to read the text about the Compensation Act carefully. The other versions (3 and 4) describe a scenario explaining that relief for flood damage by the government will no longer be granted, but that it is instead possible to purchase private flood insurance coverage. A comparison of WTP for flood insurance in these different versions of the questionnaire allows for an assessment of the charity hazard.

Next, a short text in all versions explains why flood probabilities differ across regions, providing a comparison of risks in a textual context (see Hsee et al. 1999). Moreover, the concept of expressing probabilities in terms of return periods or frequencies is introduced. Estimated flood probabilities of an area not protected by dikes (1 in 100) are compared with flood probabilities in urban areas that are protected by dikes (1 in 1,250). Subsequently, the main risk communication device is shown, which is a risk ladder on which flood risks are compared with other insurable risks commonly faced by Dutch citizens (see Appendix 10A). The linear risk ladder depicts the risks of car theft, fire in a residential building, flooding, car fire and the risk of a fatal traffic accident.[6] All adverse events are expressed as yearly probabilities.

The first contingent valuation question elicits WTP under the current flood probability of 1 in 1,250. Respondents are first asked an initial question: 'If you could get an insurance policy that completely covers flood damage to your house and home contents, would your household be willing to pay for it?' Those who answered 'yes' were asked to give a WTP amount for the insurance. A payment card is given as a visual aid (see Appendix 10B).[7] The payment card method

[6] Traffic risks are often given as an anchor in risk belief surveys (Lichtenstein et al. 1978; Viscusi and Zeckhauser 2006).

[7] A payment card shows a range of possible WTP amounts. This approach is generally regarded to be free from anchoring effects or starting point biases that are sometimes observed with other elicitation approaches as long as the range of the card is not arbitrarily cut off (Boyle 2003; Roach et al. 2002). Therefore, a zero bid and a bid larger than the highest amount shown need to be included explicitly (Rowe et al. 1996).

has already been successfully used to measure WTP for risk changes (e.g. Bhattacharya et al. 2007). The values on the payment card are derived using the method of Rowe et al. (1996). It was explicitly stated that respondents could answer with a WTP value that was shown on the payment card or any other value they preferred. Open-ended WTP questions were included during the pre-tests of the survey in order to determine the range of values shown on the payment card. The next two contingent valuation questions elicit WTP in scenarios in which flood risks rise to 1 in 600 and 1 in 400 as a result of climate change. Changes in risks are communicated by stating the probability and frequency of the new risk, as well as the proportional change relative to the baseline probability ('doubled' or 'tripled') in order to facilitate comprehension (e.g. McDaniels 1992). Risk ladders are shown only in versions 1 and 3, so that the effect of using a risk ladder on WTP can be estimated. In all questions, respondents are asked to provide a motivation if they do not want flood insurance, which allows protest responses to be identified. The questionnaire ends with the usual socio-economic questions. The survey was administered over the Internet among homeowners living in the river delta in areas with a flood safety norm of 1 in 1,250 years. The implementation of the survey is discussed in detail in Section 9.4.1.

10.4 Results of the survey

10.4.1 Descriptive statistics of the willingness-to-pay questions

Sample characteristics were discussed in Section 9.4.2. The results of the contingent valuation questions are shown in Table 10.2. The percentage of total respondents who are willing to insure (WTI), the mean willingness-to-pay of all respondents (WTP) and the mean WTP conditional on wanting insurance (CWTP) are shown for the different flood probability scenarios and versions. These three measures convey different information relevant for policy makers and insurers. *WTI* is an indicator of the percentage of homeowners in the river delta who are willing to pay a positive amount for flood insurance. *WTP* is an indicator of the mean WTP for flood insurance of all homeowners, and is the relevant measure for policy makers who are interested in the welfare effects of introducing flood insurance. *CWTP* is the mean WTP of homeowners who are willing to pay a positive amount for flood

Table 10.2 ***Descriptive statistics of willingness-to-insure (WTI), mean willingness-to-pay (WTP) and mean conditional willingness-to-pay (CWTP)***

	WTI	WTP	CWTP
Flood probability and questionnaire version	Percentage of respondents	€ per month	€ per month
Flood probability 1 in 1,250			
Government relief with risk ladder	24.2	3.07 [0.59]	12.90 [1.92]
Government relief without risk ladder	32.7	2.49 [0.39]	7.85 [0.92]
No government relief with risk ladder	33.6	3.21 [0.68]	9.62 [1.84]
No government relief without risk ladder	38.0	4.14 [0.62]	10.97 [1.41]
Flood probability 1 in 600			
Government relief with risk ladder	30.1	4.27 [0.74]	14.56 [2.00]
Government relief without risk ladder	40.0	4.47 [0.73]	11.72 [1.60]
No government relief with risk ladder	44.9	5.85 [1.14]	13.15 [2.39]
No government relief without risk ladder	44.2	5.13 [0.67]	12.04 [1.33]
Flood probability 1 in 400			
Government relief with risk ladder	35.5	4.79 [0.79]	13.62 [1.83]
Government relief without risk ladder	40.0	4.96 [0.78]	12.58 [1.64]
No government relief with risk ladder	50.8	6.32 [0.85]	12.53 [1.48]
No government relief without risk ladder	46.4	6.21 [0.80]	13.56 [1.49]

Note: Standard errors are given in square brackets.

insurance. This indicates the potential revenue per policy, excluding homeowners who do not want the insurance. Comparing CWTP with the mean expected loss (compensation provided) per policy serves as an indicator of profitability, which is interesting for insurers who are considering introducing flood insurance.

WTI is between 24 per cent and 50 per cent, depending on the flood probability, availability of compensation for flood damage by the government, and use of risk ladders to communicate risk.[8] The large proportion of respondents with a zero WTP value has been observed in other studies that estimate insurance demand for low-probability hazards (e.g. McClelland et al. 1993). It indicates that many individuals dismiss low-probability flood risk altogether (Kunreuther and Pauly 2004). The neglecting of risk contradicts expected utility theory and may be consistent with prospect theory (Kahneman and Tversky 1979). WTI is higher when the flood probability is greater in all versions of the questionnaire, which suggests that fewer homeowners neglect the risk if the probability rises. Slovic et al. (1977) argue that individuals only buy insurance against low-probability hazards if the probability exceeds a certain threshold, which varies across individuals. An increase in probability reduces probability neglect, as the probability exceeds the threshold for more individuals. The absence of government compensation increases WTI, which indicates that the government can stimulate the emergence of a private market by abolishing the current compensation scheme.

The mean WTP for flood insurance is between €2.49 and €6.32 per month. The independent effects of the risk ladder, the flood probability and the availability of government compensation for damage on WTP are examined in subsequent sections using a statistical model that controls for other explanatory variables.

Mean CWTP, which excludes zero WTP responses, is between €7.85 and €14.56 per month. The large change in the number of respondents with a positive WTP value across versions makes differences in CWTP difficult to interpret. It is relevant to compare CWTP with the expected value of the loss per policy. The expected value of the damage per household in the 1 in 1,250 norm dike-ring areas is computed as

[8] In total, 50, 67 and 77 protest responses with a zero WTP were excluded from, respectively, the contingent valuation questions with 1/1,250, 1/600 and 1/400 flood probabilities. Such responses are motivated by individuals saying that they do not believe that flood damage is not already covered, do not believe or accept the stated flood probability, do not believe or accept the change in the flood probability, or do not believe that offering flood insurance is possible. In versions 3 and 4, a protest response may result from individuals not accepting the abolition of government compensation.

in Chapter 8 and equals €4.7 per month according to price levels at the time of the survey. Mean CWTP is between 70 per cent and 175 per cent higher than the expected value of the average flood damage per house under current climate conditions, which indicates that offering flood insurance is profitable.

10.4.2 A model of WTP for flood insurance

A single model was constructed for WTP elicited with the three contingent valuation questions. WTP was transformed to a logarithmic scale. This has statistical advantages, as problems with non-normality and heteroskedasticity are likely to become less serious. In addition, estimation results become more intuitive, since coefficients can be interpreted as elasticities if the explanatory variable is also in logarithmic form, or as semi-elasticities otherwise. It is assumed that for each respondent n, WTP depends on the probability p_n, the expected flood damage d_n, the presence of government compensation g_n and individual characteristics x_n. In particular we have:

$$WTP_n = \exp(\beta_1 x_n) \times \exp(\beta_2 g_n) \times d_n^{\beta_3} \times (p_n)^{\beta_4} \times \exp(\varepsilon_n) \tag{10.1}$$

where x_n is a $1 \times k$ vector of individual characteristics, and ε_n is the error term.

10.4.3 Bayesian updating using perceived flood probabilities

Several studies show that individuals follow a Bayesian updating of probabilities when they are provided with new information about the likelihood of adverse events (e.g. Viscusi 1989; Viscusi and O'Connor 1984). If individuals follow a Bayesian updating of their flood risk, then WTP depends not only on the flood probability stated in the questionnaire but also on prior beliefs. A Bayesian model of updating risk is:

$$p_{0n}^{*} = \frac{\alpha \pi_n + \theta p_n}{\alpha + \theta} \tag{10.2}$$

where α and θ are the weights assigned to the prior belief about, respectively, the flood probability π_n and the probability presented in the survey p_n (see Viscusi 1985a, 1985b; Viscusi and O'Connor 1984). The WTP equation with the Bayesian model is:

$$WTP_n = \exp(\beta_1 x_n) \times \exp(\beta_2 g_n) \times d_n^{\beta_3} \times \left(\frac{\alpha \pi_n + \theta p_n}{\alpha + \theta} \right)^{\beta_4} \times \exp(\varepsilon_n). \tag{10.3}$$

We do not observe the prior beliefs about flood probability. Nevertheless, the survey generates an estimate of beliefs about the current return period and a broad range of indicators of risk perception. The latter can serve as proxies for prior beliefs, enabling the modelling of Bayesian updating. In particular, several variables capturing the perceived risk of flooding c_n are used here as proxy variables for π_n in Eqn 10.2, as in Bhattacharya et al. (2007). Then, taking the natural logarithm, the regression equation becomes:

$$\log WTP_n = \beta_1 x_n + \beta_2 g_n + \beta_3 \log d_n + \beta_4 \log p_n + \beta_5 c_n + \varepsilon_n. \tag{10.4}$$

10.4.4 Testing the effects of risk communication

The effect of showing the risk ladders on the sensitivity of WTP to the flood probability is tested by creating an interaction term $RL_n{}^* p_n$, where RL_n is a dummy variable that equals 1 if a risk ladder is shown and 0 otherwise. Including RL_n directly tests the effect of showing the risk ladder on the level of WTP irrespective of the probability. The model to test risk communication is then:

$$\begin{aligned}\log WTP_n = {} & \beta_1 x_n + \beta_2 g_n + \beta_3 \log d_n + \beta_4 \log p_n \\ & + \beta_5 c_n + \beta_6 \log p_n * RL_n + \beta_7 * RL_n + \varepsilon_n.\end{aligned} \tag{10.5}$$

The effect of the probability is captured by β_4 in the versions without the risk ladder and by $\beta_4 + \beta_6$ in the versions with the risk ladder. If $\beta_4 = 1$ or $\beta_4 + \beta_6 = 1$, then WTP is proportional to the size of the risk change (Hammitt and Graham 1999). Finally, this equation has to be adjusted to reflect the panel structure of the data, since each individual is asked their WTP under three different flood probabilities. This will result in the correlation of the errors for a given individual, which is taken into account by a panel data model:

$$\log WTP_{nt} = \beta_1 x_n + \beta_2 g_n + \beta_3 \log d_n + \beta_4 \log p_{nt} + \beta_5 c_n + \beta_6 \log p_{nt} * RL_n + \beta_7 * RL_n + \varepsilon_{nt} \quad (10.6)$$

where $n = 1, 2, \ldots N$ and $t = 1, 2, 3$.

10.4.5 Estimation method and coding of the explanatory variables

Censoring of the data at zero is considered by estimating a random effects Tobit model. The WTP values are treated as a positive continuous variable. This is justified since the elicitation method applied explicitly stated that respondents could answer with any value listed on the payment card or any other value they preferred. For this reason a standard Tobit model is employed instead of an interval Tobit model. Moreover, the payment card includes a bid larger than the highest amount stated (>€120), meaning that the WTP variable was not censored at the highest amount stated on the card. The panel model is:

$$y_{nt}^{*} = \alpha_n + x_{nt}^{'}\beta + \varepsilon_{nt} \quad (10.7)$$

where $\varepsilon_{nt} \sim \mathrm{N}[0, \sigma_{\varepsilon}^2]$ and $\alpha_n \sim \mathrm{N}[0, \sigma_{\alpha}^2]$. We observe $y_{nt} = y_{nt}^{*}$ if $y_{nt}^{*} > 0$, and $y_{nt} = 0$ if $y_{nt}^{*} \leq 0$. The random effects Tobit model maximizes the log likelihood $\sum_{n=1}^{N} \ln f(y_n \mid X_n, \beta, \sigma_{\varepsilon}^2, \sigma_{\alpha}^2)$ with:

$$f(y_n \mid X_n, \beta, \sigma_{\varepsilon}^2, \sigma_{\alpha}^2) =$$

$$\int \prod_{i=1}^{T} \left[\frac{1}{\sigma_{\varepsilon}} \phi_{nt} \right]^{d_{nt}} [1 - \Phi_{nt}]^{1-d_{nt}} \frac{1}{\sqrt{2\pi\sigma_{\alpha}^2}} \exp\left(\frac{-\alpha_n}{2\sigma_{\alpha}^2} \right)^2 d\alpha_n \quad (10.8)$$

and $\phi_{nt} = \phi((y_{nt} - \alpha_n - x_{nt}^{'}\beta) / \sigma_{\varepsilon})$, $\Phi_{nt} = \Phi((\alpha_n + x_{nt}^{'}\beta) / \sigma_{\varepsilon})$, here $\phi(.)$ and $\Phi(.)$ denote the standard normal probability density function (pdf) and cumulative density function (cdf), respectively.

Appendix 10C gives a detailed explanation of the variables and their descriptive statistics. Different methods of coding categorical variables were applied. Continuous variables were created from categorical variables that represent monetary classes, such as income, and the value of the house and contents (e.g. Blumenschein et al. 2008). Ordinal qualitative variables,[9] which are partitioned into J intervals, can be included using $J - 1$ dummies or can be transformed into values on the real axis using an approach proposed by Terza (1987). An advantage of dummy coding is that the interpretation of the coefficients is straightforward, but many variables are needed if J is large. In this case, the transformation of Terza (1987) can result in gains in efficiency and less estimation bias. For this reason, the latter approach has been applied in several studies (e.g. van Praag et al. 2003). We apply dummy coding for variables with a small number of categories: namely, the rating of the flood risk compared with that of an average resident, the effect of climate change on the flood probability, and education level. The Terza transformation (see Appendix 7B) is used for variables with a large number of categories: namely, the risk of suffering flood damage, and the risk-seeking index.

10.4.6 Estimation results of the panel Tobit model

Scenarios for flood probabilities, risk ladders and the availability of government relief

Table 10.3 shows the coefficients, standard errors and t statistics of the panel Tobit model (Eqn 10.7) and the marginal effects of a change in the explanatory variables on the (ln) WTP. The probability variable is expressed as a natural logarithm, which implies that the marginal effect has an elasticity interpretation. WTP is significantly and positively related to the flood probability stated in the survey. The marginal effect of the flood probability is less than proportional; it is approximately 0.11 in versions without risk ladders, and 0.16 in versions with risk ladders (computed as 0.11 + 0.05). The risk ladders have a large and statistically significant (at the 1% level) effect on the sensitivity of WTP to changes in the flood probability stated in the questionnaire. The sensitivity to probability, which is a measure of sensitivity to scope in this survey, increases by about 45 per cent in the versions where flood risk is communicated with risk ladders.[10] In addition, the risk ladders increase the level of WTP considerably, independent of the

[9] These are characterized by a continuous unobservable ordinal latent index, while each interval is ranked (1 through J) in increasing order according to its supremum (Terza 1987).

[10] Computed as (0.16 – 0.11)/0.11*100.

Table 10.3 *Estimation results of a random effects Tobit model of ln(WTP)*

Variable	Coefficient	Standard error	*t* statistic	Marginal effect
Scenarios in the survey				
Log probability	0.5025***	[0.0736]	6.83	0.1100
Log probability * Risk ladder	0.2460***	[0.0961]	2.56	0.0538
Risk ladder	2.0878***	[0.6286]	3.32	0.4570
Government compensation	-0.1788***	[0.0648]	-2.76	-0.0391
Risk perception				
Climate change causes higher flood risk	0.9448***	[0.0794]	11.90	0.2068
Risk of suffering flood damage	0.4195***	[0.0392]	10.70	0.0918
Lower flood risk than an average resident	-1.1128***	[0.0780]	-14.26	-0.2436
Zero expected flood damage	-0.9069***	[0.2174]	-4.18	-0.4085
Log expected flood damage	0.0509***	[0.0109]	4.68	0.0111
Zero expected return period flood	-1.8662***	[0.3735]	-5.00	-0.1985
Log expected return period flood	-0.0747***	[0.0161]	-4.64	-0.0163
Flooding is exogenous to human control	-0.3594***	[0.0673]	-5.34	-0.0787
Experience of flooding and evacuation	-0.1197	[0.1158]	-1.03	-0.0262
Individual risk aversion				
Insurance purchase index	0.2691***	[0.0904]	2.98	0.0589
Risk-seeking index	-0.4817***	[0.0355]	-13.54	-0.1055
Geographical characteristics				
Elevation of house relative to water level	-0.0004***	[0.0008]	-5.38	-0.00009
Area is not protected by dikes	-0.4873***	[0.1162]	-4.19	-0.1067
Distance of house to a main river	-0.0082***	[0.0030]	-2.76	-0.0017
Rural area	0.7420***	[0.1572]	4.72	0.1624
Socio-economic characteristics				
Female	-1.0866***	[0.0775]	-14.01	-0.2379
Age	-0.0089***	[0.0031]	-2.83	-0.0020
Log value of property	-0.2228***	[0.1000]	-2.23	-0.0488
University degree	-0.1597*	[0.0856]	-1.86	-0.0350
Log income	0.7768***	[0.0922]	8.43	0.1701
Constant	0.3627	[1.2364]	0.29	
σ_v	0.8604	[0.0135]		
σ_u	2.1750	[0.0532]		
Number of observations	2,645			
Log likelihood	-2,294			
Restricted log likelihood	-2,987			

Notes: *, ** and *** indicate, respectively, significance at the 10%, 5% and 1% levels. Estimations were performed with Limdep software.

change in flood probability, as indicated by the positive and significant coefficient of the dummy variable representing versions with risk ladders. The level of WTP is about 46 per cent higher when risk ladders are shown to communicate risks.

Examining the individual WTP responses reveals that in the version with (without) the risk ladder and without government relief, 29 per cent (30 per cent) of these increase less than proportionally, while about 4 per cent (7 per cent) of them increase approximately proportionally, and only 1 per cent (1 per cent) increase more than proportionally, with a probability increase from 1 in 1,250 to 1 in 600. About 57 per cent of individuals neglect the risk and 10 per cent show a WTP increase from zero to a positive amount. These responses to probability changes are very similar if flood risk increases to 1 in 400.[11] In summary, few individuals adjust their WTP proportionally or more than proportionally to probability changes. In aggregate, the response to probability is considerably less than proportional. This is in accordance with several other studies, suggesting that individuals are insufficiently sensitive to probability changes (e.g. Tversky and Wakker 1995). This behaviour may be consistent with the probability weighting functions of rank-dependent and prospect theories (Quiggin 1982; Tversky and Kahneman 1992), while it contradicts the linear processing of probabilities assumed in expected utility theory (von Neumann and Morgenstern 1947). Overall, communicating the risk with risk ladders has a large effect on the sensitivity of individual WTP adjustments to probability changes as well as on the level of WTP. Nevertheless, the effect of probability changes on WTP remains less than proportional even when risk ladders are shown to communicate flood risk.

The variable representing the version describing the scenarios where compensation for flood damage by the government may be granted has a negative and significant coefficient. This implies that demand for flood insurance is lower if individuals expect that the government might compensate for flood damage than when such compensation is

[11] In the version without (with) the risk ladder and without government relief 36% (33%) of individual WTP values increase less than proportionally, 2% (4%) increase approximately proportionally, and 6% (5%) increase more than proportionally to the probability change from 1 in 600 to 1 in 400. Probability neglect is 50% (55%) and 7% (3%) of WTP values that change from zero to a positive amount.

not available and damage can only be reimbursed by insurance. The size of the marginal effect shows that WTP is approximately 4 per cent less due to the current government compensation scheme.

Perceptions of flood risk and flood experience

The variables that represent respondents' perceptions of flood risk are all highly statistically significant and have the expected sign. The greater that individuals perceive their flood risk, the greater is their demand for flood insurance. These results suggest that individuals follow an updating process of flood probability, since prior perceptions of flood probability are a significant determinant of WTP in addition to the flood probabilities stated in the questionnaire. In particular, if respondents think that climate change increases flood risks, then their WTP is 21 per cent higher. WTP is positively related to the expected probability that the household suffers damage from flooding, as the positive coefficient of the 'risk of suffering flood damage' variable indicates. Moreover, demand for flood insurance is 25 per cent less if respondents rate their flood risk as lower than the national average compared with respondents who rate their flood risk as being equal to, or higher than, average.

The expected flood damage is included as two variables: a dummy variable that captures the respondents who answered with zero expected flood damage and a continuous variable of the expected flood damage for respondents with positive answers in logarithmic form. WTP is 41 per cent less if the respondent expects to suffer no damage during floods, and it is higher the more that the respondent expects flood damage to be. The expected return period represents the flood risk on a quantitative scale and is included in a similar way, using a separate variable that represents the zero answers. WTP is 20 per cent lower for respondents who expect a zero return period for flooding, and decreases the greater the expected return period is for respondents who indicate a non-zero return period. This effect is as expected, since a longer flood return period represents a lower (perceived) flood probability.

Respondents who indicate causes of flooding that are beyond their or the relevant water manager's direct control, such as extreme weather events or climate change, have a lower WTP value. This is consistent with studies showing that individuals have lower flood risk perceptions if they regard floods as natural phenomena or outside human control (e.g. Brilly and Polic 2005). Research in the USA suggests that demand for flood insurance is positively related to experience of flooding (see Section 10.2). However, the experience of flooding and evacuation does not have a significant influence on WTP for flood insurance in this study, which may be the case because the last flood in the Netherlands occurred in 1995, which may have been too long ago to affect demand significantly.

Individual risk aversion

Two variables are included in the model of the respondent's risk attitudes. The insurance purchase index represents individuals who have purchased many other insurance policies and is an indicator of revealed risk attitudes. The positive and significant coefficient of this variable indicates that such individuals also have a higher demand for flood insurance. In addition, a variable has been included that represents risk attitudes as stated by the respondents themselves. This risk-seeking index has been derived from answers to a question in which respondents indicated how similar they regarded themselves to a person who avoids taking financial risks and prefers to be well insured. The negative coefficient of this variable indicates that the more risk-seeking individuals have a lower WTP for flood insurance.

Geographical characteristics

Geographical characteristics are included as explanatory variables since they serve as indicators of the actual flood risk, which may influence flood insurance demand as Section 10.2 has shown. Data for these variables (computed from the respondents' zip code areas) have been obtained using geographical information system (GIS) maps, as explained in Section 9.6.2.[12] A variable has been constructed that represents the difference between the elevation of the zip code area of the individual and the height of the potential water level of a flood.[13] The results show that WTP is negatively related to the elevation of the respondent's living area relative to the potential water level. Some

[12] These data are based on zip code numbers and letters for 950 respondents, which is highly accurate because the GIS data are available at street level. The data for 32 respondents are based on zip code numbers only because the letters were incomplete.

[13] Adjustments have been made for respondents who live in flats on the first floor by adding 2.5 metres to the height of the area.

respondents live in flood-prone areas that are not protected by primary river dikes, which implies that they face a considerably higher probability of flooding. It was investigated whether respondents who live in flood-prone areas unprotected by dikes have a greater demand for flood insurance. A variable representing such respondents is statistically significant and negative. The marginal effect indicates that WTP for flood insurance is 11 per cent lower in these high-risk areas. Moreover, the analysis of individual flood risk perceptions in Chapter 9 indicates that risk perceptions are lower in unprotected flood-prone areas. Possible reasons are that houses in areas that are unprotected by dikes are generally better adapted to flooding, resulting in a lower demand for flood insurance, or that such individuals are unaware of the high risk they face.

Homeowners who live close to major rivers are more likely to suffer flood damage; for example due to higher flow velocities of water. The effect of living close to a river on WTP was examined by including a variable that represents the distance of the respondent's house to the closest major river. The coefficient of this variable is significant and negative, so that homeowners who live further away from a major river have a lower flood insurance demand than homeowners who live close to a river. Finally, it was examined whether inhabitants of rural areas place a different value on flood insurance than inhabitants of cities. The marginal effect indicates that inhabitants of rural areas have a 16 per cent higher WTP for flood insurance.

These findings about the role of geographical characteristics on insurance demand are very relevant for insurers. It has sometimes been suggested by Dutch insurers that problems with adverse selection may be severe in flood insurance markets, if it was the case that only those individuals who live in unprotected areas with high flood risks would demand flood insurance (de Vries 1998). Our findings indicate that the problem of adverse selection may be minor, because WTP may actually be lower for some homeowners in these high-risk areas.

Socio-economic characteristics

The socio-economic variables indicate that being female decreases WTP by approximately 24 per cent. Age influences demand for flood insurance negatively, suggesting that older individuals are more risk-seeking. Having a university degree decreases WTP slightly, but this effect is only significant at the 10% level. Adverse selection may still be an issue if homeowners with an expensive house and home contents have a higher demand for flood insurance, since these homeowners are more likely to suffer high flood damage. We examined this by including a variable of the logarithm of the total value of the respondent's property, i.e. house and contents, which is also an indicator of wealth. Ganderton et al. (2000) explain that the effect of higher wealth on the demand for insurance is in theory ambiguous. On the one hand, insurance costs are a smaller fraction of an individual's wealth for richer individuals, while they may be more likely to suffer high property damage, which would increase their insurance demand. On the other hand, potential losses may be more easily carried by wealthy people, which makes insurance less attractive and increases self-insurance. The latter effect seems to dominate in our study since the total value of property has a negative influence on the WTP for flood insurance. The net income level of households positively influences WTP for flood insurance, which is consistent with the international studies reviewed in Section 10.2. The estimated income elasticity is 0.17.

10.5 Conclusions

An important requirement for the adequate functioning of insurance as an instrument to spread climate change related natural disaster risk is that sufficient demand for natural disaster insurance is present among households. Several international studies have shown that many people may neglect low-probability natural disaster risk and do not purchase natural disaster insurance, which may impede the adequate functioning of insurance markets for such risks. Therefore, it is imperative to obtain insights into the determinants of flood insurance demand, and to make accurate predictions of demand before new insurance arrangements for climate change damage are introduced. This chapter started with a review of several representative international studies about factors of influence on the demand for flood insurance. In summary, previous disaster experience is found to increase insurance demand, as well as flood risk perceptions and certain socio-economic variables such as income. Moreover, several geographical variables which are indicators of objective flood risks influence flood insurance demand significantly. Evidence on the charity hazard is mixed; government relief does not appear to reduce flood insurance demand in the USA and Bangladesh, but it does so in Austria.

It is expected that climate change may increase the likelihood of certain low-probability, high-impact natural disasters, but this has not been researched in previous studies. This chapter examined the demand for flood insurance under current flood probabilities and increases in flood probability due to climate change, using a survey among homeowners living in the river delta area of the Netherlands. Particular attention was paid to the effects of availability of government relief, perceptions of flood risks, and geographical and household characteristics on WTP for flood insurance. Risk attitudes to low-probability risk and the influence of communicating probabilities with risk ladders on insurance demand were also analysed.

The results indicate that a large proportion of homeowners neglect the low probability of flood risk and do not want to purchase flood insurance. This behaviour contradicts expected utility theory and may be explained by prospect theory. The mean WTP value of homeowners who are interested in buying flood insurance (defined as CWTP) is above the average expected value of the flood damage per household, which suggests that opportunities for partly private flood insurance exist. Fewer homeowners would neglect the risk if the flood probability were to increase as a result of climate change, as would be expected. Moreover, the removal of the government compensation scheme could increase the number of homeowners purchasing insurance. Fears concerning adverse selection in flood insurance seem unfounded, as geographical characteristics of the respondents show that the inhabitants of areas more vulnerable to flooding do not necessarily have a greater demand for insurance than the inhabitants of less vulnerable areas. In fact, the results indicate that inhabitants of unprotected areas even have a lower WTP for flood insurance.

The panel Tobit estimation revealed that WTP increases less than proportionally if climate change increases flood probability. Examination of individual risk attitudes confirms that very few respondents adjust their WTP proportionally or more than proportionally to the rise in flood probability, and that WTP is adjusted less than proportionally to the probability change by most individuals. The implication of this finding is that offering insurance may become unprofitable if climate change increases flood probabilities considerably, since WTP is expected to rise less than the expected value of the loss of the insurance. During pre-tests, respondents indicated that the risk ladders were useful in increasing their understanding of the flood probability faced. Therefore, it was decided to include risk ladders in the survey, as this would allow respondents to make a more rational insurance purchase decision. Using risk ladders as a device to communicate the flood probability considerably increased the sensitivity of WTP to probability changes. The effect of the risk ladder on WTP, irrespective of probability, was also to increase WTP by a relatively large amount. In other words, the way in which low-probability risks are communicated has a large impact on the outcomes of the valuation task.

Perceptions of flood risk play an important role in the demand for flood insurance. In particular, the greater the respondents perceive their flood risk to be, the higher is their WTP value. Several measures of prior risk are statistically significant in explaining WTP; that is, in addition to the flood probability stated in the questionnaire. This suggests that it is difficult to capture the complete effect of climate change on WTP for insurance by focusing only on the effect of probabilities stated in surveys. A gradual increase in flood risk due to climate change may increase risk perceptions, which in turn could increase demand for flood insurance, so that the final effect may not be entirely captured by the coefficient of the flood probabilities. Furthermore, the results suggest that increasing awareness of the flood risk may increase demand for insurance. Future research could focus on examining the updating of prior beliefs in responses to changes in risk due to climatic conditions and the corresponding effects on insurance demand.

Appendix 10A Determination of risk using a risk ladder

The linear[14] risk ladders depict the risk of car theft, fire in a residential building, flooding, a car fire and the risk of a fatal traffic accident. The probability of a deadly traffic accident is the probability that a traffic participant dies because of an accident, and is obtained from RVWS and VROM-R (2003). The information on the other probabilities is

[14] The risk ladder has a scale that increases by 2/4,000 per step, implying a linear scale (e.g. Vassanadumrongdee and Matsuoka 2005).

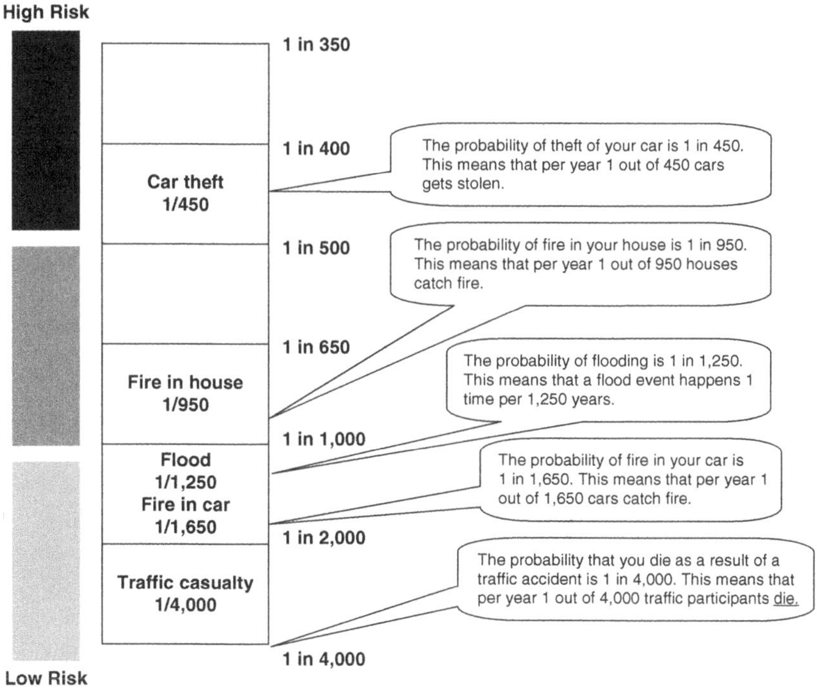

Figure 10.A1 Risk ladder that compares the current flood probability with other risks

obtained from Verbond van Verzekeraars (2005, 2006, 2007) and the Central Bureau of Statistics (CBS 2005, 2006, 2007). The probabilities are based on 5-year averages between 2002 and 2006 and are computed as the number of times an event (car fire, car theft, fire in a house) occurs per year divided by the number of private cars or residential houses, as discussed in Botzen et al. (2008b).

Appendix 10B Payment card

€0	€2	€4	€8	€15	€30	€65	€120
€1.25	€2.50	€5	€10	€20	€40	€80	More than €120
€1.50	€3	€6	€13	€25	€50	€100	Don't know

Figure 10.B1 The payment card included with the contingent valuation questions

Appendix 10C Explanatory variables and descriptive statistics

Table 10.C1 *Summary overview of the variables used in the statistical analysis*

Climate change causes higher flood risk	Dummy variable, 1 = respondent expects that climate change causes higher flood risks
Risk of suffering flood damage	Categorical variable (range 1–11), 1 = no risk and 11 = extremely high risk
Lower flood risk than an average resident	Dummy variable, 1 = respondent expects that his/her flood risk is lower than that of an average resident in the Netherlands
Zero expected flood damage	Dummy variable, 1 = respondent does not expect to suffer flood damage during a flood
Expected flood damage	Continuous variable, damage in euros
Zero expected return period	Dummy variable, 1 = respondent expects return period to be zero
Expected return period flood	Continuous variable, return period in years
Flooding is exogenous to human control	Dummy variable, 1 = respondent mentions only climate or natural conditions and not human activity or water management as causes of flooding
Experience with floods and evacuation	Dummy variable, 1 = respondent has experienced a flood and has been evacuated in the past
Insurance purchase index	Dummy variable, 1 = respondent has purchased more than eight of the following insurances: health insurance with no deductible, home, home contents, dentist, continuous travel, all-risk car, life, bike, legal assistance, disability insurance
Risk-seeking index	Categorical variable (range 1–5), 1 = very risk-averse, 5 = not risk-averse at all
Elevation of house relative to water level	Continuous variable, elevation of the living area relative to the potential water level of a flood in centimetres
Area is unprotected by dikes	Dummy variable, 1 = the area is not protected by dikes
Distance of house to a main river	Continuous variable, distance of the respondent's house from a major river in kilometres
Rural area	Dummy variable, 1 = the area is a rural area

Table 10.C1 (*cont.*)

Female	Dummy variable, 1 = respondent is female
Age	Continuous variable, age in years
Value of property[a]	Continuous variable, total market value of the respondent's house and contents in €
Educational level	Dummy variable, 1 = highest level is university degree
Income[b]	Continuous variable of the monthly after-tax household income in €

Notes:

[a] For housing value, the respondent could mark one of the following categories: <€100,000; €100,000–150,000; €150,000–200,000; €200,000–250,000; €250,000–300,000; €300,000–350,000; €350,000–400,000; €400,000–500,000; €500,000–600,000; >€600,000. For home contents value, the respondent could mark one of the following categories: <€25,000; €25,000–50,000; €50,000–75,000; €75,000–100,000; €100,000–125,000; €125,000–150,000; €150,000–175,000; €175,000–200,000; €200,000–300,000; >€300,000. Continuous value of housing and home contents variables were constructed by setting the housing and home contents value of each respondent to the midpoint of the interval (€650,000 and €350,000 were used for, respectively, the highest housing and highest home contents value categories).

[b] For monthly income, the respondent could mark one of the following categories: <€750; €751–1,000; €1,001–1,250; €1,251–1,500; €1,501–2,000; €2,001–2,500; €2,501–3,000; €3,001–3,500; €3,501–4,000; >€4,000. A continuous income variable was constructed by setting the income of each respondent to the midpoint of the interval (€4,500 was used for the highest category).

Table 10.C2 *Descriptive statistics for the explanatory variables*

Variable	Number of observations	Mean	Standard deviation
Climate change causes higher flood risk	982	0.65	0.48
Risk of suffering flood damage[a]	982	4.19	2.20
Lower flood risk than an average resident	982	0.48	0.50
Zero expected flood damage	981	0.12	0.33
Expected flood damage	981	70,383	185,400
Zero expected return period	982	0.04	0.19
Expected return period flood	982	113,780	3,190,695

Table 10.C2 (*cont.*)

Variable	Number of observations	Mean	Standard deviation
Flooding is exogenous to human control	982	0.39	0.49
Experience of floods and evacuation	982	0.03	0.17
Insurance purchase index	982	0.12	0.33
Risk-seeking index[a]	982	3.24	1.16
Elevation of house relative to water level	982	620	–1,165
Area is unprotected by dikes	982	0.11	0.32
Distance of house to a main river	982	9.40	11.68
Rural area	982	0.05	0.222
Female	982	0.42	0.49
Age	982	45.59	11.95
Value of property	982	411,947	167,669
University degree	982	0.21	0.41
Income	982	2,856	1,018

Notes: [a] The statistics of this variable are in accordance with the coding in Table 10.C1. This original coding has been transformed for the analysis according to Terza (1987), as described in Appendix 7B.

11 Market shares for insurance against flood risk under different climate change scenarios

11.1 Introduction

Climate change is projected to increase the frequency and severity of weather extremes in certain regions of the world, which is likely to increase risks that are covered by the insurance sector, as has been outlined in Chapters 2 and 3. Several studies have examined the impact of climate change on the insurance sector, and have predicted that future premiums for natural disaster insurance may increase (e.g. Dlugolecki 2008; Kunreuther and Michel-Kerjan 2007; Mills 2005). Moreover, the insurability of natural disaster risk may be hampered if insurance premiums that reflect increased risk are more than what homeowners are willing to pay.

There have only been a few empirical studies which have estimated the effect of climate change on the demand for natural catastrophe insurance. The willingness-to-pay (WTP) for insurance is expected to increase as a result of a rise in the probability of suffering weather-related damage, but it is uncertain how large this rise will be in relation to the change in the underlying risk. Insights into the influence of climate change on WTP for disaster insurance is required so that insurers can assess the future profitability of offering coverage against damage caused by extreme weather. The previous chapter presented a contingent valuation study that estimated how individual WTP for flood insurance and the neglect of low-probability risk may change if flood probabilities increase as a result of climate change. The influence of climate change on the WTP for flood insurance will be examined by means of a choice experiment in this study for the Netherlands. This case is a good example with a broader relevance for three reasons: floods are a major cause of natural disaster damage in many countries; the insurance sector has great economic importance because it is the largest industry worldwide in terms of revenues; and climate change will have implications worldwide (IFRC 2010; Mills 2009b).

A choice experiment is described in this chapter in which individuals were asked to choose one option from a set of choices, as this method is expected to result in less hypothetical bias than a contingent valuation study (Carlsson and Martinsson 2001).[1] The choice model for flood insurance demand estimated here is particularly suitable for examining how demand can be affected by climate change because it is able to simulate potential market shares and assess WTP under various scenarios of increased risk caused by climate change. In addition, the influence of socio-economic developments on WTP and potential market shares for flood insurance are analysed. Socio-economic developments, such as settlement in vulnerable areas as well as population and economic growth, are likely to increase the damage caused by natural disasters (Chapters 2 and 3), while investments in damage mitigation may restrain rising trends in disaster losses (Chapters 6 and 7).

Chapter 8 presented risk premiums for flood insurance demand in the Netherlands under different climate change scenarios using prospect and rank-dependent utility theories, with parameters obtained from existing experimental studies. The results indicate that an increase in flood probabilities from 1 in 1,250 up to 1 in 550 causes WTP to increase more than the expected value of the loss. The representative agent assumption underlying that study, which results in average WTP values, is relaxed here by estimating individual heterogeneity in WTP using a stated preference survey among homeowners. A stated preference study is appropriate here since flood insurance is not available in the Netherlands, which implies that insurance demand cannot be analysed using data on revealed preferences. It was argued in Chapter 4 that the introduction of a public–private flood

[1] This application of insurance demand may be regarded as being less susceptible to hypothetical bias than other commonly undertaken valuation studies which are concerned with intangible public goods such as environmental assets, because people have experience with insurance, and because it concerns the protection of one's private financial position (income/wealth). Moreover, problems with 'warm glow' (Andreoni 1990) and 'purchase of moral satisfaction' (Kahneman and Knetsch 1992) are likely to be less serious with the valuation of private goods.

insurance in the Netherlands may be beneficial for the spreading of climate change induced flood risk. Therefore, the results discussed in this chapter about the potential market shares of flood insurance are of practical relevance to insurers and policy makers who are considering introducing flood insurance.

This chapter, moreover, analyses the effect of the current institutional setting, characterized by availability of government compensation for flood damage, on the potential market share of flood insurance. The Dutch government can grant partial compensation for damage caused by large-scale floods via the Calamities and Compensation Act (Wet Tegemoetkoming Schade bij Rampen en Zware Ongevallen – WTS), as is also the situation in several other countries (Crichton 2008). Decisions about whether to grant compensation and how much damage will be compensated are political decisions. Experiences with flood damage in 1993 and 1995 resulted in considerable relief payments via the WTS. As a consequence, householders might expect that the government will compensate for future flood damage unconditionally, regardless of the risk they take. This may reduce the desirability of private insurance, which is often referred to as the 'charity hazard' or the 'Samaritan syndrome' (Buchanan 1975; Harrington 1999; Raschky and Weck-Hannemann 2007). Even though there is no guarantee that government compensation for disaster damage will be efficiently provided or that it will be distributed to those households who are most in need of assistance, disaster assistance is often accompanied by broad media coverage which may create expectations among householders that the government will also compensate them for their damage if they were to be affected by a future flood. The theoretical prediction is that government compensation for disaster damage reduces insurance demand (Lewis and Nickerson 1989), but the few empirical studies that exist on this topic have produced mixed results (see Section 10.2).

The remainder of this chapter is organized as follows. Section 11.2 outlines the set-up of the survey and its implementation. Section 11.3 explains the design of the choice experiment and the estimation methods. Section 11.4 presents the estimation results of logit and mixed logit models. Section 11.5 estimates market penetration and WTP for flood insurance under various climate change and socio-economic scenarios. Section 11.6 concludes the chapter.

11.2 Explanation of the questionnaires

11.2.1 The commodity valued

WTP for flood insurance was elicited by means of a choice experiment. The choice experiment evaluates insurance with different levels of coverage in situations with varying flood probabilities and damage caused by river flooding to homes and their contents. Careful consideration is given to communicating these low flood probabilities between the current safety standard of 1 in 1,250 (see Chapter 5) and increases in probabilities up to 1 in 100 as a result of climate change.

11.2.2 Pre-tests

Four pre-tests were conducted using interviews (n = 88) as explained in Chapter 10. The difficulties in communicating low probabilities are discussed in Section 10.3. The pre-tests indicated that risk ladders were effective in explaining (changes in) flood probabilities, as clarified in Chapter 10, so risk ladders were included in the final survey. Such a risk ladder illustrates the size of the change, as well as how the new probability compares with other types of risks (e.g. flooding compared with a car accident). We tested a labelled experiment (one label per insurance type with label-specific attributes) consisting of insurance options that cover flood damage to home contents, housing, both of these, and no insurance. The resulting choice experiment turned out to be overly complex. Instead, an experiment with generic (unlabelled) alternatives, which values an insurance covering both damage to home contents and buildings, was used in the final pre-tests and survey. This turned out to be easier for respondents. Both yearly and monthly premiums were provided in the choice experiment, and their levels were derived from answers to open-ended WTP questions.

11.2.3 Structure of the questionnaires

A full description of the survey and overview of the questions is given in an online report (see Botzen et al. 2008b). The questionnaire contains questions on the experience of the respondent with respect to flooding, flood damage and evacuation because of flood threats, and knowledge about the causes of flooding. In addition, several questions address the perception of flood risks. These questions and answers to these risk perception questions are discussed in detail in Chapter 9. The assessed perceptions may be important in decision making under risk, as several studies suggest (Viscusi 1989), and serve as explanatory variables. Additional questions are included about risk aversion and actual insurance purchases.

While the above-mentioned part of the questionnaire was identical in all versions, some subsequent explanations differed, giving rise to two versions. In version 1, the current regulation for compensating flood damage by the government was explained, and it also indicated that it is possible to purchase additional private insurance. In other words, it states that the government may partly compensate for damage caused by major floods, while this compensation is not granted for small flooding events. The uncertainty about receiving relief is mentioned, and several recent examples of floods where damage was partly compensated are given. The other version describes a scenario in which relief for flood damage by the government will no longer be granted, but instead it is possible to purchase insurance coverage.

Next, a short text was included in all versions about flood probabilities. It is explained why flood probabilities differ across regions, providing a comparison of risks in a textual context (see Hsee et al. 1999). Flood probabilities are expressed in terms of return periods or frequencies. This provides clearer information to respondents than just probabilities. In addition, estimated flood probabilities of an area not protected by dikes (1 in 100) were compared with flood probabilities of urban areas that are protected by dikes (1 in 1,250). This explanation preceded our main device for communicating risk, which was a risk ladder on which flood risk is compared with other insurable risks commonly faced by Dutch citizens (see Appendix 10A). Risk ladders that depict the change in flood probabilities in the climate change scenarios were shown subsequently.

The choice experiment asks the preferences of respondents with regard to flood insurance with a certain coverage level in a situation with a specified flood probability and expected damage. In particular, individuals decide whether to buy a certain degree of insurance coverage against a risk – probability and damage – for a certain price (premium). The attributes 'coverage', 'premium', 'flood probability' and 'flood damage' can take on different values in the choice options for insurance, as indicated in Figure 11.A1. An unlabelled experiment is used where respondents choose between insurance 'Situation A', 'Situation B', or an 'opt out' (an illustrative choice card is shown in Figure 11.A2). This set-up is consistent with the common practice of presenting respondents with two hypothetical choice alternatives with varying attribute levels (in our case A and B), and the status quo ('opt out') (Ferrini and Scarpa 2007). Individuals are asked to indicate their most preferred alternative, which is either one of the insurance alternatives or the opt out. The experiment is therefore embedded in the random utility maximization model of economic choice in which individuals are assumed to choose the alternative that maximizes their utility, which is specified as a function of the observed attributes of the alternatives (McFadden 2001).

The experiment starts with a 'practice choice card' which is carefully explained. It is mentioned that flood probabilities can differ between the two situations because of different impacts of climate change. Flood damage relates to the severity of the flood. Individuals are instructed to choose the 'opt out' if they do not want to purchase the flood insurance in either of the two situations A and B. If respondents choose the 'opt out', then they are not insured against the flood risk and do not have to pay an insurance premium. Subsequently, respondents are randomly assigned to three choice cards with attribute levels that are in accordance with the experimental design (see Section 11.3.1). Finally, a fixed card with a dominant option is shown in order to identify those respondents who have trouble in understanding the experiment. Follow-up questions ask for the main reasoning behind the choices made and the perceived difficulty of the experiment. Following the recommendation of the NOAA panel, we asked respondents to consider their budget constraints, which may reduce hypothetical bias (Arrow et al. 1993). The questionnaire concludes with the usual sociodemographic questions. The survey was administered over the Internet among homeowners living in the river delta in areas with a flood safety norm of 1 in 1,250 years. Section 9.4.1 explains the implementation of the survey and Section 9.4.2 provides characteristics of the sample.

11.3 Experimental design and estimation methods

11.3.1 Experimental design

The experimental design allocates combinations of values to the attributes which together make up the choice card that is shown to a respondent. All respondents are asked for their preferences for flood insurance on three such choice cards. The design was generated using Sawtooth CBC software, in accordance with the principles of level balance, moderate attribute level overlap, and orthogonality (uncorrelated attributes). This resulted in an almost optimal design that can achieve a high degree of model estimation accuracy, as the *D* efficiency statistic of 95% indicates (Ferrini and Scarpa 2007).[2] A few strictly dominant choices were excluded from the final design, but this hardly affected the design efficiency. The final design was balanced and consisted of 250 versions of three choice cards. This large number of versions implies that many of the possible combinations of the attribute levels were included in the experimental design. After removing protest responses, this resulted in a total of 2,751 completed choices.[3]

11.3.2 Estimation methods

Choice models are based on the random utility model. Here, the probability p_{ni} of an individual n choosing alternative i is set equal to the probability that the utility of alternative i is greater than, or equal to, the utility associated with an alternative j for every alternative in the choice set ($j = 1 \ldots J$). Formally:

$$p_{ni} = prob[(V_{ni} + \varepsilon_{ni}) \geq (V_{nj} + \varepsilon_{nj})\forall j \in j = 1,\ldots,J; i \neq j] \quad (11.1)$$

where V_{ni} and ε_{ni} are, respectively, the observed and unobserved components of individual n's utility associated with alternative i. The observed utility V_{ni} can be expressed as βx_{ni}, where x_{ni} are observed variables that influence the choice for alternative i with coefficients β. Different assumptions about the distribution of ε_{ni} result in different choice models. The logit model is the most commonly used. This is

[2] The design was generated using the efficient design module ('Balanced Overlap') of Sawtooth CBC software that is suitable for estimating interaction effects between the attributes (Chrzan and Orme 2000). Because of a lack of reliable priors of the coefficients of the attributes, we did not include such information in generating the design.

[3] 65 protest responses were excluded. These include individuals who said that they do not believe that flood damage is not already covered in their insurance, do not believe or accept the stated flood probability, or do not believe that offering flood insurance is possible. In version 2 of the survey, a protest response may also result from individuals not accepting the abolition of the current arrangement regarding governmental compensation.

derived under the assumption that ε_{ni} is *iid* extreme value type 1 (or Gumbel) distributed for all i (see Train 2003) for more details). The logit probability is:

$$p_{ni} = \frac{e^{\beta x_{ni}}}{\sum_j e^{\beta x_{nj}}}. \tag{11.2}$$

The independence assumption within the logit model is restrictive, because unobserved characteristics associated with alternatives in a choice situation may be similar. Moreover, unobserved factors that affect the choice in one choice situation may affect the choice in a subsequent choice situation, which implies dependence between choices over time. The more general mixed logit model can overcome these problems by allowing for random taste variation, unrestricted substitution patterns, and correlation in unobserved characteristics over different choice situations (McFadden 2001). The mixed logit probability is (Train 2003):

$$p_{ni} = \int \prod_{t=1}^{T} \left(\frac{e^{\beta x_{nit}}}{\sum_j e^{\beta x_{njt}}} \right) f(\beta) d\beta. \tag{11.3}$$

The mixed logit probability is a weighted average of the logit representation (Eqn 11.2) evaluated at different values of β, with weights given by the density $f(\beta)$. Coefficients can be specified as random in the mixed logit model, meaning that they can vary between respondents. The resulting model allows us to estimate the mean coefficients and standard deviations of the random parameters, which represent unobserved heterogeneity in preferences. Respondents answer three sequential choice cards (so that $T = 3$), which means that error terms are not independent. This panel data structure is represented by the time subscript t, and is explicitly modelled by estimating the same error for choices made by the same individual (Train 2003). Moreover, a normally distributed zero mean error component is added to the mixed logit model in order to allow for different variances of the insurance alternatives and the opt-out option, i.e. to accommodate heteroskedasticity (Scarpa et al. 2007). The parameters are estimated using maximum simulated likelihood with Monte Carlo integration using 1,000 Halton quasi-random draws (Bhat 2001). Individual specific parameters, conditional on choice for a random coefficient, can be estimated with the simulation of the following equation (Hensher et al. 2003; Train 2003):

$$\beta_n = \frac{\int_{\beta_n} \beta_n \left[\prod_{t=1}^{T} P(choice\ j \mid x_{nt}, \beta_n) g(\beta \mid \theta) \right] d\beta_n}{\int_{\beta_n} \left[\prod_{t=1}^{T} P(choice\ j \mid x_{nt}, \beta_n) g(\beta \mid \theta) \right] d\beta_n} \tag{11.4}$$

where $g(\beta|\theta)$ is the distribution of β in the population, and θ is its vector of parameters.

11.3.3 Coding of the explanatory variables

Appendix 11B describes all the variables and their descriptive statistics. Different methods of coding categorical variables were applied. Dummy variables are used for several categorical variables. Continuous variables are created from categorical variables that represent monetary classes, such as the value of the house and home contents (Blumenschein et al. 2008). A variable representing the total value of property is created by adding the home contents and house values. Ordinal qualitative variables, which are partitioned into J intervals, can be included using $J - 1$ dummies or can be transformed into values on the real axis using an approach proposed by Terza (1987).[4] Using Terza (1987), an ordered categorical variable (e.g. categories of risk perception such as 'small' or 'large') can be translated into an underlying numerical variable with a continuous scale (e.g. a latent variable of the degree of risk perception) (see also van Praag et al. 2003). An advantage of the dummy approach is that the interpretation of the coefficients is straightforward, but many variables are needed when J is large. In this case, the transformation of Terza (1987) can result in gains in efficiency and bias. The transformation (see Appendix 7B) has been applied to variables containing a large number of categories: namely, the perceived risk of suffering flood damage and risk-seeking attitude.

[4] These are characterized by a continuous unobservable ordinal latent index, where each interval is ranked (1 through J) in increasing order according to its supremum (Terza 1987).

11.4 Estimation results of the choice models for flood insurance demand

The choice experiment is unlabelled, which implies that there is no reason to expect a general preference for either of the two situations with flood insurance shown to respondents. This is supported by the data, since the two situations with insurance (A and B) were each chosen about 19 per cent of the time, while the 'opt out' or no insurance was chosen 62 per cent of the time. The choice experiment was followed by a question that asked how difficult it was for respondents to make a choice, with the answer options: 'very easy', 'easy', 'neutral', 'difficult' and 'very difficult'. Only 2.7 per cent of the respondents indicated that the choice experiment was 'very difficult' and 14 per cent indicated that it was 'difficult'. The last choice card included a dominant option to check understanding of the experiment by respondents. The dominant option was chosen by only 2 per cent of the respondents. On the basis of these answers and the pilot survey, it can be concluded that the experiment was not too difficult for the great majority of respondents, despite the inclusion of the probability attribute and the unfamiliarity of Dutch homeowners with buying flood insurance in practice.

11.4.1 Results of a model for insurance demand without heterogeneity

It is common practice in studies that evaluate insurance coverage or health risks to use a general utility function for the retained attributes, and not anchor the utility function in expected utility theory (e.g. Goldberg and Roosen 2007; Schneider and Zweifel 2004). A reason for this is that expected utility theory often provides a poor description of individual choices under risk (Camerer 1998). Common violations of expected utility theory relevant to the insurance application concerned are that individuals may ignore low probabilities or weight them in a non-linear fashion (Mason et al. 2005; Slovic et al. 1977; Schmeidler 1989; Tversky and Kahneman 1992). In addition, it is often found in insurance markets that individuals place a larger value on the level of coverage than is predicted by expected utility theory (Doherty and Eeckhoudt 1995). The following utility specification is used for the model that includes only the attributes of the experiment:

$$\begin{aligned} U_{Insurance} &= \beta_1 * low\ probability + \beta_2 * medium\ probability \\ &+ \beta_3 * high\ probability + \beta_4 * damage \\ &+ \beta_6 * coverage + \beta_7 * price \\ U_{No\ insurance} &= \beta_8 * constant. \end{aligned} \tag{11.5}$$

The utility of having insurance depends on the expected flood damage, the probability of flooding, insurance coverage and price.[5] The parameters of the attributes are the same under both scenarios because the experiment is unlabelled, i.e. there is no a priori reason to expect that the attributes have a different effect on the utility in the generic scenarios A or B. The utility of the option without insurance is modelled with a constant term. The three probability variables are dummy variables representing low (1/600), medium (1/400) and high (1/100) flood probabilities. The current flood probability (1/1,250) is excluded so that the coefficients β_1, β_2 and β_3 measure the effect relative to the current flood probability. Using dummies for the probability variable allows us to examine the non-linear effects of flood probability on the utility of insurance without restricting the functional form of this non-linearity. Subsequently, an adequate functional form for a continuous probability variable can be derived from the coefficient estimates of the dummy variables.

The two left-hand columns in Table 11.1 show the estimation results (Eqn 11.5). The pseudo-R^2 is 0.27, which indicates a good fit for this kind of model. The coefficients of all attributes are statistically significant at the 1% level, and of the expected sign. In particular, the utility of flood insurance increases with flood risk (probability and damage) and coverage level, while it decreases with price. The dummy variables of the flood probability indicate a monotonic and

[5] Individuals may value the insurance according to the monetary pay-off (*damage*coverage*), as appeared to be the case in a choice experiment for nuclear risk insurance by Schneider and Zweifel (2004). This is examined by including an interaction between the coverage and damage attribute in Eqn 11.5. The coefficient of this interaction term is non-significant (at the 10% level). Furthermore, non-linearity in the reaction to damage has been estimated by including *damage* squared as an explanatory variable in Eqn 11.5. The non-linear reaction is non-significant (at the 10% level). Coverage cannot be modelled in a non-linear fashion, as the choice experiment included no more than two levels of this attribute (75% and 100%) to limit the complexity of the choice for respondents.

Table 11.1 ***Results of logit models without heterogeneity***

	Logit model (Eqn 11.5)		Logit model (Eqn 11.6)	
Variable	Coefficient	*t* statistic	Coefficient	*t* statistic
Low probability of flooding	0.2835***	2.63	n.a.	n.a.
Medium probability of flooding	0.4270***	3.97	n.a.	n.a.
High probability of flooding	0.7750***	7.43	n.a.	n.a.
SQRT flood probability	n.a.	n.a.	9.5121***	7.50
Flood damage	0.0036***	3.31	0.0036***	3.33
Insurance coverage	0.0100***	3.56	0.0102***	3.62
Insurance premium	-0.0407***	-20.84	-0.0407***	-20.85
Constant	1.4028***	5.10	1.5638***	5.67
Number of observations	2,751		2,751	
Log likelihood	-2,217		-2,219	
Pseudo-R^2	0.27		0.27	

Notes: *, ** and *** indicate, respectively, significance at the 10%, 5% and 1% levels; n.a. stands for not applicable; SQRT = square root. Estimations were performed with NLOGIT software.

non-linear increase in utility when the probability rises. This relation is concave, which means that utility and WTP increase in a less than proportional way, with a decreasing slope in response to a probability increase.

The specification with the probability variables coded as dummy variables provided useful insights into the non-linear shape of the relation between utility of insurance and flood probability. A disadvantage of the dummy specification is that it is only possible to evaluate insurance demand for the flood probability levels captured by the dummies, that is, 1/600, 1/400 and 1/100, and not for the whole range of probabilities between 1/1,250 and 1/100. The flood probability is a continuous variable (between 0 and 1) and this property can be exploited by including it as a single variable in the utility specification. The non-linear relation between probability and utility of flood insurance observed in Eqn 11.5 can be approximated by specifying the utility function dependent on the square root of the probability.[6] This model can be written as:

$$U_{Insurance} = \beta_1 * (SQRT(probability)) + \beta_2 * damage + \beta_3 * coverage$$
$$+ \beta_4 * price$$
$$U_{No\ insurance} = \beta_5 * constant. \tag{11.6}$$

The results of this more parsimonious model (Eqn 11.6) are shown in the two right-hand columns of Table 11.1. Overall, results are very similar to the model (Eqn 11.5), apart from the relation with the flood probability that is now captured by a single variable instead of the three dummies.

11.4.2 *Results of a model for insurance demand with observed and unobserved heterogeneity*

Insights into individual heterogeneity in flood insurance demand are of interest to insurers for two main reasons. First, this provides information about what groups of customers insurers could target. Second, it is very useful to know how demand for flood insurance relates to risk characteristics of individuals in order to determine pricing strategies, i.e. premium differentiation, and to assess potential problems with adverse selection. Adverse selection could hamper the development of flood insurance markets if mainly high-risk individuals who live in unprotected areas are interested in purchasing insurance and insurers are unable to adequately distinguish low-risk from high-risk customers in order to charge high-risk customers a higher (risk-based) premium. Examining heterogeneity further provides relevant insights

[6] Non-linearity of continuous variables in choice experiments is often modelled by including a squared term of the variable in addition to its level. Estimation results do not change much if the square of the flood probability and the flood probability level are included, instead of the square root of the probability (Eqn 11.6) or the dummies (Eqn 11.5). The coefficient of the level is significant and positive, and the coefficient of the squared term is negative and significant. An unrealistic characteristic of such a specification in this application is that the utility of insurance declines if the probability increases to very large probabilities. Furthermore, the probability variable has been specified as the logarithm of the probability: overall, the results are rather similar. Specification 11.6 is preferred, since the results are similar to those obtained with specification 11.5.

into the risk characteristics of individuals faced with low-probability, high-impact climate risks.

It was examined how individual characteristics can explain heterogeneity in preferences for the attributes of the choice experiment by testing all possible interactions of all explanatory variables with the attributes of the situations with flood insurance (probability, damage, coverage and price). The final model specification includes only the significant interactions with the attributes of the experiment. Other explanatory variables are included in the utility specification of the alternative without insurance, so as to capture the utility difference between the insurance alternatives and the case without insurance. The resulting utility specifications are:

$$
\begin{aligned}
U_{Insurance} &= \beta_1 * SQRT(probability) + \beta_2 * damage \\
&+\beta_3 * damage * perceived\ flood\ probability \\
&+\beta_4 * coverage + \beta_5 * coverage * risk\ seeking\ attitude \\
&+\beta_6 * \mathrm{coverage} * age + \beta_7 * (coverage * close\ to\ main\ river) \\
&+\beta_8 * price + \beta_9 * (price * high\ income) \\
U_{No\ insurance} &= \beta_7 * constant + \beta_k * x_n.
\end{aligned}
\qquad (11.7)
$$

Here x_n represents variables which capture the availability of government relief, individual flood risk perceptions, experience with floods, insurance purchases, geographical characteristics and socio-economic characteristics. The behavioural economics literature indicates that individuals can react in very different ways to probabilistic information (e.g. Ida and Goto 2009). This is examined by specifying the coefficient of the probability variable as random, using the mixed logit model. A lognormal distribution is specified for the coefficient of the square root of the flood probability. This distribution is behaviourally plausible since it ensures that the coefficients are positive, meaning that the WTP for insurance increases for all individuals if the probability of suffering damage rises.[7] We also tested various random distributions for the parameters of the other attributes in the experiment, but these turned out to be insignificant, which means that these parameters can be specified as fixed in the final model. Elasticities and marginal effects (ME) are estimated, as they have a more intuitive interpretation

[7] The results are similar with normal and triangular distributions. The lognormal distribution has the best fit.

than coefficients (β) in choice models. Elasticities indicate the percentage change in the probability of choosing a particular alternative with respect to a 1 per cent change in the level of an attribute of that alternative or explanatory variable. The ME is the change in choice probability for an alternative, given a unit change in an explanatory variable. Both elasticities and MEs are computed for continuous variables using probability weighted sample enumeration (Louviere et al. 2000). For categorical variables, only MEs are computed, since elasticities are not defined here.

Table 11.2 provides the results of the logit and mixed logit estimations of Eqn 11.7. Elasticities are shown between brackets, and MEs without brackets. The standard deviation of the random coefficient is statistically significant. This indicates that individual preference heterogeneity exists in the coefficient of the probability attribute around the mean coefficient, and that the mixed logit model is preferred to the logit specification. The standard deviation of the error component is 4.2742 and statistically significant, which implies a considerably larger variance of the insurance alternatives than of the opt out. This finding is consistent with other studies showing that the variance of hypothetical alternatives is larger than that of the reference alternative, which has been called the 'status quo' effect (Hess and Rose 2009; Hu et al. 2009). The logit model can be rejected in favour of the mixed logit model based on the AIC (Akaike information criterion) and the Vuong test (p value <0.01). Therefore, the remainder of this chapter discusses only the results of the mixed logit model. The overall fit of this model is good, as the pseudo-R^2 statistic of 0.46 indicates.

The attributes of the choice experiment determine the utility of having insurance ($U_{Insurance}$ in Eqn 11.7), so that a positive coefficient indicates a positive relation between the attribute and the value placed on flood insurance. The coefficients of the attributes are statistically significant with the expected sign: the utility of insurance increases with flood probability (elasticity = 0.57), damage (elasticity = 0.38) and the coverage level (elasticity = 1.14), while it decreases with the insurance premium. Figure 11.1 shows the frequency of choices for flood insurance for flood probabilities between 1 in 1,000 and 1 in 100, computed using the parameters conditional on choice (Eqn 11.4), which is a concave relation. The price elasticity is −1.27 which is in line with the findings of other studies of actual flood insurance demand which observe elasticities of about −1 (Browne and Hoyt 2000; Kunreuther et al.

Table 11.2 *Coefficients, elasticities and marginal effects (ME) of the choice models*

	Logit model		Mixed logit model	
Variable	Coefficient	(Elasticity) or ME	Coefficient	(Elasticity) or ME
Attributes and interactions ($U_{Insurance}$)				
Flood probability (mean coefficient)	10.1472***	(0.39)	14.4870***	(0.57)
Standard deviation of coefficient of flood probability	n.a.	n.a.	1.4051***	n.a.
Flood damage	0.0037***	(0.21); 0.003	0.0069***	(0.38); 0.005
Flood damage*Perceived flood probability[a, b]	0.0031***	0.03	0.0050***	0.02
Insurance coverage	0.0124***	(0.56); 0.006	0.0312***	(1.14); 0.01
Insurance coverage*Close to major river[a]	0.0036***	0.06	0.0097**	0.05
Insurance coverage*Risk-seeking attitude[a, c]	−0.0024***	−0.04	−0.0076***	−0.04
Insurance coverage*Age[a]	−0.0001***	(−0.34); −0.008	−0.0004***	(−1.14); −0.03
Insurance premium	−0.0448***	(−0.76); −0.03	−0.0775***	(−1.27); −0.05
Insurance premium*High income[a]	0.0114***	0.09	0.0240***	0.07
Government compensation ($U_{No\ insurance}$)				
Government relief for damage is available	0.3707***	0.07	1.0086***	0.07
Risk perception and experience ($U_{No\ insurance}$)				
Climate change causes higher flood risk	−0.3570***	−0.07	−1.0656***	−0.07
Lower flood risk than average resident	0.4006***	0.08	1.2572***	0.08
Expected flood damage	−0.0009***	(−0.02); −0.0003	−0.0026***	(−0.06); −0.0009
Zero expected return period flood	1.2091***	0.20	3.4214***	0.20
Return period flood	0.0011	n.s.	0.0047**	(0.01); 0.002
Flooding is exogenous to human control	0.3333***	0.06	0.9543**	0.07
Experience of flooding and evacuation	−0.8026***	−0.16	−2.5239**	−0.18

Geographical characteristics ($U_{No\ insurance}$)				
Elevation of house relative to water level	0.0002**	(0.01); 0.00006	0.0005	n.s.
Area is not protected by dikes	−0.066	n.s.	0.1038	n.s.
Rural area	−0.7442***	−0.15	−1.8135**	−0.13
Socio-economic characteristics ($U_{No\ insurance}$)				
Women	0.3068***	0.06	0.8982**	0.06
Actual insurance purchases	−0.3283**	−0.06	−1.1038**	−0.08
Value of property	0.0010***	(0.13); 0.0003	0.0027**	(0.36); 0.0009
University degree	−0.2933***	−0.06	−0.7611	n.s.
Standard deviation error component	n.a.		4.2742***	
Constant	0.9164***		1.4556*	
Log likelihood	−2,059		−1,627	
AIC	1.5153		1.2027	
Pseudo-R^2	0.19		0.46	
Number of observations	2,751		2,751	

Notes: *, ** and *** indicate, respectively, significance at the 10%, 5% and 1% levels; n.a. stands for not applicable; n.s. stands for not significant; AIC = Akaike information criterion. Estimations were performed with NLOGIT software.

[a] The ME or elasticity provided for interaction variables ($x_1 * x_2$) corresponds to the second variable (x_2).

[b] The marginal effect (ME) of this ordered categorical variable (range 1–11) in the table is calculated as the change in the proportion of choices for the flood insurance alternative, while changing the value from the mode answer (2) to the subsequent higher category (3). The ME of the mixed logit model of a change from category 1 to 2 is 0.02, and the MEs of an increase in categories 3–11 are all 0.01.

[c] The ME of this ordered categorical variable (range 1–6) in the table is calculated as the change in the proportion of choices for flood insurance, while changing the value from the mode answer (4) to the subsequent higher category (5). The MEs of the mixed logit model of an increase in categories 1–4 of one category are, respectively, −0.04, −0.03 and −0.03, and the ME of a change from category 5 to 6 is −0.04.

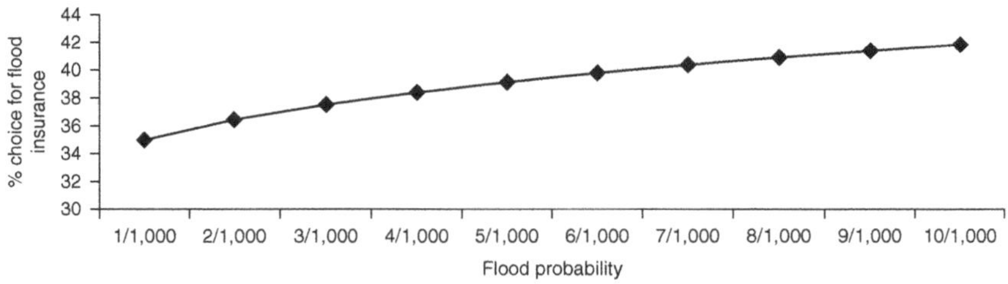

Figure 11.1 Effects of climate change on the choice of flood insurance

2011). The coefficient of the interaction variable between the insurance premium and the high-income category shows that high-income individuals worry less about the price and have a higher WTP for flood insurance, as can be expected from consumer theory. The coefficient of the interaction variable of flood damage and the perceived flood probability is positive and significant, which indicates that the positive relation between flood damage and the demand for flood insurance is stronger for individuals who have a high perception of flood probability. The coefficients of the interaction variables with insurance coverage indicate that flood insurance coverage is more valued by individuals who live close to a river, while it is less valued by individuals who are older and individuals who in general have risk-seeking attitudes. This variable of risk-seeking attitudes has been derived from the answers to a question in which respondents indicated how similar to themselves they regarded a person who avoids taking financial risks and prefers to be well insured.[8]

The other explanatory variables determine the utility of having no insurance ($U_{No\ insurance}$ in Eqn 11.7), so that a positive coefficient in Table 11.2 indicates a negative relation between the variable and the value placed on flood insurance. The socio-economic variables indicate that the probability of choosing to have flood insurance is lower for women, while it relates positively to actual (other) insurance purchases. Individuals who own properties of a higher value are more likely to self-insure and demand less flood insurance. More

[8] The question corresponds to a common format in psychological studies (e.g. Schwartz et al. 2001): 'Some people avoid financial risks as much as possible. They are also well insured. How similar are these people to you?' with answer options 'extremely similar to me', 'very similar to me', 'similar to me', 'a little bit like me', 'not similar to me' and 'not similar to me at all'.

importantly, Table 11.2 shows that the probability of choosing to have insurance is 0.07 lower if compensation for flood damage by the government is available, implying that government relief crowds out demand for private insurance.

Individual risk perceptions are important determinants in the choice of flood insurance, as has been found in other studies of decision making under risk (Viscusi 1989). In particular, the probability of choosing to have flood insurance is positively related to a respondent's perceptions that climate change increases flood risk, the expected flood damage and the expected flood return period; while demand is lower for individuals who believe that their flood risk is lower than for an average resident, and who expect that the return period of flooding equals 0. Individuals who indicate causes of flooding that are beyond their own or a water manager's direct control, such as extreme weather events or climate change, are less likely to buy insurance (see also Brilly and Polic 2005). Individuals with previous experience of flood threats place a much larger value on flood insurance coverage, as has been observed in other studies (Michel-Kerjan and Kousky 2010).

Flood risk in the Netherlands is highly dependent on the existing flood protection infrastructure of an area and the potential water level during flooding (Aerts et al. 2008b). Adverse selection could hamper the development of flood insurance markets if it is mainly high-risk individuals who live in unprotected or very low-lying flood-prone areas who are interested in purchasing insurance. The estimation results suggest that problems with adverse selection may be minor, since both a variable that represents respondents living in unprotected flood-prone areas and a variable that represents the potential water level at the respondent's home are statistically non-significant in explaining demand for flood insurance.

11.5 Market penetration and WTP for flood insurance under various climate change and socio-economic scenarios

11.5.1 Demand curves for flood insurance

Choice models can be used to predict consumer choices for a specific (product) alternative. A commonly applied method to predict aggregate demand or market shares (W) is 'sample enumeration', in which the predicted share of the sample choosing an alternative i is calculated using the following equation (Ben-Akiva and Lerman 1985):

$$W(i) = \frac{1}{N}\sum_{n=1}^{N} P(i \mid x_n) \qquad (11.8)$$

where N is the number of individuals in the sample, and $P(i|x_n)$ is the probability (Eqn 11.3) that an individual n chooses i with specific attribute values x_n. Using Eqn 11.8, market shares for flood insurance are estimated here and in Section 11.5.2 by using the mixed logit model to predict the choices for flood insurance under various levels of flood risk and other explanatory variables. These predictions are made for our sample according to the results of our utility specification of the choice alternatives (Eqn 11.7).

Potential market shares of flood insurance are approximated by predicting the percentage of choices for flood insurance in our sample under varying monthly premiums of between €5 and €80. Predicted probabilities are computed using specific premium and coverage levels, and values of the dummy that represent the availability of government compensation. Flood damage and probability are fixed at €70,000 and 1/1,250, respectively, which reflect current average expected flood damage and climate conditions, while the other explanatory variables take on their sample values. Figure 11.2 shows the estimated demand curves for flood insurance that reflect potential market penetration, and their 95% confidence intervals. The expected value of damage is about €5 per month under these conditions. The potential market share is approximately 43 per cent for flood insurance with 100 per cent coverage, which costs €5 per month in the current situation, and is about 40 per cent for flood insurance with 75 per cent coverage. These potential market shares could be 50 per cent and 46 per cent if the government did not provide compensation for flood damage. Market shares decline considerably if premiums are as high as €80 per month. Overall, these results suggest that potential market penetration could be sufficiently high for a flood insurance market, even though premiums would be considerably above the expected flood loss.

11.5.2 Market shares under various climate change and socio-economic scenarios

Market shares are now estimated under a range of climate change and socio-economic scenarios concerned with the availability of

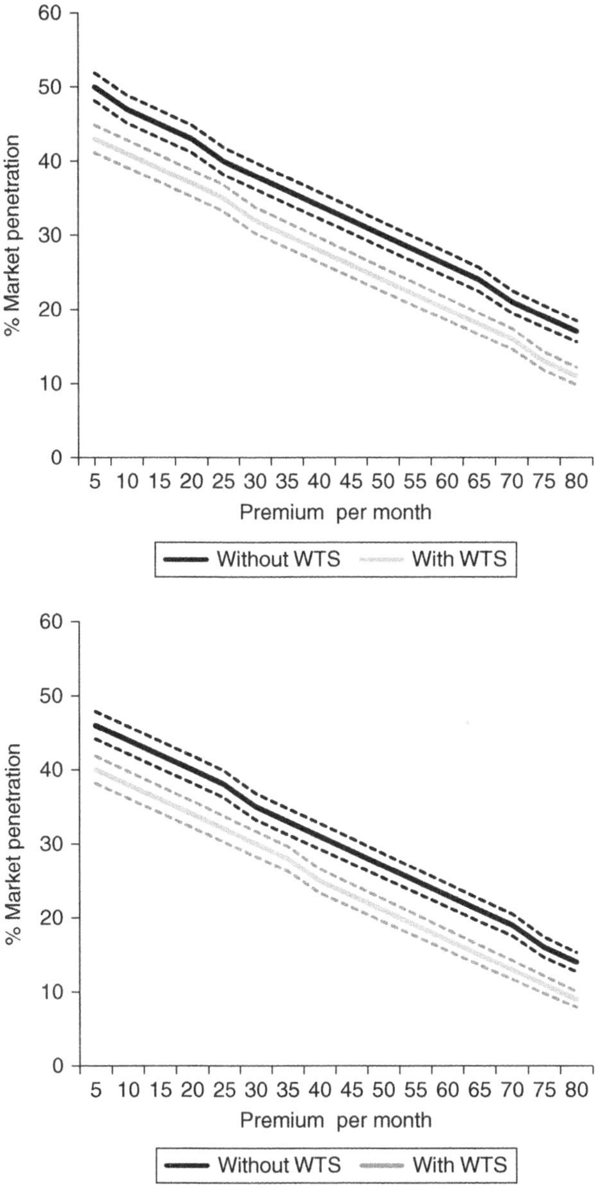

Figure 11.2 Demand curves for insurance products under 100 per cent coverage (top), 75 per cent coverage (bottom), and with or without government compensation for damage (with/without WTS)

Note: WTS stands for availability of government relief via the 'Calamities and Compensation Act'; and the dotted curves represent 95% confidence intervals.

government relief and expected flood damage. This is done for specific levels of flood damage and probability attributes in the experiment, while coverage is 100 per cent. It is likely that insurers will adjust their premiums in response to changes in flood risk. As a result, the model includes two opposing effects of climate change on the demand for flood insurance: namely, demand increases due to increased risk, and demand decreases because of higher insurance premiums. In these forecasts it is assumed that premiums are set according to risk, i.e. probability multiplied by damage, plus a surcharge for operational expenses of 35 per cent of the total premium. This surcharge is derived from an estimate of operational costs relative to the premium revenues of the National Flood Insurance Program (NFIP) in the USA (Kunreuther et al. 2011). This may be a useful approximation of such costs for a public–private flood insurance programme as has been proposed for the Netherlands, since the NFIP operates like a public–private partnership.

The results provide three main insights (Table 11.3).

1. Expected market penetration is about 40 per cent under current conditions and moderate increases in flood risk, while it might rise up to almost 50 per cent if government relief is not available.
2. The effects of higher risk and prices balance out for changes in flood probabilities of up to 1/600 if flood damage can be mitigated to €40,000, while the effect of higher premiums dominates if flood probabilities rise and expected flood damage remains at €70,000 or increases to €120,000.
3. The market will collapse if climate change results in an extreme rise in flood risk (1/100), because of the large expected rise in insurance premiums in the scenarios of middle and high flood damage, but not if expected flood damage is only €40,000. Implementing measures that reduce flood damage may be an effective means of preventing such a collapse of insurance demand because of its ameliorating influence on premiums.

11.5.3 Willingness-to-pay estimates for flood insurance using the mixed logit model

The estimation results of the choice model can be used to assess changes in an individual's utility of having flood insurance, characterized by

Table 11.3 ***Potential market penetration (as a percentage) for flood insurance under climate change and socio-economic scenarios with premiums adjusted to reflect changes in risk***

Socio-economic scenarios		Climate change scenarios			
Government relief available	Expected flood damage	Current climate 1 in 1,250	Small change 1 in 600	Medium-large change 1 in 400	Extreme change 1 in 100
No	€40,000	48	48	47	35
No	€70,000	48	46	44	22
No	€120,000	48	44	39	5
Yes	€40,000	42	42	41	30
Yes	€70,000	42	40	38	16
Yes	€120,000	42	38	33	4

specific levels of the attributes or explanatory variables, compared with having no insurance. These changes in utility can be translated into changes in consumer surplus (CS), according to:

$$\Delta E(CS_n) = \frac{1}{\alpha_n}\left[\ln\left(\sum_{j=1}^{J1} e^{V_{nj}^1}\right) - \ln\left(\sum_{j=1}^{J0} e^{V_{nj}^0}\right)\right] \tag{11.9}$$

where α_n is the marginal utility of income, which equals the negative of the coefficient of the insurance premium; and V_{nj}^1 and V_{nj}^0 represent utility after and before the change, which has been estimated using the flood probability parameters conditional on choice (Eqn 11.4). The average improvement in consumer surplus in a specific situation with flood insurance compared with the situation without flood insurance gives an estimate of average WTP for flood insurance.

Table 11.4 shows the WTP for flood insurance with high (100 per cent) or low (75 per cent) coverage in situations with varying flood probabilities, damage and availability of government relief. The mean WTP of the entire sample is computed, which is relevant as a social welfare measure of introducing flood insurance. In addition, the mean conditional willingness-to-pay (CWTP) is estimated as the average WTP for those individuals who are interested in buying flood insurance, which is an indicator of potential revenue per policy for insurance companies.[9] The risk premium is given between brackets below the WTP and the CWTP estimates, which is the difference between the willingness-to-pay measure and the expected value of the flood damage per insurance policy (*probability*damage*fraction covered*). The risk premium of CWTP can be interpreted as an indicator of the average potential profit of an insurance policy. The risk premium of the WTP can be interpreted as the average net benefit of introducing flood insurance for the population, since it represents the difference between the average improvement in welfare of introducing insurance and the average expected value of the payouts of claims.

The WTP measures in Table 11.4 indicate that there is a potentially profitable flood insurance market in the current situation. In particular, the risk premiums of CWTP are always positive with the current safety standard of flooding. Furthermore, average welfare improves for the entire population in the river delta in the current situation by more than the expected value of the flood damage covered. This is the case because the average consumer surplus or the mean WTP is larger than the expected payouts for damage, as reflected by positive risk premiums. WTP and CWTP values are about 10 per cent less if coverage is low in most scenarios compared with complete coverage. WTP and CWTP values are about 20 per cent less if damage relief from the government is available. Finally, the results in Table 11.4 suggest that offering flood insurance may remain profitable if flood probabilities rise moderately. Risk premiums of CWTP are still positive if climate change increases flood probabilities to 1 in 600 and 1 in 400. However, most risk premiums of WTP are negative if climate change were to result in an extreme rise of flood probabilities to 1 in 100, since the WTP increases less than the expected value of damage. In this scenario the risk premiums of CWTP are negative if expected flood damage is large (€120,000).

11.6 Policy implications and conclusions

This chapter presents a comprehensive study of demand for flood insurance using a stated preference survey with a choice experiment, and current state-of-the-art mixed logit estimation. Even though the focal point of this study is on demand for flood insurance in the

[9] The CWTP is computed as the WTP for individuals who choose flood insurance at least once in the choice experiment, which is the case for 52% of the sample.

Table 11.4 ***Mean willingness-to-pay (WTP), conditional willingness-to-pay (CWTP) and risk premiums for flood insurance with high or low coverage levels, under varying conditions of flood risk and availability of government relief (in € per year)***

Insurance coverage and socio-economic scenarios			Flood probabilities under climate change scenarios							
			Current climate 1 in 1,250		Small change 1 in 600		Medium-large change 1 in 400		Extreme change 1 in 100	
Coverage in percent	Government relief	Expected flood damage	WTP	CWTP	WTP	CWTP	WTP	CWTP	WTP	CWTP
100	No	€40,000	313	595	338	643	358	679	467	885
			(281)	(563)	(272)	(576)	(258)	(579)	(67)	(485)
100	No	€70,000	331	628	357	677	376	714	487	921
			(275)	(572)	(240)	(560)	(201)	(539)	(-213)	(221)
100	No	€120,000	363	686	390	736	410	774	522	983
			(267)	(590)	(190)	(536)	(110)	(474)	(−678)	(−217)
75	No	€40,000	275	525	300	572	319	607	425	808
			(251)	(501)	(250)	(522)	(244)	(532)	(125)	(508)
75	No	€70,000	293	557	318	605	336	640	444	843
			(251)	(515)	(231)	(518)	(205)	(509)	(−81)	(318)
75	No	€120,000	323	613	348	662	368	698	477	897
			(251)	(541)	(198)	(512)	(143)	(473)	(−423)	(−3)
100	Yes	€40,000	241	460	264	505	282	538	384	733
			(209)	(428)	(197)	(438)	(182)	(438)	(−16)	(333)

Table 11.4 (*cont.*)

Insurance coverage and socio-economic scenarios			Flood probabilities under climate change scenarios							
			Current climate 1 in 1,250		Small change 1 in 600		Medium-large change 1 in 400		Extreme change 1 in 100	
Coverage in percent	Government relief	Expected flood damage	WTP	CWTP	WTP	CWTP	WTP	CWTP	WTP	CWTP
100	Yes	€70,000	257	490	281	536	299	570	403	768
			(201)	(434)	(164)	(419)	(124)	(395)	(-297)	(68)
100	Yes	€120,000	285	544	310	590	328	625	435	826
			(189)	(448)	(110)	(390)	(28)	(325)	(-765)	(-374)
75	Yes	€40,000	208	399	231	441	248	474	347	662
			(184)	(375)	(181)	(391)	(173)	(399)	(47)	(362)
75	Yes	€70,000	223	427	246	471	263	503	364	695
			(181)	(385)	(159)	(384)	(132)	(372)	(-161)	(170)
75	Yes	€120,000	250	477	273	522	291	556	394	752
			(178)	(405)	(123)	(372)	(66)	(331)	(-506)	(-148)

Note: WTP and CWTP are shown without brackets, and the corresponding risk premiums are given within brackets below these values.

Netherlands, its findings may be more broadly applicable to individual behaviour with respect to other low-probability, high-impact risks. The insights are of interest to the insurance industry in assessing the current and future profitability of offering insurance against extreme weather events in a changing climate. Climate change may increase the risks of natural disasters in several regions of the world, which could result in a substantial increase in premiums for insurance arrangements that cover such risks. An important question that has been examined in this chapter is whether consumers would be willing to pay premiums that reflect increased risk and how market shares of flood insurance may develop in the face of climate change.

The results of this survey provide four main insights.

1. Offering flood insurance may be profitable in the Netherlands in the current situation.
2. The availability of government compensation reduces demand for flood insurance.
3. Problems with adverse selection seem to be minor.
4. The effect of climate change is a non-linear increase in the proportion of homeowners who take out insurance and in the willingness-to-pay (WTP) for flood insurance. Our results show that WTP increases less than the risk-based insurance premium if climate change were to cause a large increase in the flood probability, which might result in a considerable decline in the market penetration of flood insurance.

With respect to the first insight, it should be noted that under current climate conditions a considerable proportion of homeowners would be willing to pay for flood insurance, and that average WTP exceeds the actuarially fair premium.

With regard to the second insight, we find evidence that the 'charity hazard' should be avoided; that is, flood insurance demand is expected to increase if the government credibly refrains from compensating for flood damage.

In relation to the third insight, the role of perceptions of flood risk in the demand for flood insurance seems to be more important than actual measures of risk that are based on geographical characteristics. An important measure of flood risk is whether flood-prone areas are protected by dikes. The results indicate that this variable, which represents individuals who live in unprotected flood-prone areas, is insignificant, suggesting there is little need to worry about adverse selection. Our findings indicate that demand is sufficiently high in the protected areas in the Netherlands, which is important for the purpose of risk-pooling.

With respect to the fourth insight, offering flood insurance remains profitable if climate change results in a moderate increase in flood probabilities, but not if there is an extreme rise in flood probabilities. The choice model shows that demand for flood insurance decreases considerably if climate change results in large increases in flood probabilities and insurers adjust premiums in accordance with risk. The reason for this is that household WTP for flood insurance increases by less than the rise in the expected value of the loss. This could have profound implications for insurance markets in the future if the expected effects of climate change on flood risk were to materialize. The market penetration of flood insurance could decrease if insurers need to substantially revise premiums upwards to reflect increased risk. A decreased market penetration of flood insurance may impair the ability of insurers to pool risks between a large number of policyholders, and could disrupt the function of insurance to spread the societal costs of flooding.

Appendix 11A Overview of attributes and levels of the choice experiment

Attributes	Levels
Flood probability	Once per 1,250 years
	Once per 600 years
	Once per 400 years
	Once per 100 years
Damage to home contents and house	€40,000
	€70,000
	€120,000
Insurance coverage	High (100 per cent)
	Low (75 per cent)
Insurance premium	€10 per month
	€20 per month
	€35 per month
	€55 per month
	€80 per month

Figure 11.A1 Attributes and levels used in the choice experiment

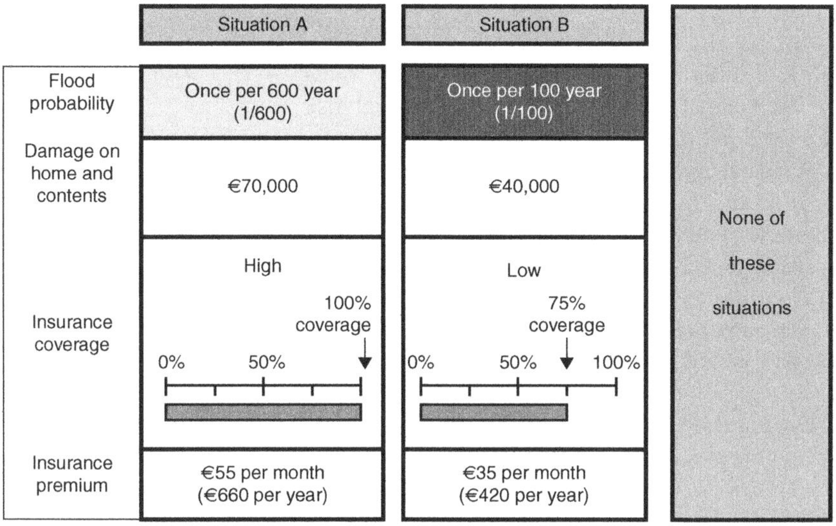

Figure 11.A2 Example of a choice card

Appendix 11B Explanatory variables and descriptive statistics

Table 11.B1 ***Summary overview of the variables used in the statistical analysis***

Flood probability	Attribute choice experiment, probability of flooding
Flood damage	Attribute choice experiment, total flood damage, in €1,000
Insurance coverage	Attribute choice experiment, coverage of flood insurance in percentage of total damage
Close to major river	Dummy variable, 1 = the respondent's zip code area is within 10 kilometres of the nearest major river
Insurance premium	Attribute choice experiment, monthly insurance premium in €
High income	Dummy variable, 1 = monthly household after-tax income is within the highest category >€4,000
Government relief for damage is available	Dummy variable, 1 = questionnaire version stating that respondents may receive partial relief for flood damage by the government
Climate change causes higher flood risk	Dummy variable, 1 = respondent expects that climate change causes higher flood risks

Table 11.B1 (*cont.*)

Risk of suffering flood damage	Categorical variable (range 1–11), 1 = no risk and 11 = extremely high risk
Lower flood risk than average resident	Dummy variable, 1 = respondent believes that his/her flood risk is lower than that of an average resident in the Netherlands
Expected flood damage	Continuous variable, damage a respondent expects if a flood occurs, in €1,000
Zero expected return period flood	Dummy variable, 1 = respondent expects return period to be zero
Return period flood	Continuous variable, flood return period that the respondent expects in 1,000 years
Flooding is exogenous to human control	Dummy variable, 1 = respondent mentions only climate or natural conditions and not human activity or water management as causes for flooding
Experience of floods and evacuation	Dummy variable, 1 = respondent has experienced a flood and has been evacuated in the past
Actual insurance purchases[a]	Dummy variable, 1 = respondent has purchased more than eight of the following insurances: health insurance with no deductible, home, home contents, dentist, continuous travel, all-risk car, life, bike, legal assistance, disability insurance
Risk-seeking attitude	Categorical variable (range 1–5), 1 = very risk-averse, 5 = not risk-averse at all
Elevation of house relative to water level[b]	Continuous variable, elevation of the living area relative to the potential water level of a flood in centimetres
Area is unprotected by dikes	Dummy variable, 1 = the area is not protected by dikes
Rural area	Dummy variable, 1 = the area is a rural area
Age	Continuous variable, age in years
Women	Dummy variable, 1 = respondent is a woman
Value of property[c]	Continuous variable, total market value of the respondent's house and its contents, in €1,000
University degree	Dummy variable, 1 = highest degree is university education

Notes: [a] Separate variables of actual insurance purchases cannot be included in the model because of multi-collinearity. These correlations arise because risk-averse individuals are more likely to purchase many insurance policies. Another advantage of including insurance purchases as a single variable is that this saves on degrees of freedom.

Notes to Table 11.B1 (*Cont.*)

[b] Adjustments have been made for respondents who live in flats on the first floor by adding 2.5 metres to the height of the area.

[c] For housing value, the respondent could mark one of the following categories: <€100,000; €100,000–150,000; €150,000–200,000; €200,000–250,000; €250,000–300,000; €300,000–350,000; €350,000–400,000; €400,000–500,000; €500,000–600,000; >€600,000. For home contents value, the respondent could mark one of the following categories: <€25,000; €25,000–50,000; €50,000–75,000; €75,000–100,000; €100,000–125,000; €125,000–150,000; €150,000–175,000; €175,000–200,000; €200,000–300,000; >€300,000. Continuous values of housing and home contents variables were constructed by setting the housing and home contents value of each respondent to the midpoint of the interval (€650,000 and €350,000 were used for, respectively, the highest housing and home contents value categories).

Table 11.B2 *Descriptive statistics for the explanatory variables*

	Number of observations	Mean	Standard deviation
Close to major river	982	0.78	0.41
High income	982	0.15	0.36
Climate change causes higher flood risk	982	0.66	0.47
Risk of suffering flood damage[a]	982	4.19	2.20
Lower flood risk than average resident	982	0.48	0.50
Expected flood damage	981	69.9	186.56
Zero expected return period flood	982	0.04	0.19
Return period flood	982	229.58	4,665.53
Flooding is exogenous to human control	982	0.40	0.49
Experience with floods and evacuation	982	0.03	0.17
Actual insurance purchases	982	0.12	0.33
Risk-seeking attitude[a]	982	3.24	1.16
Elevation of house relative to water level	982	-133.93	625.27
Area is unprotected by dikes	982	0.11	0.32
Rural area	982	0.05	0.22
Age	982	45.50	12.02
Women	982	0.42	0.49
Value of assets	982	410.29	168.94
University degree	982	0.21	0.40

Notes: [a] The statistics of this variable are in accordance with the coding in Table 11.B1. This original coding has been transformed for the analysis according to Terza (1987), as described in Appendix 7B.

12 Conclusions

12.1 Introduction

Around the world, insured and other economic losses from natural disasters have been increasing sharply during the last few decades. Several recent destructive disasters highlight the vulnerability of modern societies to destructive forces of nature, such as the 2011 floods in Australia and Pakistan, and the earthquake and tsunami in Japan which resulted in problems with the nuclear power plant at Fukushima. These events may be signalling more disasters in the future. It is projected that natural disaster risks will increase in several regions in the world as a result of climate change and socio-economic developments. This book provides a comprehensive study of the management of natural disaster risks through insurance, with a particular focus on the expected change in risks from climate change. This is done by examining the contribution that insurance can make in the management of climate change risks, the potential impacts of climate change on the insurance sector and insurers' responses to climate change, the design of comprehensive climate change adaptation strategies including insurance arrangements for natural disaster risks, and individual behaviour towards climate change related risks and demand for natural disaster insurance in a changing climate. Several main findings on each of these issues are discussed in this chapter.

12.2 Managing risks in a changing climate

The recent upswing in natural disasters and resulting damage illustrates the vulnerability of human societies to extreme weather events, such as storms, floods and droughts. The more frequent occurrence of destructive natural disasters is in line with the projections made by climate modellers, who predict that anthropogenic global warming may increase the frequency and severity of extreme weather events in many regions. Considerable research efforts have been devoted to examining whether or not trends in past damages have been influenced by climate change or socio-economic influences on risks. Chapter 2 provides an extensive review of these studies, from which it can be concluded that socio-economic developments have been the main reason behind the rapid increase in total amount of damage that has been observed, worldwide. Nevertheless, climate change projections indicate that in the future more extreme weather may further augment upward trends in natural disaster losses in several regions of the world. Moreover, socio-economic developments are likely to increase the number of people and the value of assets at risk from disasters, while an increased interconnectedness of global economies has the effect that adverse impacts of disasters are not confined to disaster-prone regions but can have ripple effects to other economies.

The projections of increased natural disaster risks require innovative adaptation policies to prevent natural disaster damage or to reduce the impacts of disasters on our societies. Regional projections of climate and socio-economic change and the modelling of future changes in natural disaster risks are needed to steer such adaptation and risk management strategies. The rapid development of catastrophe models for assessing risk and for estimating climate change impacts is promising in this respect. Indeed, the literature on the potential effects of climate change on region-specific natural disaster risks has been expanding during recent years. An understanding of householders' perceptions of risk and their behaviour is also important in this respect. Individual perceptions of risk may deviate from expert assessments and influence the risk management of households, such as investments in damage mitigation measures or insurance purchases, which may form a key component of an overall risk management and climate change adaptation strategy. Chapter 1 briefly shows that a combination of well-diversified risk management strategies, investments in damage mitigation measures by households, prevention undertaken by the public sector, and adequate arrangements to compensate victims of disaster damage is likely to result in enhanced economic resilience to natural hazards.

The experience of the insurance sector in assessing, managing and spreading risks may be useful in fostering the adaptation of modern societies to climate change. Well-designed financial compensation arrangements can speed up the recovery process after natural disasters have struck, and provide compensation to the victims of disaster damage, which ameliorates the impacts of more frequent natural disasters that are expected to arise due to climate change. Moreover, insurance with risk-based premiums can act as a price signal of risks, which provides an economic incentive for making more informed trade-offs about the benefits and risks of locating housing and industry in hazard-prone areas. Moreover, insurance may provide financial incentives that encourage policyholders to implement measures that limit damage. It is self-evident that insuring for climate change risks will not completely solve the adverse economic impacts of a higher frequency and severity of natural disasters in the future, but it may ameliorate them. In addition, sensible public policies are needed that take a long-term perspective on the management of natural disaster risk, including land-use planning and building code policies, and investments in disaster prevention infrastructure. This book explores the complementary role that insurance arrangements can play in designing comprehensive climate change adaptation policies that comprise risk prevention and reduction and efficient risk-sharing strategies.

12.3 Climate change impacts on insurers and insurers' responses to climate change

Apart from the useful societal function that insurers may fulfil in spreading, and limiting, risks from climate change, the insurers will themselves be faced with an increase in natural disaster risks because of climate change. This is the topic of Chapters 2 and 3. Insurers provide coverage against a large proportion of weather-related risks, although the exposure of insurers to natural disasters varies considerably among different insurers and between countries. This is apparent from analyses in Chapter 2 about the potential impacts of climate change on the insurance sectors in Canada, the UK, the USA and the Netherlands. The relevance of weather risks for the insurance sector becomes evident when observing past trends in insured and other economic natural disaster losses. Mainly as a result of increased growth of population and property values in hazard-prone areas, insured natural disaster losses have increased substantially during recent decades. Even though insurance is a short-term business with mostly annual insurance policies of which the policy terms can be adjusted over time, the influence on risk of long-term developments, such as climate and socio-economic changes, can pose considerable challenges for insurers. The growing exposures to natural disasters and concentrations of risks due to socio-economic developments increase the probability of extreme damages and drive up insurers' capital requirements. Climate change can result in more extreme insured damages, which complicates the insurability of low-probability, high-impact events that are already costly to insure under the present climatic conditions. An increase in extreme risks requires the insurance company to hold larger reserves, to purchase more reinsurance, or to hedge risks on capital markets: all of these strategies to increase the capacity to sustain losses are costly. Moreover, the effects of climate change, such as sea level rise or increased regional hurricane activity, have implications for several types of natural disaster risks in many regions of the world, which can increase the dependence of natural disaster risks between regions and insurers' lines of business.

Chapter 3 shows, for a variety of hazards, that climate change can result in a large increase in natural disaster damage. However, the current state of climate research is still unable to reach firm conclusions concerning the possible influence of global warming on small-scale severe weather phenomena such as hailstorms. Chapter 3 studies this in detail by reviewing international studies on this topic and by estimating the effect of global warming on hailstorm damage in the Netherlands by extrapolating the historical relations between data on insured hailstorm damage to the agricultural sector and indicators of temperature and precipitation. According to the best prediction model, annual hailstorm damage to outdoor farming may rise between 25 per cent and 48 per cent, while hailstorm damage to greenhouse horticulture in summer is expected to increase between 116 per cent and 219 per cent depending on the scenario of climate change. These results are in line with several other studies showing that climate change impacts on small-scale weather phenomena may result in a considerable increase in risks for insurance companies. Moreover, this chapter provides a review of 27 studies that have estimated the potential impacts of climate change on (insured) natural disaster damage for different kinds of storm and flood hazards in a variety of countries. These studies show that both climate change and socio-economic change are expected to have a non-negligible impact on future natural disaster damage for most of the hazards and countries considered. It is difficult to draw conclusions on whether the influences of climate or socio-economic change on risk are likely to dominate, because their relative effects on risk differ considerably by hazard and by region. The large differences between some projected minimum and maximum changes in damage indicate that there is great uncertainty associated with these projections of future damage. These uncertainties appear to be especially large for projections of tropical storm damage and flood damage, and seem to be slightly smaller for extratropical storm damage.

Insurers should assess how risks have changed and should account for changes in natural disaster risk as a result of climate change in setting premiums and in their financial risk management, such as reinsurance purchases. In this respect, a distinction should be made between the persistent effects of a changed climate on risk and short-term natural variability in weather which affects natural disaster losses. However, disentangling the effects of climate change and natural weather variability on risks is challenging, given the large uncertainties associated with the effects of climate change on natural hazards and considerable short-term weather variability. An increased understanding is required both of the processes by which the climate system and extreme weather changes and of the timing of changes in risks. Chapter 2 presented several studies on the impacts of climate change on insurance in various countries. These studies show that a broad range of insurance arrangements can be identified that will be affected by climate change. Although some effects of climate change on insurance may be positive, the majority of the reported effects are negative and indicate increased risks for insurers. However, climate change can also bring new business opportunities for the insurance sector. In particular, insurers, in collaboration with the public sector, could explore the development of new insurance products for weather-related risks that are currently not yet covered by insurance, such as flooding in the Netherlands. The increasing weather-related risks of insurers require the implementation of risk management strategies to handle these risks and the undertaking of strategies to reduce insurers' vulnerability. This could involve traditional measures, such as limiting risks, raising premiums to reflect changed risks, transferring damage or damage control. Problems with reduced insurance availability are more likely to arise for risks that are already difficult to insure privately, such as drought risk and flood risk in certain regions. Other strategies that reduce risk are also beneficial for insurance customers in the sense that they limit climate change risks and premiums of the associated insurance products and do not impair insurance availability. These strategies involve the promotion of emission reduction and adaptation measures in cooperation with public agencies. In addition, insurance companies could stimulate the undertaking of cost-effective risk-reducing adaptation measures at the policyholder level. International responses of insurers to climate change show that many insurance companies have indeed been very active in this respect, but that much more can be done.

The issue of climate change and insurability of risks is different in developing countries. Climate change impacts in the developing world are expected to be large because these countries have fewer resources to spend on effective adaptation policies. Insurance penetration is very low in many developing countries, which implies that most natural disaster losses are not covered by insurance. Increasing the coverage of natural disaster insurance could provide an important safety net for the poor and reduce their reliance on uncertain assistance from their social environment, governments or international donors after a major disaster has occurred. The lack of affordability of insurance is a major obstacle for increasing insurance coverage in the developing world. A solution is to develop innovative micro-insurance schemes and/or index-based insurance with support from governments or international funding, in order to facilitate the provision of insurance at low cost. These schemes should be embedded in broader programmes in which public investments in risk reduction and insurance incentives for risk reduction by policyholders are combined, with the objective of limiting the projected risks posed by climate change.

12.4 Insurance arrangements for natural disaster risks and a comprehensive climate change adaptation strategy

The projected increases in the future risks posed by natural hazards can be limited, and the impacts of natural disasters on societies can be ameliorated, by undertaking climate change adaptation measures. Insurance arrangements can play an important function in an adaptation strategy by providing incentives to limit natural disaster risk and by spreading these risks so that victims of natural disasters receive compensation for damage. Chapter 4 studies the potential benefits and challenges associated with insurance against natural disaster risk, in particular for the case of flooding. As an exemplary case, Chapter 4 examines opportunities for introducing flood insurance in the Netherlands, which is currently not available. In the Netherlands, flood damage may be compensated for by the Dutch government by an *ex post* public compensation scheme for which no funds have been established nor premiums charged. Moreover, this scheme lacks clear and transparent rules about which cases of flood damage will be compensated and how much compensation will be provided. The Netherlands is an interesting case study area where floods have a very low probability of occurring, due to the high safety standards of flood protection infrastructure, but at the same time have very high potential damage because of a large concentration of economic values in low-lying areas. Many of the issues discussed are also of relevance for flood insurance markets in other countries, as is apparent from the international comparison in Chapter 4 of flood insurance arrangements that exist in the USA, the UK, Germany and France.

Worldwide, there is a large gap in natural disaster insurance coverage and a considerable growth in these markets is necessary if insurance is to be used as an instrument to spread climate change induced natural disaster risks. In general, the ability of insurance arrangements to efficiently spread and segregate risks, reduce damage by providing loss-reducing incentives, and monitor and control policyholders could be a rationale for extending private flood insurance coverage with risk-based premiums. However, this is not an easy task, as Chapter 4 reveals that there are several challenges associated with providing insurance against natural disaster risks. Natural disaster risks often have a low probability of occurring, which implies that few observations and data about damage exist to serve as a basis for determining risk-based premiums. The uncertainty associated with assessing natural disaster risks can be partly overcome by catastrophe models, which can be used for assessing risk and estimating insurance premiums. Nevertheless, uncertainties cannot be resolved completely by modelling, and large uncertainties, especially of extreme risks, will often result in higher premiums than for insurance against risks that can be more accurately estimated. Moreover, adverse selection and moral hazard problems can impede the insurability of natural hazards. Adverse selection may arise when individuals who face a high natural disaster risk are more likely to purchase insurance than individuals who face a low risk, while insurers have difficulty with distinguishing high- from low-risk individuals in order to charge the former a higher premium. In practice, this problem can be overcome by an adequate screening of policyholders by the insurer and by implementing premium differentiation according to risk classes. Moreover, this book shows that individuals, in general, have difficulty in assessing low-probability and high-impact climate change-related risks (see Chapter 9) and that, for example, individuals with a high flood risk may insufficiently realize this and therefore do not have a high enough demand for flood insurance (Chapters 10 and 11). This implies that problems with adverse selection may be minor, in practice. Moral hazard occurs when insured individuals behave less carefully when they have insurance coverage, which increases risk for the insurer, while the insurer may not realize this and is therefore unable to charge a premium that reflects the higher risk. In practice, the problem of moral hazard can be overcome by an adequate monitoring of the behaviour of the insured and by making the policyholder pay for part of the damage, via deductibles or co-insurance, if a disaster occurs.

A more fundamental challenge in providing affordable natural disaster insurance is the correlated nature of natural disaster risks and the potential for catastrophic damage, which can result in huge numbers of large claims following a disaster. These characteristics of natural disaster risks imply that insurers need to hold large reserves, purchase large amounts of reinsurance, or attract capital directly from capital markets by hedging risks with alternative risk transfer (ART) instruments, which all come at a cost. A solution to this problem, which Chapter 4 proposes for insuring flood risk in the Netherlands, is the introduction of a multilayered insurance programme that operates like a public–private partnership. In such a programme, the first layer of small losses is paid by the policyholders themselves, via deductibles or co-insurance arrangements, in order to provide incentives to policyholders to act prudently and to prevent damage during floods, which attenuates the moral hazard problem. Insurers cover the layer of middle-sized losses and charge risk-based premiums that act as a price signal of risk. The government covers the layer of very large losses, for example through a public reinsurance scheme. The advantage of keeping the government responsible for extreme damage is that this reduces capital and reinsurance costs for insurers and reduces their uncertainty in assessing these extreme risks, which has the effect of keeping the insurance affordable. Such public–private partnerships may be useful to investigate for covering natural disaster risks in other countries. Public–private partnerships can be an efficient mechanism for spreading climate change risks if they make optimal use of the expertise of the insurance sector to assess, cover and provide incentives to limit these risks. Governments who have the ability to spread risks intertemporally by borrowing are well suited to covering the extreme portion of risks, which provides a solution for the insurability of correlated and catastrophic natural disaster risks at affordable premiums.

Insurance arrangements for natural disasters should be combined with broader adaptation programmes that limit the risks posed by climate change. Given the uncertainty associated with climate change risks, it is advisable to seek solutions in diversified strategies for managing these risks. Chapter 5 illustrates this by presenting modern portfolio theory (MPT) as a framework for evaluating portfolios of flood management options in terms of their return (avoided damage) and risk (variance in this return). The application of MPT to flood risk management was illustrated for a case study area in the Netherlands, but obviously this framework has a broader relevance and could be applied to other countries and natural hazards. The application indicates that diversification of different water management investments, such as public investments in flood protection infrastructure and implementing measures at the household level that mitigate damage, has the potential to increase the expected returns of water management investments and/or reduce risks. For example, combining investment in the strengthening of dikes with flood-proofing of houses results in lower variances in return (risk) compared with only investing in dikes, which means that the probability of suffering catastrophic damage decreases. It appears that combining risk prevention measures (e.g. dike infrastructure) with damage mitigation measures (e.g. elevating buildings) can be especially attractive because these investments have the lowest correlation of returns. The results indicate that diversification of water management investments is necessary in the face of climate change if an increase in risk is regarded as undesirable. These findings suggest that the implementation of a diversified portfolio of adaptation measures that are able to limit risk in different flooding conditions is likely to be an effective strategy to increase resilience to climate change.

Chapter 6 reviews international studies on the effectiveness of measures taken at the household level that mitigate flood damage. The results of these studies suggest that substantial reductions in flood damage can be achieved through mitigation. However, a challenge is how to motivate households to invest in such measures, since experience in several countries shows that individuals rarely voluntarily invest in mitigation of natural disaster risks. Several reasons for this have been put forward, such as underestimation of risk by households, a lack of information on mitigation strategies, and high perceptions of the cost of mitigation compared with the potential benefits in terms of risk reduction. A solution may be to design measures that completely eliminate natural disaster risks, since individuals appear to have a stronger preference for these, compared with measures that only partly reduce risks (see Chapter 6). Alternatively, Chapter 7 demonstrates that natural disaster insurance with risk-based premiums can be effective in providing incentives for householders to invest in damage mitigation measures, for example by rewarding mitigation with premium discounts or higher levels of coverage or by requiring certain minimal mitigation activities to be carried out before coverage is provided. The ability of insurance companies to provide incentives to mitigate damage is very relevant in the context of adaptation to climate change. In this respect, it should be noted that efforts of insurers to encourage mitigation will be targeted at current levels of risk and not future risk because most insurers write annual policies. Nevertheless, if insurers are focused on reducing the risk during the time frame in which they provide insurance coverage, then the resulting reduction in vulnerability and exposure to natural hazards may be effective in limiting climate change impacts as long as the influence of climate change on risks follows a gradual process. In addition, public investments in risk reduction and land-use planning and building code policies should be geared towards achieving the reduction of natural disaster risks over a long time horizon.

12.5 Individual behaviour under risk and demand for natural disaster insurance in a changing climate

Insurance arrangements for natural disaster risk can only fulfil a meaningful role in spreading climate change related risks if sufficient demand for natural disaster insurance exists, resulting in a high market penetration of disaster insurance. High market penetration of insurance implies that fixed costs can be spread over many policyholders, that risks can be diversified in a large pool of insured hazards, and that many individuals who reside in hazard-prone areas will be subject to incentives or requirements provided by insurance to limit risks. Moreover, high market penetration of insurance in combination with adequate insurance coverage reduces the need for ad hoc government compensation of (uninsured) damage after a major disaster. The first three chapters of this book show that natural disaster risks may increases substantially in some regions as a result of climate change, which implies that premiums for insurance against such risks are also expected to increase. Important questions for the design of insurance arrangements for climate change losses are whether or not individual insurance demand is an impediment to the adequate functioning of current natural disaster insurance markets and whether or not the willingness-to-pay (WTP) for natural disaster insurance is likely to increase sufficiently if climate change increases natural disaster risks and associated insurance premiums in the future. This book examines these questions by estimating potential demand for flood insurance in the Netherlands using theories of decision making under risk (Chapter 8) and by conducting a contingent valuation survey (Chapter 10) and a choice modelling survey (Chapter 11) among approximately 1,000 Dutch households. These surveys elicited the WTP for flood insurance under the present climatic conditions and under scenarios where flood probabilities have increased because of climate change. An advantage of the survey methods is that they allow the provision of an in-depth analysis of the driving forces of insurance demand at the individual level. Moreover, Chapter 10 presents a review of international studies of flood insurance demand, in order to arrive at a complete picture and to place the results of the Dutch surveys in a broader perspective.

The traditional economic theory of individual decision making under risk – expected utility theory – has been unable to accurately describe individual decision making regarding the purchase of natural disaster insurance. Therefore, Chapter 8 examines the demand side of flood insurance arrangements in the Netherlands by estimating WTP and risk premiums, which are the difference between WTP and the expected value of the loss, using alternative theories of individual decision making under risk: namely, rank-dependent utility and prospect theories. These theories model individual attitudes towards both probabilities and monetary outcomes and are generally better in calibrating individual behaviour than expected utility, which models risk attitudes solely through the processing of monetary outcomes. The study delivers two main insights. First, estimation results suggest that a profitable flood insurance market could be feasible. Second, climate change has the potential to increase the profitability of offering flood insurance if individuals overweight low-probability flood risks. Under current flood probabilities, individuals are willing to purchase flood insurance as long as the premiums charged by insurance companies are not higher than approximately 70 per cent to 500 per cent of the expected value of the insurance. The estimation of risk premiums for three climate change scenarios shows that WTP increases more than the expected value of the insurance when the probability of flooding rises. This result is due to overweighting of low probabilities. However, several international studies have shown not only that not all individuals overweight low-probability risks but also that many people neglect low-probability natural disaster risks and do not purchase insurance against these risks. Therefore, it is imperative to obtain insights into individual attitudes towards natural disaster risks and into the determinants of natural disaster insurance demand, and to make accurate predictions of demand, before new insurance arrangements for climate change damage are introduced. This is the topic of Chapters 9, 10 and 11 for the case of flood insurance.

A review of several representative international studies about factors of influence on demand in Chapter 10 shows that previous disaster experience is a major driver of insurance demand; in addition, flood risk perceptions and certain socio-economic variables, such as income, significantly influence the demand for flood insurance. Moreover, several geographical variables which are indicators of objective flood risks influence flood insurance demand significantly. In several countries the possibility exists of government compensation for disaster damage, which may reduce individual demand for private natural disaster insurance. This has been called the 'charity hazard', although evidence on the charity hazard is mixed. For example, government relief does not appear to reduce flood insurance demand in the USA and Bangladesh, but it does so in Austria. An explanation for the absence of an empirical proof of the charity hazard in the USA may be that the aggregate indicators of flood insurance (e.g. at a state or county level) used in most of these studies inaccurately capture the influence of individual experiences with, or expectations of, federal disaster relief on the decision of the individual to (not) purchase flood insurance.

Chapter 10 also examines the demand for flood insurance under current flood probabilities, and increases in flood probability due to climate change, using a survey among homeowners living in the river delta area of the Netherlands. Particular attention has been paid to the effects of availability of government relief, perceptions of flood risks, and geographical and household characteristics on WTP for flood insurance. Risk attitudes to low-probability risk and the influence of communicating probabilities with risk ladders on insurance demand were analysed. This survey method is particularly useful for providing an in-depth analysis of the factors of influence on insurance demand at the individual (respondent) level. The results indicate that a large proportion of homeowners neglect the low probability of flood risk and do not want to purchase flood insurance. Fewer homeowners would neglect the risk if the flood probability were to increase as a result of climate change, as would be expected. Moreover, the removal of the government compensation scheme could increase the number of homeowners purchasing insurance, which provides evidence of the charity hazard. The mean WTP value of homeowners who are interested in buying flood insurance (defined as CWTP) is above the average expected value of the flood damage per household, which suggests that opportunities for partly private flood insurance do exist. Concerns about adverse selection with regard to flood insurance seem unfounded, as the geographical characteristics of the respondents show that the inhabitants of areas more vulnerable to flooding do not necessarily have a greater demand for insurance than the inhabitants of less vulnerable areas. A panel Tobit estimation reveals that WTP increases less than proportionally if climate change increases the flood probability. Using risk ladders as a device to communicate the flood probability considerably increases the sensitivity of WTP to probability changes. The effect of the risk ladder on WTP, irrespective of probability, is also to increase WTP by a relatively large amount. In other words, the way in which low-probability risks are communicated has a large impact on flood insurance demand.

Perceptions of flood risk play an important role in the demand for flood insurance. In particular, the greater the respondents perceive their flood risk to be, the higher is their WTP value for insurance. An analysis of the factors determining a range of indicators of risk perceptions in Chapter 9 provides several relevant insights for policy makers and insurers. In general, individuals have difficulty with assessing their flood risks and individuals' perceptions of risk may deviate considerably from expert judgements. However, differences in expected risk are, in part, consistently related to actual risk levels, since individuals in the vicinity of a major river and in low-lying areas generally have elevated risk perceptions. An explanation is that flood risk may be more salient for such individuals. However, individuals in areas unprotected by dikes tend to underestimate their risk of flooding. It appears that these people are not aware of the higher risk they face. Moreover, it is observed that individuals with little knowledge of the causes of flood events have a lower perception of flood risk. The findings indicate that improving the knowledge of citizens about the causes of flooding may increase flood risk awareness and associated flood insurance demand. It is especially important to target individuals who live in areas unprotected by dike infrastructure, since they tend to be unaware of, or ignore, their high exposure to flood risk.

Chapter 11 examines the development of WTP and market shares of flood insurance under various scenarios of climate change, using a choice model. The insights are of interest to the insurance industry in assessing the current and future profitability of offering insurance against extreme weather events in a changing climate. An important question that this chapter deals with is whether or not consumers would be willing to pay premiums that reflect increased risk as a result of climate change, and how market shares of flood insurance may develop under scenarios of increased risk-based premiums. The results of this survey show that offering flood insurance may be profitable in the Netherlands in the current situation, but the availability of government compensation reduces the demand for flood insurance. The effect of climate change is a non-linear increase in the proportion of homeowners who take out insurance and in the willingness-to-pay (WTP) for flood insurance. WTP increases less than the risk-based insurance premium if climate change were to cause a large increase in the probability of flooding, which might result in a considerable decline in the market penetration of flood insurance. Offering flood insurance remains profitable if climate change results in a moderate increase in flood probability, but not if there is an extreme rise in flood probability. The choice model shows that demand for flood insurance decreases considerably if climate change results in large increases in flood probability and insurers adjust premiums in accordance with risk. The reason for this is that household WTP for flood insurance increases by less than the rise in the expected value of the loss. This could have profound implications for insurance markets in the future if the expected effects of climate change on flood risk materialized. The market penetration of flood insurance could decrease if insurers need to substantially revise premiums upwards to reflect increased risk. A decreased market penetration of flood insurance may impair the ability of insurers to pool risks between a large number of policyholders and could disrupt the function of insurance to spread the societal costs of flooding. These findings highlight the necessity of designing comprehensive climate change adaptation programmes that limit the expected increase in natural disaster risks, in combination with efficient insurance arrangements that spread these losses at low cost and provide adequate incentives to policyholders to limit natural disaster risks.

12.6 An agenda for future research

The first part of this book examines the possible effects of climate change on a variety of natural disaster risks in several countries, as well as the potential impacts of climate change on the insurance sector. There are several studies that have estimated the potential effects of climate change on future natural disaster damage, but only a few have also assessed the influence of socio-economic factors on risks. However, these have a large effect on the exposure and vulnerability of societies to natural hazards and thereby determine, to a large extent, the damage caused by natural disasters. Further research on potential future developments of natural disaster risks should examine in more detail how both climatic and socio-economic influences on risks interact; ignoring one or the other will always result in an incomplete risk assessment.

Considerable uncertainty is associated with projections of future developments of risks as a result of climate change, which complicates the assessment of the impacts of climate change for insurers, in particular, and the adequate design of climate change adaptation strategies, in general. Further research is needed to quantify the prevailing uncertainty, and to refine assessments of the influence of climate change on natural disaster risks. In this respect, regional risk assessments are required, since the influence of climate change on extreme weather events is very region-specific, while risks posed by natural hazards are also strongly determined by local geographical characteristics and local economic exposure to the hazard. Future research should focus on improving projections of the effects of global warming on the frequency and intensity of extreme weather events. An important step forward would be to quantify the uncertainty associated with climate change projections that, for example, result from differences in climate models, and derive probabilistic scenarios of climate change. Moreover, a further refinement of catastrophe models could reduce uncertainty about the impacts of (changes in) natural hazards on risks. For this purpose, the insurance sector could work more closely with the academic research community and provide detailed historical data on claims for damage caused by extreme weather. The availability of high-quality data on extreme weather conditions and natural disaster losses is very valuable for calibrating climate and risk models and for improving their accuracy. Given the wider societal benefits associated with the recording and sharing of such data, governments or other public organizations could encourage the establishment of coordinated programmes for improved data recording and dissemination. This is especially important for developing countries, where reliable data on weather conditions and natural disaster damage are often lacking.

A major challenge is how to increase the global capacity of coverage of insurance arrangements against natural disaster risk. Extreme natural disaster risks are already challenging to insure privately in several regions around the world, and the availability of affordable natural disaster insurance may become more problematic if the risks posed by extreme weather increase as a result of climate change. Chapter 4 explores how public–private insurance can be implemented to cover flood risk in the Netherlands, which is currently not covered by private insurance. Further research could examine whether or not similar solutions could be feasible for covering extreme natural disaster risks in other countries where private insurance coverage is either limited or not available. A main focus of such research should be on how to design insurance programmes that have adequate incentives and strategies in place to achieve risk reduction. Adequate implementation of cost-effective risk reduction measures is essential for keeping insurance affordable in the face of climate change because, otherwise, risk-based premiums may exceed policyholders' WTP for insurance.

A combination of public investment in the prevention and damage limitation of natural disaster risks with measures implemented at the household level that reduce damage once a disaster occurs is likely to be a promising way forward to increase resilience to climate change impacts, as Chapter 5 shows in the case of flooding. Although

Chapters 6 and 7 reveal that damage mitigation measures can be very effective in limiting flood damage, practical experience suggests that few households voluntarily implement such measures. Financial incentives provided by insurance, such as premium discounts, could be a promising solution for this problem, but future research is needed to examine how effective such incentives would be in practice. Moreover, the recent literature proposes the introduction of long-term insurance contracts as an instrument to provide better incentives for promoting (long-term) risk reduction. Even though the arguments in favour of long-term insurance are compelling, the effectiveness of such incentives has scarcely been researched. In addition, insights are needed into the potential demand for such policies. In general, this book has shown that more research is needed on how to ensure that sufficient demand exists for natural disaster insurance. Risk perceptions appear to be a major factor of influence on insurance demand, which suggests that future research should focus on how the risk awareness of individuals can be improved, for example by risk communication campaigns, and how this translates into increased disaster preparedness and demand for natural disaster insurance.

Overall, this book provides a comprehensive study of the potential role that insurance can play in reducing the vulnerability of societies to natural disasters and climate change impacts. This provides a solid basis for further research on how climate change adaptation programmes that are currently being designed, or being implemented, around the world can be integrated with financial arrangements for natural disaster losses, in such a way that the projected impacts of future disasters on our societies can be limited and less damage and hardship needs to be suffered. My expectations are that this will remain an active area of research for several decades to come.

References

Abdellaoui, M. 2000. Parameter-free elicitation of utility and probability weighting functions. *Management Science*, 46(11): 1497–1512.

ABI 2005a. *Financial Risks of Climate Change: Summary Report*. Association of British Insurers (ABI), London.

2005b. *Financial Risks of Climate Change: Technical Annexes*. Association of British Insurers (ABI), London.

2009. *Financial Risks of Climate Change*. Research Paper No. 19. Association of British Insurers (ABI), London.

Ackerman, F., Stanton, E.A. and Bueno, R. 2010. Fat tails, exponents, extreme uncertainty: Simulating catastrophe in DICE. *Ecological Economics* 69(8): 1657–1665.

Adger, W.N., Brooks, N., Kelly, M., Bentham, G., Agnew, M. and Eriksen, S. 2004. *New Indicators of Vulnerability and Adaptive Capacity*. Tyndall Centre Technical Report 7. University of East Anglia, Norwich, UK.

Aerts, J.C.J.H. and Botzen, W.J.W. 2011a. Climate change impacts on pricing long-term flood insurance: A comprehensive study for the Netherlands. *Global Environmental Change*, 2(1): 1–7.

2011b. Climate-resilient waterfront development in New York City: Bridging flood insurance, building codes, and flood zoning. *Annals of the New York Academy of Sciences*, 1227: 1–82.

2012a. Hurricane Irene: A wake up call for New York City? *Natural Hazards and Earth System Sciences*, 2012: 1837–1840.

2012b. The Delta Plan of the Netherlands: Past and future perspectives on flood risk protection. In D. Hill, M.J. Bowman and J.S. Khinda (eds). *Against the Deluge: Storm Surge Barriers to Protect New York City*. American Society of Civil Engineers (ASCE), Reston, VA.

Aerts, J.C.J.H., Botzen, W.J.W., van der Veen, A., Krywkow, J. and Werners, S. 2008a. Dealing with uncertainty in flood management through diversification. *Ecology and Society*, 13(1): 41.

Aerts, J.C.J.H. and Droogers, P. 2004. *Climate Change in Contrasting River Basins: Adaptation Strategies for Water, Food and Environment*. CABI, Wallingford, UK.

Aerts, J.C.J.H., Renssen, H., Ward, P.J., de Moel, H., Odada, E., Bouwer, L.M. and Goosse, H. 2006. Sensitivity of global river discharges under

Holocene and future climate conditions. *Geophysical Research Letters*, 33(19): L19401.

Aerts, J.C.J.H., Sprong, T.A. and Bannink, B. (eds) 2008b. *Aandacht voor Veiligheid.* Rapport BSIK programma Klimaat voor Ruimte no. 009/2008. ISBN 978–90–8815–004–3. VU University Press, Amsterdam.

Akter, S., Brouwer, R., Choudhury, S. and Aziz, S. 2009. Is there a commercially viable market for crop insurance in rural Bangladesh? *Mitigation and Adaptation Strategies for Global Change*, 14(3): 215–229.

Albala-Bertrand, J.M. 2006. *The Unlikeliness of an Economic Catastrophe: Localization and Globalization.* Working Paper No. 576. Queen Mary University of London.

Alberini, A., Cropper, M., Krupnick, A. and Simon, N.B. 2004. Does the value of a statistical life vary with age and health status? Evidence from the US and Canada. *Journal of Environmental Economics and Management*, 48(1): 769–792.

2006. Willingness to pay for mortality risk reductions: Does latency matter? *Journal of Risk and Uncertainty*, 32(2): 231–245.

Alberini, A., Tonin, S., Turvani, M. and Chiabai, A. 2007. Paying for permanence: Public preferences for contaminated site cleanup. *Journal of Risk and Uncertainty*, 34(2): 155–178.

Allen, M. 2003. Liability for climate change. *Nature*, 421: 891.

Alley, R.B., Clark, P.U., Huybrechts, P. and Joughin, I. 2005. Ice-sheet and sea-level changes. *Science*, 310(5747): 456–460.

Andreoni, J. 1990. Impure altruism and donations to public goods: A theory of warm glow giving. *Economic Journal*, 100: 464–477.

Arrow, K.J. 1971. *Essays in the Theory of Risk-Bearing.* North-Holland, Amsterdam.

Arrow, K., Solow, R., Portney, P., Leamer, E., Radner, R. and Schuman, H. 1993. Report on the NOAA Panel on Contingent Valuation. *Federal Register*, 58: 4601–4614.

Asseldonk van, M.A.P.M., Meuwissen, M.P.M. and Huirne, R.B.M. 2001. Stochastic simulation of catastrophic hail and windstorm indemnities in the Dutch greenhouse sector. *Risk Analysis*, 21(4): 761–769.

Baan, P.J.A. 2004. *Overstromingsrisico's Langs Rivieren Meer dan Kans X Gevolg: Beleving en Implicaties voor Veiligheidsstrategie.* Research Report Q3673.20. WL Delft Hydraulics, Delft.

Baan, P.J.A. and Klijn, F. 2004. Flood risk perception and implications for flood risk management in the Netherlands. *International Journal of River Basin Management*, 2(2): 113–122.

Baker, E.J. 1991. Hurricane evacuation behavior. *International Journal of Mass Emergencies and Disasters*, 9(2): 287–310.

Bargiacchi, R. 2003. Climate change scenarios and the precautionary principle. In J. Wesseler, H.P. Weikard and R.D. Weaver (eds). *Risk and Uncertainty in Environmental and Resource Economics*. Edward Elgar Publishing, London.

Barnett, B.J., Barrett, C.B. and Skees, J.R. 2008. Poverty traps and index-based risk transfer products. *World Development*, 36(10): 1766–1785.

Barredo, J.I. 2009. Normalised flood losses in Europe: 1970–2006. *Natural Hazards and Earth System Sciences*, 9: 97–104.

2010. No upward trend in normalized windstorm losses in Europe: 1970–2008. *Natural Hazards and Earth System Sciences*, 10: 97–104.

Ben-Akiva, M. and Lerman, S.R. 1985. *Discrete Choice Analysis: Theory and Application to Travel Demand*. MIT Press, Cambridge, MA.

Bender, M.A., Knutson, T.R., Tuleya, R.E., Sirutis, J.J., Vecchi, G.A., Garner, S.T. and Held, I.M. 2010. Modeled impact of anthropogenic warming on the frequency of Atlantic hurricanes. *Science*, 327: 454–458.

Bergh van den, J.C.J.M. 2004. Optimal climate policy is a utopia: From quantitative to qualitative cost–benefit analysis. *Ecological Economics*, 48(4): 385–393.

Bergh van den, R. and Faure, M. 2006. Compulsory insurance of loss to property caused by natural disasters: Competition or solidarity? *World Competition*, 29(1): 25–54.

Berz, G.A. 1999. Catastrophes and climate change: Concerns and possible countermeasures of the insurance industry. *Mitigation and Adaptation Strategies for Global Change*, 4(3–4): 283–293.

Bhat, C.R. 2001. Quasi-random maximum simulated likelihood estimation of the mixed multinomial logit model. *Transportation Research B*, 35(7): 677–695.

Bhattacharya, S., Alberini, A. and Cropper, M.L. 2007. The value of mortality risk reductions in Delhi, India. *Journal of Risk and Uncertainty*, 34(1): 21–47.

Bleichrodt, H., Pinto, J.L. and Wakker, P.P. 2001. Making descriptive use of prospect theory to improve the prescriptive use of expected utility theory. *Management Science*, 47(11): 1498–1514.

Bliss, R.R. and Panigirtzoglou, N. 2004. Option-implied risk aversion estimates. *Journal of Finance*, 9(1): 407–446.

Blumenschein, K., Blomquist, G.C., Johannesson, M., Horn, N. and Freeman, P. 2008. Eliciting willingness to pay without bias: Evidence from a field experiment. *Economic Journal*, 118(525): 114–137.

Board on Natural Disasters 1999. Mitigation emerges as major strategy for reducing losses caused by natural disasters. *Science*, 284(5422): 1943–1947.

Bočkarjova, M. 2007. *Major disasters in modern economies: An input-output based approach at modelling imbalances and disproportions*. Doctoral dissertation, University of Twente, Enschede.

Booij, A.S. and van de Kuilen, G. 2006. *A Parameter-free Analysis of the Utility of Money for the General Population under Prospect Theory*. CREED Working Paper. Faculty of Economics and Econometrics, University of Amsterdam.

Boruff, B.J., Easoz, J.A., Jones, S.D., Landry, H.R. Mitchem, J.D. and Cutter, S.L. 2003. Tornado hazards in the United States. *Climate Research*, 24: 103–117.

Botzen, W.J.W., Aerts, J.C.J.H. and van den Bergh, J.C.J.M. 2009a. Willingness of homeowners to mitigate climate risk through insurance. *Ecological Economics*, 68(8–9): 2265–2277.

2009b. Dependence of flood risk perceptions on socioeconomic and objective risk factors. *Water Resources Research*, 45: W10440.

2012. Individual preferences for reducing flood risk to near zero through elevation. *Mitigation and Adaption Strategies for Global Environmental Change*. DOI: 10.1007/s11027-012-9359-5.

Botzen, W.J.W., Bouwer, L.M. and van den Bergh, J.C.J.M. 2010a. Climate change and hailstorm damage: Empirical evidence and implications for agriculture and insurance. *Resource and Energy Economics*, 32(3): 341–362.

Botzen, W.J.W., Gowdy, J.M. and van den Bergh, J.C.J.M. 2008a. Cumulative CO_2 emissions: Shifting international responsibilities for climate debt. *Climate Policy*, 8: 569–576.

Botzen, W.J.W. and van den Bergh, J.C.J.M. 2008. Insurance against climate change and flooding in the Netherlands: Present, future and comparison with other countries. *Risk Analysis*, 28(2): 413–426.

2009a. Managing natural disaster risk in a changing climate. *Environmental Hazards*, 8: 209–225.

2009b. Bounded rationality, climate risks, and insurance: Is there a market for natural disasters? *Land Economics*, 85(2): 265–278.

2012a. How sensitive is Nordhaus to Weitzman? Climate policy in DICE with an alternative damage function. *Economics Letters*, 117: 372–374.

2012b. Risk attitudes to low-probability climate change risks: WTP for flood insurance. *Journal of Economic Behavior and Organization*, 82(1): 151–166.

Botzen, W.J.W., van den Bergh, J.C.J.M. and Aerts, J.C.J.H. 2008b. *Report on a Survey about Perceptions of Flood Risk, Willingness to pay for Flood Insurance, and Willingness to Undertake Mitigation Measures: Explanation of the Survey Instrument*. Netherlands Climate Insurance

Program (NCIP) September, 2008. Institute for Environmental Studies, Vrije Universiteit, Amsterdam.

Botzen, W.J.W., van den Bergh, J.C.J.M. and Bouwer, L.M. 2010b. Climate change and increased risk for the insurance sector: A global perspective and an assessment for the Netherlands. *Natural Hazards*, 52(3): 577–598.

Bouwer, L.M. 2011. Have disaster losses increased due to anthropogenic climate change? *Bulletin of the American Meteorological Society*, January: 39–46.

2012. Projections of future extreme weather losses under changes in climate and exposure. *Risk Analysis*, doi:10.1111/j.1539–6924.2012.01880.x.

Bouwer, L.M. and Botzen, W.J.W. 2011. How sensitive are US hurricane damages to climate? Comment on a paper by W.D. Nordhaus. *Climate Change Economics*, 2(1): 1–7.

Bouwer, L.M., Bubeck, P. and Aerts, J.C.J.H. 2010. Changes in future flood risk due to climate and development in a Dutch polder area. *Global Environmental Change*, 20: 463–471.

Bouwer, L.M., Crompton, R.P., Faust, E., Höppe, P. and Pielke, R.A. 2007. Confronting disaster losses. *Science*, 318(5851): 753.

Bouwer, L.M. and Vellinga, P. 2005. Some rationales for risk sharing and financing adaptation. *Water Science and Technology*, 5(51): 89–95.

2007. On the flood risk in the Netherlands. In S. Begum, M.J.F. Stive and J.W. Hall (eds). *Flood Risk Management in Europe: Innovation in Policy and Practice*. Springer, Berlin: 469–484.

Boyle, K.J. 2003. Contingent valuation in practice. In P.A. Champ, K.J. Boyle and T.C. Browne (eds). *A Primer on Nonmarket Valuation*. Kluwer, Dordrecht.

Brainbridge, J. 1952. *Biography of an Idea: The Story of Mutual Fire and Casualty Insurance*. Doubleday & Co, Garden City, NY.

Brilly, M. and Polic, M. 2005. Public perceptions of flood risk, flood forecasting and mitigation. *Natural Hazards and Earth System Sciences*, 5: 345–355.

Brinke ten, W.B.M. and Bannink, B.A. 2004. *Risico's in Bedijkte Termen*. RIVM Rapport 500799002. Rijksinstituut voor Volksgezondheid en Milieu (RIVM), Milieu- en Natuurplanbureau (MNP), Bilthoven (in Dutch).

Brooks, H.E. (2006). *Tornado and Severe Thunderstorm Damage*. Climate change and disaster losses workshop, Hohenkammer, Germany: 30–34.

Brooks, H.E. and Doswell, C.A. 2001. Normalized damage from major tornadoes in the United States: 1890–1999. *Weather Forecasting*, 16: 168–176.

Brookshire, D.S., Thayer, M.A., Tschirhart, J. and Schulze, W.D. 1985. A test of the expected utility model: Evidence from earthquake risks. *Journal of Political Economy*, 93(2): 369–89.

Brouwer, R. and Akter, S. 2010. Informing micro insurance contract design to mitigate climate change catastrophe risks using choice experiments. *Environmental Hazards*, 9: 74–88.

Brouwer, R., Akter, S., Brander, L. and Haque, E. 2007. Socioeconomic vulnerability and adaptation to environmental risk: A case study of climate change and flooding in Bangladesh. *Risk Analysis*, 27(2): 313–326.

2008. Economic valuation of flood risk exposure and reduction in a severely flood-prone developing country. *Environment and Development Economics*, 14: 397–417.

Browne, M.J. and Hoyt, R.E. 2000. The demand for flood insurance: Empirical evidence. *Journal of Risk and Uncertainty*, 20(3): 291–306.

Bruijn de, K.M., Green, C., Johnson, C. and McFadden, L. 2007. Evolving concepts in flood risk management: Searching for a common language. In S. Begum, M.J.F. Stive and J.W. Hall (eds). *Flood Risk Management in Europe: Innovation in Policy and Practice*. Springer, Netherlands.

Bruijn de, K.M. and Klijn, F. 2001. Resilient flood risk management strategies. In L. Guifen and L. Wenxue (eds). *Proceedings of the IAHR Congress*. Tsinghua University Press, Beijing.

Bubeck, P., Botzen, W.J.W. and Aerts, J.C.J.H. 2012. A review of risk perceptions and other factors that influence mitigation behaviour? *Risk Analysis*, doi:10.1111/j.1539-6924.2011.01783.x.

Buchanan, J.M. 1975. The Samaritan's dilemma. In E.S. Phelps (ed). *Altruism, Morality and Economic Theory*. Russell Sage Foundation, New York: 71–85.

Buishand, T.A. and Klein Tank, A.M.G. 1996. Regression model for generating time series of daily precipitation amounts for climate change impacts. *Stochastic Hydrology and Hydraulics*, 10(2): 87–106.

Burby, R.J. 2001. Flood insurance and floodplain management: The US experience. *Environmental Hazards*, 3(3–4): 111–122.

2006. Hurricane Katrina and the paradoxes of government disaster policy: Bringing about wise governmental decisions for hazardous areas. *Annals of the American Academy of Political and Social Science*, 604(1): 171–191.

Burby, R.J., Beatley, T., Berke, P.R., Deyle, R.E., French, S.P., Godschalk, D.R., Kaiser, E.J., Kartez, J.D., May, P.J., Olshansky, R., Paterson, R.G. and Platt, R.H. 1999. Unleashing the power of planning to create disaster resistant communities. *Journal of the American Planning Association*, 65: 247–258.

Burn, D.H. 1999. Perceptions of flood risk: A case study of the Red River flood of 1997. *Water Resources Research*, 35(11): 3451–3458.

Camerer, C. 1998. Bounded rationality in individual decision making. *Experimental Economics*, 1(2): 163–83.

Camerer, C.F. and Kunreuther, H.C. 1989. Decision processes for low probability events: Policy implications. *Journal of Policy Analysis and Management*, 8(4): 565–592.

Cameron, T.A. 2005. Updating subjective risks in the presence of conflicting information: An application to climate change. *Journal of Risk and Uncertainty*, 30(1): 63–97.

Carlsson, F. and Martinsson, P. 2001. Do hypothetical and actual marginal willingness to pay differ in choice experiments? Application to the valuation of the environment. *Journal of Environmental Economics and Management*, 41: 179–192.

CBS 2005. *Assurantiejaarboek 2005*. Centraal Bureau voor de Statistiek. Kluwer, The Netherlands (in Dutch).

2006. *Assurantiejaarboek 2006*. Centraal Bureau voor de Statistiek. Kluwer, The Netherlands (in Dutch).

2007. *Assurantiejaarboek 2007*. Centraal Bureau voor de Statistiek. Kluwer, The Netherlands (in Dutch).

Chang, H., Franczyk, J. and Kim, C. 2009. What is responsible for increasing flood risk? The case of Gangwon Province, Korea. *Natural Hazards*, 48: 339–354.

Changnon, S.A. 2001. Damaging thunderstorm activity in the United States. *Bulletin of the American Meteorological Society*, 82(4): 597–608.

2003a. Shifting economic impacts from weather extremes in the United States: A result of societal changes, not global warming. *Natural Hazards*, 29(2): 273–290.

2003b. Measures of economic impacts of weather extremes: Getting better but far from what is needed – a call for action. *Bulletin of the American Meteorological Society*, 84(9): 1231–1235.

2007. Catastrophic winter storms: An escalating problem. *Climatic Change*, 84(2): 131–139.

2009a. Increasing major hail losses in the US. *Climatic Change*, 96: 161 166.

2009b. Temporal and spatial distributions of wind storm damages in the United States. *Climatic Change*, 94: 473–483.

Changnon, S.A. and Changnon, J.M. 1992. Temporal fluctuations in weather disasters: 1950–1989. *Climatic Change*, 22(3): 191–208.

Changnon, S.A., Pielke, R.A., Changnon, D., Sylves, R.T. and Pulwarty, R. 2000. Human factors explain the increased losses from weather and climate extremes. *Bulletin of the American Meteorological Society*, 81(3): 437–482.

Charpentier, A. 2008. Insurability of climate risks. *Geneva Papers of Risk and Insurance – Issues and Practice*, 33(1): 91–109.

Chilton, S., Jones-Lee, M., Kiraly, F., Metcalf, H. and Pang, W. 2006. Dread risk. *Journal of Risk and Uncertainty*, 33(3): 165–182.

Choi, O. and Fisher, A. 2003. The impacts of socioeconomic development and climate change on severe weather catastrophe losses: Mid-Atlantic region (MAR) and the US. *Climatic Change*, 58: 149–170.

Chrzan, K. and Orme, B. 2000. *An Overview of Design Strategies for Choice-Based Conjoint Analysis*. Sawtooth Software Research Paper Series, www.sawtoothsoftware.com/download/techpap/desgncbc.pdf.

Clark, M.J. 1998. Flood insurance as a management strategy for UK coastal resilience. *Geographical Journal*, 164(3): 333–343.

Cline, W.R. 1992. *The Economics of Global Warming*. Institute for International Economics, Washington, DC.

Cohen, M.D. 1995. Risk-aversion concepts in expected- and non-expected-utility models. *Geneva Papers on Risk and Insurance Theory*, 20(1): 73–91.

Connelly, N.A. and Knuth, B. 1998. Evaluating risk communication: Examining target audience perceptions about four presentation formats for fish consumption health advisory information. *Risk Analysis*, 18(5): 649–659.

Coomber, J.R. 2006. Natural and large catastrophes: Changing risk characteristics and challenges for the insurance industry. *Geneva Papers on Risk and Insurance – Issues and Practice*, 31: 88–95.

Corso, P.S., Hammitt, J.K. and Graham, J.D. 2001. Valuing mortality-risk reduction: Using visual aids to improve the validity of contingent valuation. *Journal of Risk and Uncertainty*, 23(2): 165–184.

Costanza, R., Daly, H., Folke, C., Hawken, P., Holling, C.S., McMichael, A.J., Pimentel, D. and Rapport, D. 2000. Managing our environmental portfolio. *BioScience*, 50(2): 149–155.

Crichton, D. 2005. *Flood Risk and Insurance in England and Wales: Are There Lessons to be Learned from Scotland?* Technical Paper 1. Benfield Greig Hazard Research Centre, London.

2008. Role of insurance in reducing flood risk. *Geneva Papers on Risk and Insurance – Issues and Practice*, 33(1): 117–132.

Crompton, R.P. and McAneney, K.J. 2008. Normalised Australian insured losses from meteorological hazards: 1967–2006. *Environmental Science and Policy*, 11(5): 371–378.

Crompton, R.P., McAneney, K.J., Chen, K., Pielke, R.A. and Haynes, K. 2010. Influence of location, population, and climate on building damage and fatalities due to Australian bushfire: 1925–2009. *Weather and Climate Society*, 2: 300–310.

Cubitt, R.P., Starmer, C. and Sugden, R. 2001. Discovered preferences and the experimental evidence of violations of expected utility theory. *Journal of Economic Methodology*, 8(3): 385–414.

Cutler, D.M. and Zeckhauser, R.J. 1997. *Reinsurance for Catastrophes and Cataclysms*. Working Paper 5913. NBER Working Paper Series, Cambridge, MA.

Dantzig van, D. 1956. Economic decision problems for flood prevention. *Econometrica*, 24(3): 276–287.

Daun, M.C. 2000. *Flood Risk and Contingent Valuation Willingness to Pay Studies: A Methodological Review and Applied Analysis*. Technical Report 6. Institute for Urban Environmental Risk Management, Marquette University, Milwaukee, WI.

Dawson, R.J., Ball, T., Werritty, J., Werritty, A., Hall, J.W. and Roche, N. 2011. Assessing the effectiveness of non-structural flood management measures in the Thames Estuary under conditions of socio-economic and environmental change. *Global Environmental Change*, 21: 628–646.

Dawson, R.J., Dickson, M.E., Nicholls, R.J., Hall, J.W., Walkden, M.J.A., Stansby, P.K., Mokrech, M., Richards, J., Zhou, J., Milligan, J., Jordan, A., Pearson, S., Rees, J., Bates, P.D., Koukoulas, S. and Watkinson, A.R. 2009. Integrated analysis of risks of coastal flooding and cliff erosion under scenarios of long-term change. *Climatic Change*, 95: 249–288.

Defra 2008. *Developing the Evidence Base for Flood Resistance and Resilience: Summary Report*. R&D Technical Report FD2607/TR1. Environment Agency and the Department for Environment Food and Rural Affairs (Defra), London.

Delavande, A. 2008. Measuring revisions to subjective expectations. *Journal of Risk and Uncertainty*, 36(1): 43–82.

Dessai, S. and Hulme, M. 2007. Assessing the robustness of adaptation decisions to climate change uncertainties: A case study on water resources management in the East of England. *Global Environmental Change*, 17(1): 59–72.

Dessens, J. 1995. Severe convective weather in the context of a nighttime global warming. *Geophysical Research Letters*, 22(10): 1241–1244.

Diecidue, E. and Wakker, P.P. 2001. On the intuition of rank-dependent utility. *Journal of Risk and Uncertainty*, 23(3): 281–298.

Dixon, L., Clancy, N., Seabury, S.A. and Overton, A. 2006. *The National Flood Insurance Program's Market Penetration Rate: Estimates and Policy Implications*. American Institutes for Research, Washington, DC.

Dlugolecki, A.F. 2000. Climate change and the insurance industry. *Geneva Papers on Risk and Insurance – Issues and Practice*, 25(4): 582–601.

2006. Thoughts about the impact of climate change on insurance claims. In P. Höppe and R.A. Pielke (eds). *Workshop on Climate Change and Disaster Losses*. Hohenkammer, Germany.

2008. Climate change and the insurance sector. *Geneva Papers on Risk and Insurance – Issues and Practice*, 33(1): 71–90.

Doherty, N. 1997. Insurance markets and climate change. *Geneva Papers on Risk and Insurance*, 83: 223–237.

Doherty, N.A. and Eeckhoudt, L. 1995. Optimal insurance without expected utility theory: The dual theory and the linearity of insurance contracts. *Journal of Risk and Uncertainty*, 10(2): 157–179.

Domencich, T. and McFadden, D. 1975. *Urban Travel Demand: A Behavioural Analysis*. North-Holland, Amsterdam.

Donat, M.G., Leckebusch, G.C., Wild, S. and Ulbrich, U. 2011. Future changes in European winter storm losses and extreme wind speeds inferred from GCM and RCM multi-model simulations. *Natural Hazards and Earth System Sciences*, 11: 1351–1370.

Donkers, B., Melenberg, B. and van Soest, A. 2001. Estimating risk attitudes using lotteries: A large sample approach. *Journal of Risk and Uncertainty*, 22(2): 165–195.

Dorland, C., Tol, R.S.J. and Palutikof, J.P. 1999. Vulnerability of the Netherlands and northwest Europe to storm damage under climate change: A model approach based on storm damage in the Netherlands. *Climatic Change*, 43(3): 513–535.

Dowdeswell, J.A. 2006. The Greenland Ice Sheet and global sea level rise. *Science*, 311(5763): 963–964.

Downing, T., Bharwani, S., Warwick, C. and Ziervogel, G. 2005. *Climate Adaptation: Actions, Strategies and Capacity from an Actor Oriented Perspective*. SEI Working Paper. March 2005, Stockholm Environment Institute (SEI), Oxford.

Downton, M.W., Miller, J.Z.B. and Pielke, R.A. 2005. Reanalysis of U.S. National Weather Service flood loss database. *Natural Hazards Review*, 6: 13–22.

Downton, M.W. and Pielke, R.A. 2001. Discretion without accountability: Politics, flood damage and climate. *Natural Hazards Review*, 2(4): 157–166.

DWW 2001. *Case Study Naar Effecten en Consequenties van Noodoverloopgebieden*. Report Q2902. Dienst Weg- en Waterbouw kunde (DWW), Delft (in Dutch).

2005a. *Flood Risks and Safety in the Netherlands (Floris)*. Interim Report No. DWW-2006–13. Ministry of Transport, Public Works and Water Management, November 2005, The Hague.

2005b. *Veiligheid Nederland in Kaart Globale Schadeberekening.* September 2005, Dienst Weg- en Waterbouwkunde (DWW), Delft (in Dutch).

2005c. *HIS Hoogwater Informatie Systeem voor Rampenbestrijding bij Overstromingen.* Rijkswaterstaat Dienst Weg- en Waterbouwkunde (DWW), Delft (in Dutch).

2005d. *Veiligheid Nederland in Kaart: Hoofdrapport Onderzoek Overstromingsrisico's.* Report DWW-2008–081. Rijkswaterstaat, Dienst Weg- en Waterbouwkunde (DWW), Den Haag (in Dutch).

EIA 2007. *International Energy Outlook 2007.* Energy Information Administration (EIA), Washington, DC.

Elsner, J.B., Kossin, J.P. and Jagger, T.H. 2008. The increasing intensity of the strongest tropical cyclones. *Nature*, 455(72091): 92–95.

Elton, E.J. and Gruber, J.M. 1995. *Modern Portfolio Theory and Investment Analysis.* John Wiley, New York.

Emanuel, K. 2005. Increasing destructiveness of tropical cyclones over the past 30 years. *Nature*, 436(7051): 686–688.

Epstein, P.R. and McCarthy, J.J. 2004. Assessing climate stability. *Bulletin of the American Meteorological Society*, 85(12): 1863–1870.

Epstein, S. 1994. Integration of the cognitive and the psychodynamic unconscious. *American Psychologist*, 49(8): 709–724.

Ernst & Young 2008. *Strategic Business Risk 2008: Insurance.* EYG no. EG0015.

Etchart-Vincent, N. 2004. Is probability weighting sensitive to the magnitude of consequences? An experimental investigation on losses. *Journal of Risk and Uncertainty*, 28(3): 217–235.

Faure, M. 2007. Insurability of damage caused by climate change: A commentary. *University of Pennsylvania Law Review*, 155: 1875–1899.

FEMA 1999. *Protecting Building Utilities from Flood Damage: Principles and Practices for the Design and Construction of Flood Resistant Building Utility Systems.* Federal Emergency Management Agency (FEMA) Publications, Washington, DC.

Ferrini, S. and Scarpa, R. 2007. Designs with *a priori* information for non market valuation with choice experiments: A Monte Carlo study. *Journal of Environmental Economics and Management*, 53(3): 342–363.

Feyen, L., Barredo, J.I. and Dankers, R. 2009. Implications of global warming and urban land-use change on flooding in Europe. In J. Feyen, K. Shannon and M. Neville (eds). *Water and Urban Development Paradigms.* Taylor & Francis, London: 217–225.

Figge, F. 2004. Bio-folio: Applying portfolio theory to biodiversity. *Biodiversity and Conservation*, 13(4): 827–849.

Finucane, M.L., Alhakami, A., Slovic, P. and Johnson, S.M. 2000. The affect heuristic in judgments of risks and benefits. *Journal of Behavioral Decision Making*, 13: 1–17.

Flynn, J., Slovic, P., Mertz, C.K. and Carlisle, C. 1999. Public support for earthquake risk mitigation in Portland, Oregon. *Risk Analysis*, 19(2): 205–216.

Folke, C., Carpenter, S., Elmqvist, T., Gunderson, L., Holling, C.S. and Walker, B. 2002. Resilience and sustainable development: Building adaptive capacity in a world of transformations. *Ambio*, 31(5): 437–440.

Fothergill, A. 1996. Gender, risk, and disaster. *International Journal of Mass Emergencies and Disasters*, 14(1): 33–56.

Fraser, E.D.G., Mabee, W. and Figge, F. 2005. A framework for assessing the vulnerability of food systems to future shocks. *Futures*, 37(6): 465–479.

Freeman, P.K., Keen, M. and Mani, M. 2003. *Dealing with Increased Risk of Natural Disasters: Challenges and Options*. IMF Working Paper WP/03/197. International Monetary Fund, Washington, DC.

Freeman, P.K. and Kunreuther, H. 2003. Managing environmental risk through insurance. In H. Folmer and T. Tietenberg (eds). *The International Yearbook of Environmental and Resource Economics 2003/2004*. Edward Elgar Publishing, Northampton, MA.

Froot, K.A. 2001. The market for catastrophe risk: A clinical examination. *Journal of Financial Economics*, 60(2–3): 529–571.

Füssel, H.M. and Klein, R.J.T. 2006. Climate change vulnerability assessments: An evolution of conceptual thinking. *Climatic Change*, 75(3): 301–329.

Gaillard, J.-C., D'Ercole, R. and Leone, F. 2001. Cartography of population vulnerability to volcanic hazards and lahars of Mount Pinatubo (Philippines): A case study in Pasig-Potrero river basin (province of Pampanga). *Géomorphologie: Relief, Processes, Environment*, 7(3): 209–222.

Ganderton, P.T., Brookshire, D.S., McKee, D.S., Stewart, S. and Thurston, H. 2000. Buying insurance for disaster-type risks: Experimental evidence. *Journal of Risk and Uncertainty*, 20(3): 271–289.

Garrett, T.A. and Sobel, R.S. 2003. The political economy of FEMA disaster payments. *Economic Inquiry*, 41(3): 496–509.

Gaslikova, L., Schwerzmann, A., Raible, C.C. and Stocker, T.F. 2011. Future storm surge impacts on insurable losses for the North Sea region. *Natural Hazards and Earth System Sciences*, 11: 1205–1216.

Geneva Association 2009. *The Insurance Industry and Climate Change*. The Geneva Association, Geneva.

Gigerenzer, G. and Hoffrage, U. 1995. How to improve Bayesian reasoning without instruction: Frequency formats. *Psychological Review*, 102(4): 684–704.

Gleick, P.H. 2003. Global freshwater resources: Soft-path solutions for the 21st century. *Science*, 302(5650): 1524–1528.

Godschalk, D. 2003. Urban hazard mitigation: Creating resilient cities. *Natural Hazards Review*, 4(3): 136–143.

Goklany, I.M. 2007. Death and death rates due to extreme weather events: Global and U.S. trends, 1900–2004. In *The Civil Society Report on Climate Change*. International Policy Press, London.

Goldberg, J. and Roosen, J. 2007. Scope insensitivity in health risk reduction studies: A comparison of choice experiments and the contingent valuation method for valuing safer food. *Journal of Risk and Uncertainty*, 34(2): 123–144.

Goldstein, W.M. and Einhorn, H.J. 1987. Expression theory and the preference reversal phenomena. *Psychological Review*, 94(2): 236–254.

Gollier, C. 2005. *Some Aspects of the Economics of Catastrophe Risk Insurance*. CESifo Working Paper No. 1409. Centre for Economic Studies and Ifo Institute for Economic Research (CESifo), Munich.

Groen, G. 2007. *Extreme Zomerneerslag 2006 en Klimaatscenario's*. KNMI Scientific Report 215. Koninklijk Nederlands Meteorologisch Instituut (KNMI), De Bilt.

Grossi, P. and Kunreuther, H.C. 2005. *Catastrophe Modeling: A New Approach to Managing Risk*. Springer, New York.

Grossman, P.Z., Cearley, R.W. and Cole, D.H. 2006. Uncertainty, insurance and the Learned Hand Formula. *Law, Probability and Risk*, 5(1): 1–18.

Grothmann, T. and Reusswig, F. 2006. People at risk of flooding: Why some residents take precautionary action while others do not. *Natural Hazards*, 38: 101–120.

Gustafson, P.E. 1998. Gender differences in risk perception: Theoretical and methodological perspectives. *Risk Analysis*, 18(6): 805–811.

Hakes, J.K. and Viscusi, W.K. 1997. Mortality risk perceptions: A Bayesian reassessment. *Journal of Risk and Uncertainty*, 15(2): 135–150.

Hall, J.W., Sayers, P.B. and Dawson, R.J. 2005. National scale flood risk assessment of current and future flood risk in England and Wales. *Natural Hazards*, 36: 147–164.

Hallegatte, S. 2007. The use of synthetic hurricane tracks in risk analysis and climate change assessment. *Journal of Applied Meteorology and Climatology*, 46: 1956–1966.

Hammitt, J.K. 1990. Risk perceptions and food choice: An exploratory analysis of organic- versus conventional-produce buyers. *Risk Analysis*, 10(3): 367–374.

Hammitt, J.K. and Graham, J.D. 1999. Willingness to pay for health protection: Inadequate sensitivity to probability? *Journal of Risk and Uncertainty*, 18(1): 33–62.

Handmer, J.W. and Dovers, S.R. 1996. A typology of resilience: Rethinking institutions for sustainable development. *Organization and Environment*, 9(4): 482–511.

Harrington, S.E. 1999. Rethinking disaster policy: Breaking the cycle of "free" disaster assistance, subsidized insurance, and risky behaviour. *Regulation*, 23(1): 40–46.

Heal, G. and Kriström, B. 2002. Uncertainty and climate change. *Environmental and Resource Economics*, 22(1–2): 3–39.

Heerkens, R. 2003. *Inventarisatie van de Verzekerbaarheid van Schade als Gevolg van Extreme Regenval en Wateroverlast*. RIZA. Werkdocument 2003.219X. Ministerie van Verkeer en Waterstaat, Lelystad (in Dutch).

Heij, C., de Boer, P., Franses, P.H., Kloek, T. and van Dijk, H.K. 2004. *Econometric Methods with Applications in Business and Economics*. Oxford University Press, Oxford.

Heimann, M. and Reichstein, M. 2008. Terrestrial ecosystem carbon dynamics and climate feedbacks. *Nature*, 451(7176): 289–292.

Henriet, D. and Michel-Kerjan, E.O. 2006. *Optimal Risk-sharing under Dual Asymmetry of both Information and Market Power: A Unifying Approach*. Working Paper. Center for Risk Management and Decision Processes, The Wharton School, Philadelphia, PA.

Hensher, D.A., Greene, W.H. and Rose, J.M. 2003. *Using Classical Inference Methods to Reveal Individual-specific Parameter Estimates to Avoid the Potential Complexities of WTP Derived from Population Moments*. Working Paper ITS-WP-03–18. Institute of Transport Studies, University of Sydney and Monash University, Sydney.

Hensher, D.A., Rose, J.M. and Greene, W.H. 2005. *Applied Choice Analysis: A Primer*. Cambridge University Press.

Hershey, J.C., Kunreuther, H.C. and Schoemaker, P.J.H. 1982. Sources of bias in assessment procedures for utility functions. *Management Science*, 28(8): 936–954.

Hess, S. and Rose, J.M. 2009. Should reference alternatives in pivot design SC surveys be treated differently? *Environmental and Resource Economics*, 42: 297–317.

Hilferink, M. and Rietveld, P. 1999. Land use scanner: An integrated GIS model for long-term projections of land use in urban and rural areas. *Journal of Geographical Systems*, 1: 155–177.

Hilker, N., Badoux, A. and Hegg, C. 2009. The Swiss flood and landslide database 1972–2007. *Natural Hazards and Earth System Sciences*, 9: 913–925.

Hochrainer, S., Mechler, R. and Pflug, G. 2008. Climate change and financial adaptation in Africa: Investigating the impacts of climate change on the robustness of index-based microinsurance in Malawi. *Mitigation and Adaptation Strategies for Global Change*, 14: 231–250.

Hoetker, G. 2007. The use of logit and probit models in strategic management research: Critical issues. *Strategic Management Journal*, 28(4): 331–343.

Hoes, O.A.C. 2007. *Aanpak wateroverlast in polders op basis van risicobeheer*. Doctoral thesis, Delft University of Technology, Delft (in Dutch).

Hoff, H., Bouwer, L.M., Berz, G.A., Kron, W. and Loster, T. 2003. *Risk Management in Water and Climate: The Role of Insurance and Other Financial Services*. Dialogue on Water and Climate, Delft, and Munich Reinsurance Company, Munich.

Hoff, H., Warner, K. and Bouwer, L.M. 2005. The role of financial services in climate adaptation in developing countries. *Vierteljahrshefte zur Wirtschaftsforschung*, 74(2): 196–207.

Holling, C.S. 1973. Resilience and stability of ecological systems. *Annual Review of Ecology and Systematics*, 4(1): 1–23.

1986. The resilience of terrestrial ecosystems: Local surprise and global change. In W.C. Clark and R.E. Munn (eds). *Sustainable Development of the Biosphere*. Cambridge University Press.

Höppe, P. and Pielke, R. 2006. *Report of the Workshop on Climate Change and Disaster Losses*, October. Hohenkammer, Germany.

Hoyos, C.D., Agudelo, P.A., Webster, P.J. and Curry, J.A. 2006. Deconvolution of the factors contributing to the increase in global hurricane intensity. *Science*, 312(5769): 94–97.

Hsee, C.K., Blount, S., Loewenstein, G.F. and Bazerman, M.H. 1999. Preference reversals between joint and separate evaluations of options: A review and theoretical analysis. *Psychological Review*, 125(5): 576–590.

Hu, W., Cox, K.B.K. and Pan, M. 2009. Economic values of dolphin excursions in Hawaii: A stated choice analysis. *Marine Resource Economics*, 24: 61–76.

Huber, P.J. 1967. The behavior of maximum likelihood estimates under non-standard conditions. In L.M. Le Cam and J. Neyman (eds). *Proceedings of the Fifth Berkeley Symposium on Mathematical Statistics and Probability*. University of California Press, Berkeley, CA: 221–233.

Hubert, M. 2004. Insurability and regulatory reform: Is the English flood insurance regime able to adapt to climate change? *Geneva Papers on Risk and Insurance – Issues and Practice*, 29(2): 169–182.

Hurd, M.D. and McGarry, K. 1995. Evaluation of the subjective probabilities of survival in the health and retirement study. *Journal of Human Resources*, 30: 268–292.

Hurk van den, B., Klein Tank, A., Lenderink, G., van Ulden, A., van Oldenborgh, G.J., Katsman, C., van den Brink, H., Keller, F., Bessembinder, J., Burgers, G., Komen, G., Hazeleger, W. and Drijfhout, S. 2006. *KNMI Climate Change Scenarios 2006 for the Netherlands*. KNMI Scientific Report WR 2006–01. Koninklijk Nederlands Meteorologisch Instituut (KNMI), De Bilt.

Hurly, T.M. and Shogren, J.F. 2005. An experimental comparison of induced and elicited beliefs. *Journal of Risk and Uncertainty*, 30(2): 169–188.

ICPR 2002. *Non Structural Flood Plain Management: Measures and their Effectiveness*. International Commission for the Protection of the Rhine (ICPR), Koblenz, Germany.

Ida, T. and Goto, R. 2009. Simultaneous measurement of time and risk preferences: Stated preference discrete choice modeling analysis depending on smoking behavior. *International Economic Review*, 50: 1169–1182.

IFRC 2001. *World Disasters Report*. International Federation of Red Cross and Red Crescent Societies, Geneva.

2003. *World Disasters Report 2002: Focus on Reducing Risk*. International Federation of Red Cross and Red Crescent Societies, Geneva.

2010. *World Disasters Report 2010: Focus on Urban Risk*. International Federation of Red Cross (IFRC), Geneva.

IHDP 2007. The implications of global environmental change for human security in coastal urban areas. *Magazine of International Human Dimensions Programme on Global Environmental Change*, 2, October 2007.

IPCC 2001. *Climate Change 2001: Impacts, Adaptation and Vulnerability*. Contribution of Working Group II to the Third Assessment Report of the Intergovernmental Panel on Climate Change. J.J. McCarthy, O.F. Canziani, N.A. Leary, D.J. Dokken and K.S. White (eds). Cambridge University Press.

2007. *Climate Change 2007: The Physical Science Basis*. Contribution of Working Group I to the Fourth Assessment Report of the Intergovernmental Panel on Climate Change. S. Solomon, D. Qin, M. Manning, Z. Chen, M. Marquis, K. Averyt, M.M.B. Tignor and H.L. Miller (eds). Cambridge University Press, Cambridge and New York.

2012. *Managing the Risks of Extreme Events and Disasters to Advance Climate Change Adaptation*. A Special Report of Working Groups I and II of the Intergovernmental Panel on Climate Change. C.B. Field, V. Barros, T.F. Stocker, D. Qin, D.J. Dokken, K.L. Ebi, M.D. Mastrandrea, K.J. Mach, G.-K. Plattner, S.K. Allen, M. Tignor and P.M. Midgley (eds). Cambridge University Press, Cambridge and New York.

Jaffee, D., Kunreuther, H.C. and Michel-Kerjan, E. 2008. *Long Term Insurance (LTI) for Addressing Catastrophe Risk*. NBER Working Paper Series, Working Paper 14210. National Bureau of Economic Research, Cambridge, MA.

Jaffee, D.M. and Russel, T. 1997. Catastrophe insurance, capital markets, and uninsurable risks. *Journal of Risk and Insurance*, 64(2): 205–230.

Janssen, J.P.F.M. and Jorissen, R.E. 1997. Flood management in the Netherlands: Recent development and research needs. In R. Casale, K. Havno and P. Samuels (eds). *RIBAMOD: River Basin Modelling, Management and Flood Mitigation*. HR Wallingford, UK: 89–104.

Jiang, F.-Q., Zhu, C., Mu, G.-J., Hu, R.-J. and Meng, Q.-X. 2005. Magnification of flood disasters and its relation to regional precipitation and local human activities since the 1980s in Xinxiang, northwestern China. *Natural Hazards*, 36: 307–330.

Johannesson, M., Johansson P.-O. and Löfgren, K.G. 1997. On the value of change in life expectancy: Blips versus parametric changes. *Journal of Risk and Uncertainty*, 15(3): 221–239.

Johannesson, M., Johansson, P.-O. and O'Conor, R.M. 1996. The value of private safety versus the value of public safety. *Journal of Risk and Uncertainty*, 13(3): 263–275.

Johnson, E.J., Gächter, S. and Herrman, A. 2006. *Exploring the Nature of Loss Aversion*. Le Dex Discussion Papers No. 2006–02. Centre for Decision Research and Experimental Economics, University of Nottingham, UK.

Jones, C.P., Coulborne, W.L., Marshall, J. and Rogers, S.M. 2006. *Evaluation of the National Flood Insurance Program's Building Standards*. American Institutes for Research, Washington, DC.

Jones-Lee, M.W., Hammerton, M. and Philips, P.R. 1985. The value of safety: Results of a national sample survey. *Economic Journal*, 95(377): 49–72.

Jonkeren, O., Rietveld, P. and van Ommeren, J. 2007. Climate change and inland waterway transport: Welfare effects of low water levels on the River Rhine. *Journal of Transport Economics and Policy*, 41(3): 387–411.

Jonkman, S.N., Brinkhuis-Jak, M. and Kok, M. 2004. Cost–benefit analysis and flood damage mitigation in the Netherlands. *HERON*, 49(1): 95–111.

Jonkman, S.N. and Cappendijk, P. 2006. *Veiligheid Nederland in Kaart: Inschatting van het Aantal Slachtoffers ten Gevolge van Overstroming*. Rapport DWW-2006–012. DWW, Ministerie van Verkeer en Waterstaat, Den Haag (in Dutch).

Kaas, R., Goovaerts, M., Dhaene, J. and Denuit, M. 2001. *Modern Actuarial Risk Theory*. Kluwer Academic Publishers, Dordrecht.

Kabat, P., van Vierssen, W., Veraart, J., Vellinga, P. and Aerts, J. 2005. Climate proofing the Netherlands. *Nature*, 438(7066): 283–284.

Kahneman, D. 2003. Maps of bounded rationality: Psychology for behavioral economics. *American Economic Review*, 93(5): 1449–1475.

Kahneman, D. and Frederick, S. 2002. Representativeness revisited: Attribute substitution in intuitive judgment. In T. Gilovich, D. Griffin and D. Kahneman (eds). *Heuristics and Biases: The Psychology of Intuitive Judgment*. Cambridge University Press.

Kahneman, D. and Knetsch, J. 1992. Valuing public goods: The purchase of moral satisfaction. *Journal of Environmental Economics and Management*, 22: 57–70.

Kahneman, D., Slovic, P. and Tversky, A. 1982. *Judgment under Uncertainty: Heuristics and Biases*. Cambridge University Press.

Kahneman, D. and Tversky, A. 1979. Prospect theory: An analysis of decision under risk. *Econometrica*, 47(2): 263–291.

Kaplan, S. 1973. Cognitive maps in perception and thought. In R.M. Downs, D. Stea and K.E. Boulding (eds). *Image and Environment*. Aldine, Chicago.

Kaplow, L. 1991. Incentives and government relief for risk. *Journal of Risk and Uncertainty*, 4(2): 167–175.

Keijzer, X. 2008. *The effectiveness of flood mitigation measures*. MSc thesis, June 2008. Vrije Universiteit, Amsterdam.

Keller, C., Siegrist, M. and Gutscher, H. 2006. The role of the affect and availability heuristics in risk communication. *Risk Analysis*, 26(3): 631–639.

Kelly, M. and Kleffner, A.E. 2003. Optimal loss mitigation and contract design. *Journal of Risk and Insurance*, 70(1): 53–72.

Kelman, I. and Spence, R. 2003. A limit analysis of unreinforced masonry failing under flood water pressures. *Masonry International*, 16(2): 51–61.

Kennedy, M., Mrofka, D. and von der Borch, C. 2008. Snowball Earth termination by destabilization of equatorial permafrost methane clathrate. *Nature*, 453(7195): 642–645.

Kerr, R.A. 2006a. A tempestuous birth of hurricane climatology. *Science*, 312(5774): 676–678.

2006b. A worrying trend of less ice, higher seas. *Science*, 311(5768): 1698–1701.

Kind, J., Gauderis, J., van Velzen, E. and Silva, W. 2008. *Waterveiligheid 21e eeuw: Kengetallen Kosten-batenanalyse*. Rapport WD 2008.044. Ministerie van Verkeer en Waterstaat, Den Haag (in Dutch).

Klein, R.J.T., Nicholls, R.J. and Thomalla, F. 2003. Resilience to natural hazards: How useful is this concept? *Environmental Hazards*, 5(1–2): 35–45.

Kleindorfer, P.R. 2009. *Climate Change and Insurance*. Working Paper #2009–08–01. The Wharton School, University of Pennsylvania.

Kleindorfer, P.R. and Kunreuther, H.C. 1999. The complementary roles of mitigation and insurance in managing catastrophic risks. *Risk Analysis*, 19(4): 727–738.

Klein Tank, A.M.G. and Können, G.P. 1994. The dependence of precipitation on temperature. In *Proceedings of the Eighteenth Annual Climate Diagnostics Workshop*, 1–5 November 1993, Boulder, CO.

Klijn, F., Knoop, J.M., Ligtvoet, W. and Mens, M.J.P. 2012. In search of robust flood risk management alternatives for the Netherlands. *Natural Hazards and Earth System Sciences*, 12: 1469–1479.

KNMI 2009. *Klimaatverandering in Nederland: Aanvullingen op de KNMI '06 Scenarios*. Koninklijk Nederlands Meteorologisch Instituut (KNMI), De Bilt (in Dutch).

Koetse, M.J. and Rietveld, P. 2009. The impacts of climate change and weather on transport: An overview of empirical findings. *Transportation Research Part D*, 14(3): 205–221.

Kok, M. 2005. *Een Waterverzekering in Nederland: Mogelijk en Wenselijk?* Opdrachgever: Adviescommissie Water. HKV Lijn in Water. Rapport PR1038, Lelystad (in Dutch).

2006. Schade na een overstroming: Taak van de overheid of eigen verantwoordelijkheid? *Externe Veiligheid*, 4: 15–17 (in Dutch).

Kok, M. and Barendregt, A. 2004. *Verantwoordelijkheid en Aansprakelijkheid bij Wateroverlastschade*. Directoraat Generaal Water, HKV Lijn in Water, Lelystad (in Dutch).

Kok, M., van Gelder, P.H.A.J.M., Vrijling, J.K. and Vogelsang, M.P. 2002. Risk of flooding and insurance in the Netherlands. In B. Wu, Z.H. Wang, G. Wang, G.G.H. Huang, H. Fang and J. Huong (eds). *Flood Defence*. Science Press, New York.

Koomen, E., Loonen, W. and Hilferink, M. 2008. *Climate-Change Adaptations in Land-Use Planning: A Scenario-Based Approach*. Lecture Notes in Geoinformation and Cartography Series, Springer, Berlin.

Kousky, C. 2011. Managing natural catastrophe risk: State insurance programs in the United States. *Review of Environmental Economics and Policy*, doi:10.1093/reep/req020.

Kousky, C. and Cooke, R.M. 2009. *Climate Change and Risk Management: Challenges for Insurance, Adaptation, and Loss Estimation*. Discussion Paper No. RFF DP 09–03. Resources for the Future, Washington, DC.

Krantz, D.H. and Kunreuther, H.C. 2007. Goals and plans in decision making. *Judgment and Decision Making*, 2(3): 137–168.

Kreibich, H. 2011. Do perceptions of climate change influence precautionary measures? *International Journal of Climate Change Strategies and Management*, 3(2): 189–199.

Kreibich, H., Christenberger, S. and Schwarze, R. 2011. Economic motivation of households to undertake private precautionary measures against floods. *Natural Hazards and Earth System Sciences*, 11: 309–321.

Kreibich, H. and Thieken, A.H. 2009. Coping with floods in the city of Dresden, Germany. *Natural Hazards*, 51(3): 423–436.

Kreibich, H., Thieken, A.H., Petrow, T., Müller, M. and Merz, B. 2005. Flood loss reduction of private households due to building precautionary measures: Lessons learned from the Elbe flood in August 2002. *Natural Hazards and Earth System Sciences*, 5(1): 117–126.

Kriesel, W. and Landry, C. 2004. Participation in the National Flood Insurance Program: An empirical analysis for coastal properties. *Journal of Risk and Insurance*, 71(3): 405–420.

Krupnick, A., Alberini, A., Cropper, M., Simon, N., O'Brien, B., Goeree, R. and Heintzelman, M. 2002. Age, health and the willingness to pay for mortality risk reductions: A contingent valuation survey of Ontario residents. *Journal of Risk and Uncertainty*, 24(2): 161–186.

Kunreuther, H.C. 1978. *Disaster Insurance Protection: Public Policy Lessons*. Wiley & Sons, New York.

1984. Causes of underinsurance against natural disasters. *Geneva Papers on Risk and Insurance*, 9(31): 206–220.

1996. Mitigating disaster losses through insurance. *Journal of Risk and Uncertainty*, 12(2–3): 171–187.

2006a. Disaster mitigation and insurance: Learning from Katrina. *Annals of the American Academy of Political and Social Science*, 604(1): 208–227.

2006b. Has the time come for comprehensive natural disaster insurance? In R.J. Daniels, D.F. Kettl and H.C. Kunreuther (eds). *On Risk and Disaster: Lessons from Hurricane Katrina*. University of Pennsylvania Press, Philadelphia.

2008. Reducing losses from catastrophe risks through long-term insurance and mitigation. *Social Research*, 75: 905–932.

Kunreuther, H.C., Desvouges, W.H. and Slovic, P. 1988. Nevada's predicament: Public perceptions of risk from the proposed nuclear waste repository. *Environment*, 30(8): 17–33.

Kunreuther, H. and Hogarth, R. 1992. How does ambiguity affect insurance decisions? In G. Dionne (ed). *Contributions to Insurance Economics*. Kluwer Academic Publishers, Dordrecht.

Kunreuther, H.C., Meszaros, J., Hogarth, R.M. and Spranca, M. 1995. Ambiguity and underwriter decision processes. *Journal of Economic Behavior and Organization*, 26(3): 337–352.

Kunreuther, H.C. and Michel-Kerjan, E.O. 2007. Climate change, insurability of large-scale disasters and the emerging liability challenge. *University of Pennsylvania Law Review*, 155(6): 1795–1842.

Kunreuther, H.C., Michel-Kerjan, E.O., Doherty, N.A., Grace, M.F., Klein, R.W. and Pauly, M.V. 2011. *At War with the Weather: Managing Large-Scale Risks in a New Era of Catastrophes*. Paperback edition. MIT Press, Cambridge, MA.

Kunreuther, H.C., Novemsky, N. and Kahneman, D. 2001. Making low probabilities useful. *Journal of Risk and Uncertainty*, 23(2): 161–186.

Kunreuther, H.C. and Pauly, M. 2004. Neglecting disaster: Why don't people insure against large losses? *Journal of Risk and Uncertainty*, 28(1): 5–21.

2006. Rules rather than discretion: Lessons from Hurricane Katrina. *Journal of Risk and Uncertainty*, 33(1): 101–116.

Kunreuther, H.C. and Rose, A. 2004. *The Economics of Natural Hazards*. Edward Elgar Publishing, Cheltenham, UK.

Kunreuther, H.C., Sanderson, W. and Vetschera, R. 1985. A behavioral model of the adoption of protective activities. *Journal of Economic Behavior and Organization*, 6(1): 1–15.

Kunz, M., Sander, J. and Kottmeier, C.H. 2009. Recent trends of thunderstorm and hailstorm frequency and their relation to atmospheric characteristics in southwest Germany. *International Journal of Climatology*, 29: 2283–2297.

Lamond, J.E., Proverbs, D.G. and Hammond, F.N. 2009. Accessibility of flood risk insurance in the UK: Confusion, competition and complacency. *Journal of Risk Research*, 12(6): 825–841.

Landsea, C.W., Harper, B.A., Hoarau, K. and Knaff, J.A. 2006. Can we detect trends in extreme tropical cyclones? *Science*, 313(5786): 452–454.

Lattimore, P.K., Baker J.R. and Witte, A.D. 1992. The influence of probability on risky choice. *Journal of Economic Behavior and Organization*, 17(3): 377–400.

Laury, S.K., Morgen-McInnes, M. and Swarthout, J.T. 2009. Insurance decisions for low probability losses. *Journal of Risk and Uncertainty*, 39: 17–44.

Leckebusch, G.C., Ulbrich, U., Fröhlich, J.G. and Pinto, J.G. 2007. Property loss potentials for European midlatitude storms in a changing climate. *Geophysical Research Letters*, 34: L05703.

Leiter, A.M. 2009. The sense of snow: Individuals' perceptions of fatal avalanche events. Working Manuscript, March 2009. Department of Economics and Statistics, University of Innsbruck.

Lenton, T. 2011. Early warning of climate tipping points. *Nature Climate Change*, 1: 201–209.

Lenton, T.M., Footitt, A. and Dlugolecki, A. 2009. *Major Tipping Points in the Earth Climate System and Consequences for the Insurance Sector*. World Wide Fund for Nature (WWF) and Allianz SE, Gland, Switzerland.

Lenton, T.M., Held, H., Kriegler, E., Hall, J.W., Lucht, W., Rahmstorf, S. and Schellnhuber, H.J. 2008. Tipping elements in the earth's climate system. *Proceedings of the National Academy of Sciences of the USA*, 105(6): 1786–1793.

Levy, J.K. and Hall, J. 2005. Advances in flood risk management under uncertainty. *Stochastic Environmental Research and Risk Assessment*, 19(6): 375–377.

Lewis, C.M. and Murdock, K.C. 1996. The role of government contracts in discretionary reinsurance markets for natural disasters. *Journal of Risk and Insurance*, 63(4): 567–597.

Lewis, T. and Nickerson, D. 1989. Self-insurance against natural disasters. *Journal of Environmental Economics and Management*, 16: 209–223.

Lichtenstein, S., Slovic, P., Fischhoff, B., Layman, M. and Combs, B. 1978. Judged frequency of lethal events. *Journal of Experimental Psychology: Human Learning and Memory*, 4(6): 551–578.

Linde te, A., Bubeck, P., Dekkers, J.E.C., de Moel, H. and Aerts, J.C.J.H. 2011. Future flood risk estimates along the River Rhine. *Natural Hazards and Earth System Sciences*, 11: 459–473.

Lindell, M.K. 1994. Perceived characteristics of environmental hazards. *International Journal of Mass Emergencies and Disasters*, 12(3): 303–326.

Lindell, M.K. and Prater, C.S. 2000. Household adoption of seismic hazard adjustments: A comparison of residents of two states. *International Journal of Mass Emergencies and Disasters*, 18(2): 317–338.

Linnerooth-Bayer, J. and Mechler, R. 2006. Insurance for assisting adaptation to climate change in developing countries: A proposed strategy. *Climate Policy*, 6: 1–17.

Linnerooth-Bayer, J., Warner, K.. Bals, C., Höppe, P., Burton, I., Loster, T. and Haas, A. 2009. Insurance, developing countries and climate change. *Geneva Papers on Risk and Insurance – Issues and Practice*, 34: 381–400.

Litan, R.E. 2006. *Sharing and Reducing the Financial Risks of Future "Mega-catastrophes"*. Issues in Economic Policy No. 4. Brookings Institution, Washington, DC.

Lloyd's of London 2008. *Directors in the Dock: Is Business Facing a Liability Crisis?* Lloyds, London.

Loewenstein, G.F., Hsee, C.K., Weber, E.U. and Welch, N. 2001. Risk as feelings. *Psychological Bulletin*, 127(2): 267–286.

Loewenstein, G. and Mather, J. 1990. Dynamic processes in risk perception. *Journal of Risk and Uncertainty*, 3(2): 155–175.

Loonen, W. and Koomen, E. 2009. *Calibration and Validation of the Land Use Scanner Allocation Algorithms.* Netherlands Environmental Agency, Bilthoven.

Louviere, J.J., Hensher, D.A. and Swait, J.D. 2000. *Stated Choice Methods: Analysis and Application*. Cambridge University Press.

Luechinger, S. and Raschky, P.A. 2009. Valuing flood disasters using the life satisfaction approach. *Journal of Public Economics*, 93(3–4): 620–632.

Marcellis-Warin de, N. and Michel-Kerjan, E. 2001. *The Public-Private Sector Risk-Sharing in the French Insurance "CatNat System"*. Scientific Series, Centre Interuniversitaire de Recherche en Analyse des Organisations (CIRANO), Montréal.

Markowitz, H. 1952. Portfolio selection. *Journal of Finance*, 17(1): 77–91.

Mason, C.F., Shogren, J.F., Settle, C. and List, J.A. 2005. Investigating risky choices over losses using experimental data. *Journal of Risk and Uncertainty*, 31(2): 187–215.

Matthews, H.D. and Caldeira, K. 2008. Stabilizing climate requires near-zero emissions. *Geophysical Research Letters*, 35: 1–5.

McClelland, G.H., Schulze, W.D. and Coursey, D.L. 1993. Insurance for low-probability hazards: A bimodal response to unlikely events. *Journal of Risk and Uncertainty*, 7(1): 95–116.

McDaniels, T.L. 1992. Reference points, loss aversion, and contingent values for auto safety. *Journal of Risk and Uncertainty*, 5(2): 187–200.

McDonald, J.F. and Moffitt, R.A. 1980. The uses of Tobit analysis. *Review of Economics and Statistics*, 62(2): 318–321.

McFadden, D. 2001. Economic choices. *American Economic Review*, 91: 351–378.

McLeman, R. and Smit, B. 2006. Vulnerability to climate change hazards and risks: Crop and flood insurance. *Canadian Geographer*, 50(2): 217–226.

McMaster, H.J. 1999. The potential impact of global warming on hail losses to winter cereal crops in New South Wales. *Climatic Change*, 43(2): 455–476.

2001. Hailstorm risk assessment in rural New South Wales. *Natural Hazards*, 24(2): 187–196.

Meehl, G.A., Stocker, T.F., Collins, W.D., Friedlingstein, P., Gaye, A.T., Gregory, J.M., Kitoh, A., Knutti, R., Murphy, J.M., Noda, A., Raper, S.C.B., Watterson, I.G., Weaver, A.J. and Zhao, Z.C. 2007. Global climate projections. In S. Solomon, D. Qin, M. Manning, Z. Chen, M. Marquis, K.B. Averyt, M. Tignor and H.L. Miller (eds). *Climate Change*

2007: The Physical Science Basis. Contribution of Working Group I to the Fourth Assessment Report of the Intergovernmental Panel on Climate Change. Cambridge University Press.

Meer van der, R.W., van Asseldonk, M.A.P.M. and van der Meulen, H.A.B. 2007. *Klimaat voor Verzekeren? Oogstschade Verzekering in de Akkerbouw*. Rapport 6.07.07. Landbouw Economisch Instituut (LEI), Den Haag (in Dutch).

Merz, B., Kreibich, H., Thieken, A.H. and Schmidtke, R. 2004. Estimation uncertainty of direct monetary flood damage to buildings. *Natural Hazards and Earth System Sciences*, 4(1): 153–163.

Meulen van der, H., van Asseldonk, M., Buurma, J. and Nienhuis, J. 2006. *Mogelijkheden van een Brede Weersverzekering*. Rapport 6.06.15. Landbouw Economisch Instituut (LEI), Den Haag (in Dutch).

Meulen van der, M.J., van der Spek, A.J.F., de Lange, G., Gruijters, S.H.L.L., van Gessel, S.F., Nguyen, B.L., Maljers, D., Schokker, J., Mulder, J.P.M. and van der Krogt, R.A.A. 2007. Regional sediment deficits in the Dutch lowlands: Implications for long-term land-use options. *Journal of Soils and Sediments*, 7(1): 9–16.

Michaels, P. 2006. Is the sky really falling? A review of recent global warming scare stories. *Policy Analysis*, 576: 1–25.

Michel-Kerjan, E.O. 2010. Catastrophe economics: The National Flood Insurance Program. *Journal of Economic Perspectives*, 24(4): 165–186.

Michel-Kerjan, E.O. and Kousky, C. 2010. Come rain or shine: Evidence on flood insurance purchases in Florida. *Journal of Risk and Insurance*, 77(2): 369–397.

Michel-Kerjan, E.O. and Kunreuther, H.C. 2011. Redesigning flood insurance. *Science*, 333: 408–409.

Michel-Kerjan, E.O., Lemoyne de Forges, S.L. and Kunreuther, H. 2012. Policy tenure under the US National Flood Insurance Program (NFIP). *Risk Analysis*, 32(4): 644–658.

Michel-Kerjan, E. and de Marcellis-Warin, N. 2006. Public-private programs for covering extreme events: The impact of information distribution on risk-sharing. *Asia-Pacific Journal of Risk and Insurance*, 1(2): 21–49.

Michel-Kerjan, E. and Morlaye, F. 2008. Extreme events, global warming, and insurance-linked securities: How to trigger the "tipping point". *Geneva Papers on Risk and Insurance – Issues and Practice*, 33(1): 153–176.

Middelkoop, H., Daamen, K., Gellens, D., Grabs, W., Kwadijk, J.C.J., Lang, H., Parmet, B.W.A.H., Schädler, B., Schulla, J. and Wilke, K. 2001. Impact of climate change on hydrological regimes and water resources management in the Rhine basin. *Climatic Change*, 49(1–2): 105–128.

Mileti, D.S. and Darlington, J.D. 1997. The role of searching in shaping reactions to earthquake risk information. *Social Problems*, 44(1): 89–103.

Miller, S., Muir-Wood, R. and Boissonnade, A. 2008. An exploration of trends in normalized weather-related catastrophe losses. In H.F. Diaz and R.J. Murnane (eds). *Climate Extremes and Society*. Cambridge University Press: 225–347.

Mills, E. 1996. Energy efficiency: No-regrets climate change insurance for the insurance industry. *Journal of the Society of Insurance Research*, 9(3): 21–58.

2005. Insurance in a climate of change. *Science*, 309(5737): 1040–1044.

2007. *From Risk to Opportunity 2007: Insurer Responses to Climate Change*. Ceres Report, October 2007. Ceres, Boston, MA.

2009a. A global review of insurance industry responses to climate change. *Geneva Papers on Risk and Insurance – Issues and Practice*, 34: 323–359.

2009b. *From Risk to Opportunity 2008: Insurer Responses to Climate Change*. Ceres Report. Ceres, Boston, MA.

Mills, E. and Lecomte, E. 2006. *From Risk to Opportunity: How Insurers Can Proactively and Profitably Manage Climate Change*. Ceres Report, August, 2006. Ceres, Boston, MA.

Mills, E., Lecomte, E. and Peara, A. 2002. Insurers in the greenhouse. *Journal of Insurance Regulation*, 21(1): 43–78.

Mills, E., Roth, R.J. and Lecomte, E. 2005. *Availability and Affordability of Insurance under Climate Change: A Growing Challenge for the U.S.* Ceres Report, September 2005. Ceres, Boston, MA.

Mitchell, R.C. and Carson, R.T. 1989. *Using Surveys to Value Public Goods: The Contingent Valuation Method*. Resources for the Future, Washington, DC.

MNP 2007. *Nederland Later: Tweede Duurzaamheidsverkenning, Deel Fysieke Leefomgeving Nederland*. Milieu- en Planbureau (MNP), Bilthoven (in Dutch).

Moel de, H. and Aerts, J.C.J.H. 2011. Effect of uncertainty in land use: Damage models and inundation depth on flood damage estimates. *Natural Hazards*, 58: 407–425.

Moel de, H., Aerts, J.C.J.H. and Koomen, E. 2011. Development of flood exposure in the Netherlands during the 20th and 21st century. *Global Environmental Change*, 21: 620–627.

Moel de, H., van Alphen, J. and Aerts, J.C.J.H. 2009. Flood maps in Europe: Methods, availability and use. *Natural Hazards and Earth System Sciences*, 9(2): 289–301.

Mokrech, M., Nicholls, R.J., Richards, J.A., Henriques, C., Holman, I.P. and Shackley, S. 2008. Regional impact assessment of flooding under future climate and socioeconomic scenarios for East Anglia and North West England. *Climatic Change*, 90: 31–55.

Montenegro, A., Brovkin, V., Eby, M., Archer, D. and Weaver, A.J. 2007. Long-term fate of anthropogenic carbon. *Geophysical Research Letters*, 34: 1–5.

Muir-Wood, R., Miller, S. and Boissonade, A. 2006. The search for trends in a global catalogue of normalized weather-related catastrophe losses. In P. Höppe and R.A. Pielke (eds). *Workshop on Climate Change and Disaster Losses*, Hohenkammer. Germany.

Munich Re 1993. *Winter Storms in Europe: Analysis of 1990 Losses and Future Loss Potential*. Report 2042-E-e, Munich Reinsurance Group, Munich.

2005a. *Claims Management Following Natural Catastrophes: Experience, Analysis, Action Plans*. Munich Reinsurance Group, Munich.

2005b. *Natural Disasters According to Country Income Groups 1980–2004*. Munich Re NatCatService, Munich.

2006. *Annual Review: Natural Catastrophes 2005*. Munich Reinsurance Group, Munich.

2009. *Natural Catastrophes 2008: Analyses, Assessments, Positions*. Munich Re Group, Munich.

2011. *Natural Catastrophes 2010: Analyses, Assessments, Positions*. Munich Re Group, Munich.

2012. *Natural Catastrophes 2011: Analyses, Assessments, Positions*. Munich Re Group, Munich.

Munich Re NatCatService 2012. Downloadcenter for statistics on natural catastrophes. http://www.munichre.com/en/reinsurance/business/non-life/georisks/natcatservice/default.aspx

Narita, D., Tol, R.S.J. and Anthoff, D. 2009. Damage costs of climate change through intensification of tropical cyclone activities: An application of FUND. *Climate Research*, 39: 87–97.

2010. Economic costs of extratropical storms under climate change: An application of FUND. *Journal of Environmental Planning and Management*, 53: 371–384.

National Institute of Building Sciences 2005. *Natural Hazard Mitigation Saves: An Independent Study to Assess the Future Savings from Mitigation Activities*. The Multihazard Mitigation Council and the National Institute of Building Sciences, Washington, DC.

Netherlands Environmental Assessment Agency 2005. *The Effects of Climate Change in the Netherlands*. MNP Report 773001037. MNP, KNMI, RWS-RIZA, RWS-RIKZ, Alterra, IVM ICIS, Bilthoven.

Neumann von, J. and Morgenstern, O. 1947. *The Theory of Games and Economic Behavior*. 2nd edn. Princeton University Press, Princeton, NJ.

Niall, S. and Walsh, K. 2005. The impact of climate change on hailstorms in southeastern Australia. *International Journal of Climatology*, 25(14): 1933–1952.

Nielen-Kiezebrink, M.F., Bennink-Heinen, C.A., Groen, M., Meursing, L., van Reeken, H.N.M. and Rommel, M.C. 2005. *Risicobeheersing in Kustplaatsen: Beheersing van Kansen en Gevolgen van Kustafslag en Overstromig tijdens Zware Storm in Buitendijks Gebied*. Rapport RIKZ/2005.022. Ministerie van Verkeer en Waterstaat. Den Haag (in Dutch).

Nordhaus, W.D. 2008. *A Question of Balance: Weighting the Options of Global Warming Policies*. Yale University Press, New Haven, CT.

2010. The economics of hurricanes and implications of global warming. *Climate Change Economics*, 1, 1–20.

Nordhaus, W.D. and Yang, Z. 1996. A regional dynamic general-equilibrium model of alternative climate-change strategies. *American Economic Review*, 86(4): 741–765.

Norris, F.H., Smith, T. and Kaniasty, K. 1999. Revisiting the experience–behavior hypothesis: The effects of Hurricane Hugo on hazard preparedness and other self-protective acts. *Basic and Applied Social Psychology*, 21(1): 37–47.

O'Brien, B.J. and Sculpher, M.J. 2000. Building uncertainty into cost-effectiveness rankings: Portfolio risk-return tradeoffs and implications for decision rules. *Medical Care*, 38(5): 460–468.

O'Brien, K., Eriksen, S., Schjolden, A. and Nygaard, L. 2004. *What's in a Word? Conflicting Interpretations of Vulnerability in Climate Change Research*. CICERO Working paper, Oslo.

Olsthoorn, X., van der Werff, P., Bouwer, L.M. and Huitema, D. 2008. Neo-Atlantis: The Netherlands under a five-meter sea level rise. *Climatic Change*, 91(1–2): 103–122.

Overpeck, J.T., Otto-Bliesner, B.L., Miller, G.H., Muhs, D.R., Alley, R.B. and Kiehl, J.T. 2006. Paleoclimatic evidence for future ice-sheet instability and rapid sea-level rise. *Science*, 311(5768): 1747–1750.

Pahl Wostl, C., Downing, T., Kabat, P., Magnuszewski, P., Meigh, J., Schlueter, M., Sendzimir, J. and Werners, S. 2005. Transition to Adaptive Water Management: The NeWATER Project. NeWater Working Paper 1. Institute of Environmental Systems Research, University of Osnabrück, Germany.

Palm, R.I., Hodgson, M.E., Blanchard, R.D. and Lyons, D.I. 1990. *Earthquake Insurance in California*. Westview Press, Boulder, CO.

Pasterick, E.T. 1998. The National Flood Insurance Program. In H.C. Kunreuther and R.J. Roth (eds). *Paying the Price: The Status and Role of Insurance against Natural Disasters in the United States*. Joseph Henry/National Academy Press, Washington, DC.

Peacock, W.G., Brody, S.D. and Highfield, W. 2005. Hurricane risk perceptions among Florida's single family homeowners. *Landscape and Urban Planning*, 73(2–3): 120–135.

Pearce, D.W. and Smale, R. 2005. Appraising flood control investments in the UK. In R. Brouwer and D.W. Pearce (eds). *Cost–Benefit Analysis and Water Resources Management*. Edward Elgar Publishing, Cheltenham, UK.

Pelling, M. 2003. *The Vulnerability of Cities: Natural Disasters and Social Resilience*. Earthscan, London.

Penning-Rowsell, E.C., Johnson, C., Tunsall, S., Tapsell, S., Morris, J., Chatterton, J. and Green, C.H. 2005. *The Benefits of Flood and Coastal Risk Management: A Manual of Assessment Techniques*. Defra, London.

Perry, R.W. and Lindell, M.K. 1990. Predicting long-term adjustment to volcano hazard. *International Journal of Mass Emergencies and Disasters*, 8(2): 117–36.

Piani, F., Crisci, A., De Chiara, G., Maracchi, G. and Meneguzzo, F. 2005. Recent trends and climatic perspectives of hailstorm frequency and intensity in Tuscany and Central Italy. *Natural Hazards and Earth System Sciences*, 5: 217–224.

Pielke, R.A. 2005. Are there trends in hurricane destruction? *Nature*, 438(7071): E11.

2007. Future economic damage from tropical cyclones: Sensitivities to societal and climate changes. *Philosophical Transactions of the Royal Society A*, 365: 2717–2729.

Pielke, R.A., Agrawala, S., Bouwer, L.M., Burton, I., Changnon, S., Glantz, M.H., Hooke, W.H., Klein, R.J.T., Kunkel, K., Mileti, D., Sarewitz, D., Thompkins, E.L., Stehr, N. and von Storch, H. 2005. Clarifying the attribution of recent disaster losses: A response to Epstein and McCarthy. *Bulletin of the American Meteorological Society*, 86(10): 1481–1483.

Pielke, R.A., Gratz, J., Landsea, C.W., Collins, D., Saunders, M.A. and Musulin, R. 2008. Normalized hurricane damage in the United States: 1900–2005. *Natural Hazards Review*, 9(1): 29–42.

Pielke, R.A. and Landsea, C.W. 1998. Normalized hurricane damages in the United States: 1925–95. *Weather and Forecasting*, 13(3): 621–631.

Pielke, R.A., Prins, G., Rayner, S. and Sarewitz, D. 2007. Lifting the taboo on adaptation. *Nature*, 445(7128): 597–598.

Pielke, R.A., Rubiera, J., Landsea, C., Fernandez, M.L. and Klein, R. 2003. Hurricane vulnerability in Latin America and the Caribbean: Normalized damage and loss potentials. *Natural Hazards Review*, 4: 101–114.

Pierrehumbert, R.T. 2004. Warming the world. *Nature*, 432(7018): 677.

Pimm, S.L. 1984. The complexity and stability of ecosystems. *Nature*, 307(5949): 321–326.
Pinto, J.G., Frölich, E.L., Leckebusch, G.C. and Ulbrich, U. 2007. Changing European storm loss potentials under modified climate conditions according to ensemble simulations of the ECHAM5/MPI-OM1 GCM. *Natural Hazards and Earth System Sciences*, 7: 165–175.
Porfiriev, B. 2009. Community resilience and vulnerability to disasters: Qualitative models and megacities – A comparison with small towns. *Environmental Hazards*, 8(1): 23–37.
Poussin, J., Botzen, W.J.W. and Aerts, J.C.J.H. 2011. Stimulating flood damage mitigation through insurance: An assessment of the French CatNat system. Working Manuscript. Institute for Environmental Studies, VU University Amsterdam.
Praag van, B.M.S., Frijters, P. and Ferrer-i-Carbonell, A. 2003. The anatomy of subjective well-being. *Journal of Economic Behavior and Organization*, 51(1): 29–49.
Pratt, J.W. 1964. Risk aversion in the small and in the large. *Econometrica*, 32(1/2): 122–36.
Prelec, D. 1998. The probability weighting function. *Econometrica*, 66(3): 497–527.
PricewaterhouseCoopers 2007. "Too much regulation" tops insurance risks. Centre for the Study of Financial Information and PricewaterhouseCoopers, News Release, 25 May.
Priest, G.L. 1996. The government, the market, and the problem of catastrophic loss. *Journal of Risk and Uncertainty*, 12(2–3): 219–237.
Quiggin, J. 1982. A note on the existence of a competitive optimum. *Economic Record*, 55(161):174–176.
2005. *The Precautionary Principle in Environmental Policy and the Theory of Choice under Uncertainty*. Murray–Darling Program Working Paper M05–3. University of Queensland, Brisbane.
Raghavan, S. and Rajesh, S. 2003. Trends in tropical cyclone impact: A study in Andhra Pradesh, India. *Bulletin of the American Meteorological Society*, 84: 635–644.
Raid, J.K., Norris, F.H. and Ruback, R.B. 1999. Predicting evacuation in two major disasters: Risk perception, social influence, and access to resources. *Journal of Applied Social Psychology*, 29(5): 918–934.
Rapley, C. 2006. The Antarctic ice sheet and sea level rise. In H.J. Schellnhuber, W. Cramer, N. Nakicenovic, T. Wigley and G. Yohe (eds). *Avoiding Dangerous Climate Change*. Cambridge University Press.
Raschky, P.A., Schwarze, R., Schwindt, M. and Zahn, F. 2012. Uncertainty of government relief and the crowding out of insurance. *Environmental and Resource Economics*, doi:10.1007/s10640–012–9586-y.

Raschky, P.A. and Weck-Hannemann, H. 2007. Charity hazard: A real hazard to natural disaster insurance? *Environmental Hazards*, 7: 321–329.

Rijkswaterstaat 2005a. *Flood Risks and Safety in the Netherlands (Floris)*. Report DWW-2006–014. Rijkswaterstaat, Dienst Weg- en Waterbouwkunde (DWW), Delft.

2005b. *Veiligheid Nederland in Kaart: Overstromingsrisico Dijkring 36 Land van Heusden/De Maaskant en 36a Keent*. Rijkswaterstaat, Dienst Weg- en Waterbouwkunde (DWW), Delft (in Dutch).

RIZA 2003. *Hoeveel Hoogwater Kan ons Land Binnenkomen via de Rijn bij Lobith, Nu en in de Toekomst*. RIZA Report 2003.015. Rijkswaterstaat RIZA, Lelystad (in Dutch).

RIZA, HKV, Arcadis, KIWA, Korbee & Hovelunck, Klopstra, D., Versteeg, R. and Kroon, T. 2005. *Droogtestudie Nederland: Aard, Ernst en Omvang van Watertekorten in Nederland*. RIZA Rapport 2005.016, Lelystad (in Dutch).

Roach, B., Boyle, K.J. and Welsh, M.P. 2002. Testing bid design effects in multiple-bounded contingent-valuation questions. *Land Economics*, 78(1): 121–131.

Rooij de, A. and van den Bergh, J.C.J.M. 2005. Stilte voor de storm: Verzekeren tegen klimaatverandering. ESB, 17 juni 2005, jaargang 90 nr. 44463 (in Dutch).

Rose, A. 2003. A typology of economic disruptions. In *50th Annual North American Meetings of the Regional Science Association International*, Philadelphia, PA.

2004a. Defining and measuring economic resilience to disasters. *Disaster Prevention and Management*, 13(4): 307–314.

2004b. Economic principles, issues, and research priorities in hazard loss estimation. In Y. Okuyama and S.E. Chang (eds). *Modelling Spatial and Economic Impacts of Disasters*. Springer, Berlin and New York.

2006. Economic resilience to disasters: Toward a consistent and comprehensive formulation. In D. Paton and D. M. Johnston (eds). *Disaster Resilience: An Integrated Approach*. Charles C. Thomas, New York.

2007. Economic resilience to natural and man-made disasters: Multidisciplinary origins and contextual dimensions. *Environmental Hazards*, 7(4): 383–398.

2011. Resilience and sustainability in the face of disasters. *Environmental Innovation and Societal Transitions*, 1: 96–100.

Rosenman, R., Fort, R. and Budd, W. 1988. Perceptions, fear, and economic loss: An application of prospect theory to environmental decision making. *Policy Sciences*, 21(4): 327–350.

Rowe, R.D., Schulze, W.D. and Breffle, W.S. 1996. A test of payment card biases. *Journal of Environmental Economics and Management*, 31(2): 178–185.

Ruin, I., Gaillard, J.-C. and Lutoff, C. 2007. How to get there? Assessing motorists' flash flood risk perception on daily itineraries. *Environmental Hazards*, 7(3): 235–244.

RVWS and VROM-R 2003. *Verantwoorde Risico's, Veilige Ruimte*. Raad voor Verkeer en Waterstaat en VROM-Raad, Advice 037, Den Haag (in Dutch).

Sanchez-Rodriguez, R., Pacyna, J., O'Brien, K., Fragkias, M., Rosentrater, L. and Weichselgartner, J. 2007. Introduction to the "global environmental change, natural disasters, vulnerability and their implications for human security in coastal urban areas" issue. *Magazine of the International Human Dimensions Programme on Global Environmental Change*, 2: 4–5.

Sarmiento, C. and Miller, T.R. 2006. *Costs and Consequences of Flooding and the Impact of the National Flood Insurance Program*. Pacific Institute of Research and Evaluation, Claverton, MD.

Saunders, M.A. and Lea, A.S. 2008. Large contribution of sea surface warming to recent increase in Atlantic hurricane activity. *Nature*, 451(7178): 557–560.

Scarpa, R., Willis, K.G. and Acutt, M. 2007. Valuing externalities from water supply: Status quo, choice complexity and individual random effects in panel kernel logit analysis of choice experiments. *Journal of Environmental Planning and Management*, 50: 449–466.

Schiermeier, Q. 2006. Insurers' disaster files suggest climate is culprit. *Nature*, 441(7094): 674–675.

Schmeidler, D. 1989. Subjective probability and expected utility without additivity. *Econometrica*, 57(3): 571–587.

Schmidt, S., Kemfert, C. and Faust, E. 2009a. *Simulation of Economic Losses from Tropical Cyclones in the Years 2015 and 2050: The Effects of Anthropogenic Climate Change and Growing Wealth*. Discussion Paper 914. German Institute for Economic Research, Berlin.

Schmidt, S., Kemfert, C. and Höppe, P. 2009b. Tropical cyclone losses in the USA and the impact of climate change: A trend analysis based on data from a new approach for adjusting storm losses. *Environmental Impact Assessment Review*, 29: 359–369.

Schneider, Y. and Zweifel, P. 2004. How much internalization of nuclear risk through liability insurance? *Journal of Risk and Uncertainty*, 29(3): 219–240.

Schoubroeck van, C. 1997. Legislation and practice concerning natural disasters and insurance in a number of European countries. *Geneva Papers on Risk and Insurance – Issues and Practice*, 83: 238–267.

Schreider, S.Y., Smith, D.I. and Jakeman, A.J. 2000. Climate change impacts on urban flooding. *Climatic Change*, 47: 91–115.

Schwartz, S.H., Melech, G., Lehmann, A., Burgess, S., Harris, M. and Owens, V. 2001. Extending the cross-cultural validity of the theory of basic human values with a different method of measurement. *Journal of Cross-Cultural Psychology*, 32: 519–542.

Schwarze, R. and Wagner, G.G. 2007. The political economy of natural disaster insurance: Lessons from the failure of a proposed compulsory insurance scheme in Germany. *European Environment*, 17: 403–415.

Schwierz, C., Köllner-Heck, P., Zenklusen Mutter, E., Bresch, D.N., Vidale, P.L., Wild, M. and Schär, C. 2010. Modelling European winter wind storm losses in current and future climate. *Climatic Change*, 101: 485–514.

Seifert, I., Botzen, W.J.W., Aerts, J.C.J.H. and Krebich, H. 2012. The influence of flood risk characteristics on flood insurance demand. Working Manuscript. VU University, Amsterdam.

Sherbinin de, A., Schiller, A. and Pulsipher, A. 2007. Three cities and their vulnerabilities to climate hazards. *Magazine of the International Human Dimensions Programme on Global Environmental Change*, 2: 5–7.

Siegrist, M. and Gutscher, H. 2006. Flooding risks: A comparison of lay people's perceptions and experts' assessments in Switzerland. *Risk Analysis*, 26(4): 971–979.

Sitkin, S.B. and Pablo, A.L. 1992. Reconceptualizing the determinants of risk behavior. *Academy of Management Review*, 17(1): 9–38.

Sjöberg, L. 2000. Factors in risk perception. *Risk Analysis*, 20(1): 1–11.

Sloman, S.A. 1996. The empirical case for two systems of reasoning. *Psychological Bulletin*, 119(1): 3–22.

Slovic, P. 1987. Perception of risk. *Science*, 236(4799): 280–285.

1997. The public perception of risk. *Journal of Environmental Health*, 59(9): 22–29.

2000. *Perceptions of Risk*. Earthscan, London.

Slovic, P., Finucane, M.L., Peters, E. and MacGregor, D.G. 2004. Risk as analysis and risk as feelings: Some thoughts about affect, reason, risk, and rationality. *Risk Analysis*, 24(2): 311–322.

Slovic, P., Fischhoff, B., Lichtenstein, S., Corrigan, B. and Combs, B. 1977. Preference for insuring against probable small losses: Insurance implications. *Journal of Risk and Insurance*, 44(2): 237–258.

Smit, B., Burton, I., Klein, R.J.T. and Wandel, J. 2000. An anatomy of adaptation to climate change and variability. *Climatic Change*, 45(1): 223–251.

Smith, V.K. and Desvouges, W.H. 1987. An empirical analysis of the economic value of risk changes. *Journal of Political Economy*, 95(1): 89–114.

Starmer, C. 2000. Developments in non-expected utility theory: The hunt for a descriptive theory of choice under risk. *Journal of Economic Literature*, 38(2): 332–382.

Statistics Netherlands 2008. *StatLine Database*. Centraal Bureau voor de Statistiek (CBS). http://www.cbs.nl.

Stern, N. 2007. *The Economics of Climate Change: The Stern Review*. Cambridge University Press.

Storch, H. and Downes, N.K. 2011. A scenario-based approach to assess Ho Chi Minh City's urban development strategies against the impact of climate change. *Cities*, 26(6): 517–526.

Stuyt, L.C.P.M., van Bakel, P.J.T., Kroes, J.G., Bos, E.J., van der Elst, M., Pronk, B., Rijk, P.J., Clevering, O.A., Dekking, A.J.G., van der Voort, M.P.J., de Wolf, M. and Brandenburg, W.A. 2006. *Transitie en Toekomst van Deltalandbouw: Indicatoren voor de Ontwikkeling van de Land- en Tuinbouw in de Zuidwestelijke Delta van Nederland*. Alterra-Rapport 1132. Alterra, Wageningen (in Dutch).

Swiss Re 2005. *Natural Catastrophes and Man-Made Disasters in 2004: More than 300,000 Fatalities, Record in Insured Losses*. Sigma Report 1/2005. Swiss Re Reinsurance Company, Zurich.

Terpstra, T. 2011. Emotions, trust, and perceived risk: Affective and cognitive routes to flood preparedness behaviour. *Risk Analysis*, doi:10.1111/j.1539–6924.2011.01616.x.

Terpstra, T. and Gutteling, J.M. 2008. Households' perceived responsibilities in flood risk management in the Netherlands. *Water Resources Development*, 24(4): 555–565.

Terpstra, T., Gutteling. J.M., Geldof, G.D. and Kappe, L.J. 2006. The perception of flood risk and water nuisance. *Water Science and Technology*, 54(6–7): 431–439.

Terpstra, T., Lindell, M.K. and Gutteling, J. M. 2009. Does communicating (flood) risk affect (flood) risk perceptions? Results of a quasi-experimental study. *Risk Analysis*, 29(8): 1141–1155.

Terza, J.V. 1987. Estimating linear models with ordinal qualitative regressors. *Journal of Econometrics*, 34(3): 275–291.

Thieken, A.H., Müller, M., Kreibich, H. and Merz, B. 2005. Flood damage and influencing factors: New insights from the August 2002 flood in Germany. *Water Resources Research*, 41(12): 1–16.

Thieken, A.H., Petrow, T., Kreibich, H. and Merz, B. 2006. Insurability and mitigation of flood losses in private households in Germany. *Risk Analysis*, 26(2): 383–395.

Timmerman, P. 1981. *Vulnerability, Resilience and the Collapse of Society: A Review of Models and Possible Climatic Applications*. Institute for Environmental Studies, University of Toronto, Toronto.

Tobin, G.A. 1999. Sustainability and community resilience: The Holy Grail of natural hazards planning? *Environmental Hazards*, 1(1): 13–25.

Tobin, J. 1958. Estimation of relationships for limited dependent variables. *Econometrica*, 26(1): 24–36.

Tol, R.S.J. 1997. On the optimal control of carbon dioxide emissions: An application of FUND. *Environmental Modeling and Assessment*, 2: 151–163.

1998. Climate change and insurance: A critical appraisal. *Energy Policy*, 26(3): 257–262.

1999. Time discounting and optimal control of climate change: An application of FUND. *Climatic Change*, 41: 351–362.

2002a. New estimates of the damage costs of climate change. Part I. Benchmark estimates. *Environmental and Resource Economics*, 21(1): 47–73.

2002b. New estimates of the damage costs of climate change. Part II. Dynamic estimates. *Environmental and Resource Economics*, 21(2): 135–160.

2005. The marginal damage costs of carbon dioxide emissions: An assessment of the uncertainties. *Energy Policy*, 33(16): 2064–2074.

Tompkins, E.L. and Adger, W.N. 2005. Defining response capacity to enhance climate change policy. *Environmental Science and Policy*, 8(6): 562–571.

Tonhasca, A. and Byrne, D.N. 1994. The effects of crop diversification on herbivorous insects: A meta-analysis approach. *Ecological Entomology*, 19(3): 239–244.

Train, K.E. 2003. *Discrete Choice Models with Simulation*. Cambridge University Press.

Trapp, R.J., Diffenbaugh, N.S. and Gluhovsky, A. 2009. Transient response of severe thunderstorm forcing to elevated greenhouse gas concentrations. *Geophysical Research Letters*, 36: L01703.

Trenberth, K.E., Jones, P.D., Ambenje, P., Bojariu, R., Easterling, D., Klein Tank, A., Parker, D., Rahimzadeh, F., Renwick, J.A., Rusticucci, M., Soden, B. and Zhai, P. 2007. Observations: Surface and atmospheric climate change. In S. Solomon, D. Qin, M. Manning, Z. Chen, M. Marquis, K.B. Averyt, M. Tignor and H.L. Miller (eds). *Climate Change 2007: The Physical Science Basis*. Contribution of Working Group I to the Fourth Assessment Report of the Intergovernmental Panel on Climate Change. Cambridge University Press: 235–336.

Tucker, M. 1997. Climate change and the insurance industry: The cost of increased risk and the impetus for action. *Ecological Economics*, 22(2): 85–96.

Turner, R.H., Nigg, J.M. and Paz, D.H. 1986. *Waiting for Disaster: Earthquake Watch in California*. University of California Press, Berkeley, CA.

Tversky, A. 1967. Utility theory and additivity analysis of risky choices. *Journal of Experimental Psychology*, 75(1): 27–36.

Tversky, A. and Fox, C.R. 1995. Weighting risk and uncertainty. *Psychological Review*, 102(2): 269–283.

Tversky, A. and Kahneman, D. 1973. Availability: A heuristic for judging frequency and probability. *Cognitive Psychology*, 5(2): 207–232.

1974. Judgment under uncertainty: Heuristics and biases. *Science*, 185(4157): 1124–1131.

1981. The framing of decisions and the psychology of choice. *Science*, 211(4481): 453–458.

1983. Extensional versus intuitive reasoning: The conjunction fallacy in probability judgment. *Psychological Review*, 90(4): 293–315.

1986. Rational choice and the framing of decisions. *Journal of Business*, 59(4): 251–278.

1992. Advances in prospect theory: Cumulative representation of uncertainty. *Journal of Risk and Uncertainty*, 5(4): 297–323.

Tversky, A. and Wakker, P.P. 1995. Risk attitudes and decision weights. *Econometrica*, 63(6): 1255–1280.

UNEP 2007. *Insuring for Sustainability: Why and How the Leaders Are Doing It*. Inaugural Report of the Insurance Working Group of the United Nations Environmental Programme Finance Initiative. UNEP Finance Initiative, Switzerland.

UNEP-FI 2011. *Advancing Adaptation through Climate Information Services: Results of Global Survey on the Information Requirements of the Financial Sector*. Sustainable Business Institute and UNEP Finance Initiative (UNEP-FI), Geneva.

US Climate Change Science Program 2008. *Weather and Climate Extremes in a Changing Climate*. U.S. Climate Change Science Program, Washington, DC.

Van Houtven, G., Sullivan, M.B. and Dockins, C. 2008. Cancer premiums and latency effects: A risk tradeoff approach for valuing reductions in fatal cancer risks. *Journal of Risk and Uncertainty*, 36(2): 179–199.

Vassanadumrongdee, S. and Matsuoka, S. 2005. Risk perceptions and value of a statistical life for air pollution and traffic accidents: Evidence from Bangkok, Thailand. *Journal of Risk and Uncertainty*, 30(3): 261–287.

Vellinga, P. 2003. *Klimaatverandering en de veiligheid van Nederland*. Erasmus Liga, Den Haag (in Dutch).

Vellinga, P., Mills, E., Berz, G., Bouwer, L.M., Huq, S., Kozak, L.A., Palutikof, J., Schanzenbächer, B. and Soler, G. 2001. Insurance and other financial services. In J.J. McCarthy, O.F. Canziani, N.A. Leary, D.J. Dokken and K.S. White (eds). *Climate Change 2001: Impacts, Adaptation, and Vulnerability*. Cambridge University Press: 417–450.

Verbeek, K. (ed.) 2003. *De Toestand van het Klimaat in Nederland 2003*. Koninklijk Nederlands Meteorologisch Instituut (KNMI), De Bilt (in Dutch).

Verbond van Verzekeraars 2005. *Verzekerd van Cijfers 2005*. Verbond van Verzekeraars, Lelystad (in Dutch).

2006. *Verzekerd van Cijfers 2006*. Verbond van Verzekeraars, Lelystad (in Dutch).

2007. *Verzekerd van Cijfers 2007*. Verbond van Verzekeraars, Lelystad (in Dutch).

2010a. *Klimaat voor Verzekeraars: Rapport van de Issuecommissie Klimaat*. Verbond van Verzekeraars, Den Haag (in Dutch).

2010b. *Neerslag en Schade: Onderzoek naar het Verband tussen Neerslag en de Schadelast voor Brandverzekeraars met het Oog op de Klimaatverandering*. Verbond van Verzekeraars, Den Haag (in Dutch).

Vinet, F. 2001. Climatology of hail in France. *Atmospheric Research*, 56(1–4): 309–323.

Vis, M., Klijn, F., de Bruijn, K.M. and van Buuren, M. 2003. Resilience strategies for flood risk management in the Netherlands. *International Journal of River Basin Management*, 1(1): 33–40.

Viscusi, W.K. 1985a. A Bayesian perspective on biases in risk perception. *Economic Letters*, 17(1–2): 59–62.

1985b. Are individuals Bayesian decision makers? *American Economic Review*, 75(2): 381–385.

1989. Prospective reference theory: Toward an explanation of the paradoxes. *Journal of Risk and Uncertainty*, 2(3): 235–263.

1998. *Rational Risk Policy*. Oxford University Press, Oxford.

Viscusi, W.K. and Hamilton, J.T. 1999. Are risk regulators rational? Evidence from hazardous waste cleanup decisions. *American Economic Review*, 89(4): 1010–1027.

Viscusi, W.K. and O'Connor, C.J.O. 1984. Adaptive responses to chemical labelling: Are workers Bayesian decision makers? *American Economic Review*, 74(5): 942–956.

Viscusi, W.K. and Zeckhauser, R.J. 2006. National survey evidence on disasters and relief: Risk beliefs, self-interest, and compassion. *Journal of Risk and Uncertainty*, 33(1–2): 13–36.

Vranes, K. and Pielke, R.A. 2009. Normalized earthquake damage and fatalities in the United States: 1900–2005. *Natural Hazards Review*, 10: 84–101.

Vries de, F.J. 1998. Vergoeding van rampschade. *Nederlands Juristenblad*, 42: 1908–1915 (in Dutch).

Wakker, P.P. 2001. Testing and characterizing properties of nonadditive measures through violations of the sure-thing principle. *Econometrica*, 69(4): 1039–1059.

2008. Explaining the characteristics of the power (CRRA) utility family. *Health Economics*, 17(12): 1329–1344.

Wakker, P.P. and Deneffe, D. 1996. Eliciting von Neumann–Morgenstern utilities when probabilities are distorted or unknown. *Management Science*, 42(8): 1131–1150.

Wakker, P.P., Timmermans, D.R.M. and Machielse, I. 2007. The effects of statistical information on risk and ambiguity attitudes, and on rational insurance decisions. *Management Science*, 53(11): 1770–1784.

Ward, P.J., Renssen, H., Aerts, J.C.J.H., van Balen, R.T. and Vandenberghe, J. 2008. Strong increases in flood frequency and discharge of the River Meuse over the late Holocene: Impacts of long-term anthropogenic land use change and climate variability. *Hydrology and Earth System Science*, 12(1): 159–175.

Ward, R.E.T., Herweijer, C., Patmore, N. and Muir-Wood, R. 2008. The role of insurers in promoting adaptation to the impacts of climate change. *Geneva Papers on Risk and Insurance – Issues and Practice*, 33: 133–139.

Warner, K., Ranger, N., Surminski, S., Arnold, M., Linnerooth-Bayer, J., Michel-Kerjan, E., Kovacs, P. and Herweijer, C. 2009. *Adaptation to Climate Change: Linking Disaster Reduction and Insurance*. United Nations International Strategy for Disaster Reduction Secretariat (UNISDR), Geneva.

Weber, E.U. 1994. From subjective probabilities to decision weights: The effect of asymmetric loss functions on the evaluation of uncertain outcomes and events. *Psychological Bulletin*, 115(2): 228–242.

Webster, P.J., Holland, G.J., Curry, J.A. and Chang, H.R. 2005. Changes in tropical cyclone number, duration, and intensity in a warmer environment. *Science*, 309(5742): 1844–1846.

Weiss, M.A. and Chung, J.H. 2004. U.S. reinsurance prices, financial quality and global capacity. *Journal of Risk and Insurance*, 71(3): 347–467.

Weitzman, M. 2009. On modeling and interpreting the economics of catastrophic climate change. *Review of Economics and Statistics*, 91(1): 1–19.

West, J.J., Small, M.J. and Dowlatabadi, H. 2001. Storms, investor decisions, and the economic impacts of sea level rise. *Climatic Change*, 48(2): 317–342.

Wetmore, F., Bernstein, G., Conrad, D., DiVicenti, C., Larson, L., Plasencia, D. and Riggs, R. 2006. *The Evaluation of the National Flood Insurance Program: Final Report*. American Institutes for Research, Washington, DC.

White, H. 1982. Maximum likelihood estimation of misspecified models. *Econometrica*, 50(1): 1–25.

White, R. and Etkin, D. 1997. Climate change, extreme events and the Canadian insurance industry. *Natural Hazards*, 16(2–3): 135–163.

Wijngaard, J.B., Kok, M., Smits, I. and Talsma, M. 2005. Nieuwe statistiek voor extreme neerslag. *H2O*, 6: 35–37 (in Dutch).

Wilkinson, L. 2005. *The Grammar of Graphics*. Springer, Berlin.

Willemse, S. 1995. *A statistical analysis and climatological interpretation of hailstorms in Switzerland*. Doctor of Natural Sciences thesis dissertation No. 11137. Swiss Federal Institute of Technology, Zurich.

Wind, H.G., Nierop, T.M., de Blois, C.J. and de Kok, J.L. 1999. Analysis of flood damages from the 1993 and 1995 Meuse floods. *Water Resources Research*, 35(11): 3459–3465.

Windham, G.O., Possey, E.I., Ross, P. and Spencer, B. 1977. *Reactions to Storm Threat during Hurricane Eloise*. Mississippi State University, Mississippi.

Wood, R., Collins, M., Gregory, J., Harris, G. and Vellinga, M. 2006. Towards a risk assessment for the shutdown of the Atlantic thermohaline circulation. In H.J. Schellnhuber, W. Cramer, N. Nakicenovic, T. Wigley and G. Yohe (eds). *Avoiding Dangerous Climate Change*. Cambridge University Press.

Wooldridge, J.M. 2002. *Econometric Analysis of Cross Section and Panel Data*. MIT Press, Cambridge, MA.

Wouters, K. 2005. *Veiligheid Nederland in Kaart Globale Schadeberekening: Achtergronddocument*. Rijkswaterstaat DWW, Delft (in Dutch).

Wu, G. and Gonzalez, R. 1996. Curvature of the probability weighting function. *Management Science*, 42(12): 1676–1690.

Xu, J. and Long, J.S. 2005. Confidence intervals for predicted outcomes in regression models for categorical outcomes. *Stata Journal*, 5(4): 537–559.

Yeo, S.W. 2002. Flooding in Australia: A review of events in 1998. *Natural Hazards*, 25(2): 177–191.

Zahran, S., Weiler, S., Brody, S.D., Lindell, M.K. and Highfield, W.E. 2009. Modeling national flood insurance policy holding at the county scale in Florida, 1999–2005. *Ecological Economics*, 68: 2627–2636.

Zhang, Q., Wu, L. and Liu, Q. 2009. Tropical cyclone damages in China: 1983–2006. *Bulletin of the American Meteorological Society*, 90: 489–495.

Zhang, X., Zwiers, F.W., Hegerl, G.C., Lambert, F.H., Gillett, N.P., Solomon, S., Stott, P.A. and Nozawa, T. 2007. Detection of human influence on twentieth-century precipitation trends. *Nature*, 448(7152): 461–465.

Index

Lightning Source UK Ltd.
Milton Keynes UK
UKOW06f2312100316

269952UK00018B/634/P

9 781316 600887